ORIGIN, EVOLUTION AND GEOLOGICAL HISTORY OF THE EARTH

SCIENCE IN SHORT CHAPTERS

ORIGIN, EVOLUTION AND GEOLOGICAL HISTORY OF THE EARTH

Additional books and e-books in this series can be found on Nova's website under the Series tab.

Origin, Evolution and Geological History of the Earth

Science in Short Chapters

William Mattieu Williams

Copyright © 2019 by Nova Science Publishers, Inc.

All rights reserved. No part of this book may be reproduced, stored in a retrieval system or transmitted in any form or by any means: electronic, electrostatic, magnetic, tape, mechanical photocopying, recording or otherwise without the written permission of the Publisher.

We have partnered with Copyright Clearance Center to make it easy for you to obtain permissions to reuse content from this publication. Simply navigate to this publication's page on Nova's website and locate the "Get Permission" button below the title description. This button is linked directly to the title's permission page on copyright.com. Alternatively, you can visit copyright.com and search by title, ISBN, or ISSN.

For further questions about using the service on copyright.com, please contact:
Copyright Clearance Center
Phone: +1-(978) 750-8400 Fax: +1-(978) 750-4470 E-mail: info@copyright.com.

NOTICE TO THE READER

The Publisher has taken reasonable care in the preparation of this book, but makes no expressed or implied warranty of any kind and assumes no responsibility for any errors or omissions. No liability is assumed for incidental or consequential damages in connection with or arising out of information contained in this book. The Publisher shall not be liable for any special, consequential, or exemplary damages resulting, in whole or in part, from the readers' use of, or reliance upon, this material. Any parts of this book based on government reports are so indicated and copyright is claimed for those parts to the extent applicable to compilations of such works.

Independent verification should be sought for any data, advice or recommendations contained in this book. In addition, no responsibility is assumed by the publisher for any injury and/or damage to persons or property arising from any methods, products, instructions, ideas or otherwise contained in this publication.

This publication is designed to provide accurate and authoritative information with regard to the subject matter covered herein. It is sold with the clear understanding that the Publisher is not engaged in rendering legal or any other professional services. If legal or any other expert assistance is required, the services of a competent person should be sought. FROM A DECLARATION OF PARTICIPANTS JOINTLY ADOPTED BY A COMMITTEE OF THE AMERICAN BAR ASSOCIATION AND A COMMITTEE OF PUBLISHERS.

Additional color graphics may be available in the e-book version of this book.

Library of Congress Cataloging-in-Publication Data

ISBN: 978-1-53614-926-5

Published by Nova Science Publishers, Inc. † New York

CONTENTS

Preface		**vii**
Chapter 1	The Fuel of the Sun	1
Chapter 2	Dr. Siemens' Theory of the Sun	19
Chapter 3	Another World Down Here	23
Chapter 4	The Origin of Lunar Volcanoes	29
Chapter 5	Note on the Direct Effect of Sun-Spots on Terrestrial Climates	33
Chapter 6	The Philosophy of the Radiometer and Its Cosmical Revelations	35
Chapter 7	On the Social Benefits of Paraffin	39
Chapter 8	The Solidity of the Earth	43
Chapter 9	A Contribution to the History of Electric Lighting	47
Chapter 10	The Formation of Coal	53
Chapter 11	The Solar Eclipse of 1871	59
Chapter 12	Meteoric Astronomy	67
Chapter 13	The "Great Ice Age" and the Origin of the "Till"	73
Chapter 14	The Barometer and the Weather	89
Chapter 15	The Chemistry of Bog Reclamation	103
Chapter 16	Aerial Exploration of the Arctic Regions	109
Chapter 17	The Limits of Our Coal Supply	121
Chapter 18	"The Englishman's Fireside"	135
Chapter 19	"Baily's Beads"	141
Chapter 20	The Coloring of Green Tea	143
Chapter 21	"Iron Filings" in Tea	147
Chapter 22	Concert-Room Acoustics	151

Chapter 23	Science and Spiritualism	**155**
Chapter 24	Mathematical Fictions	**165**
Chapter 25	World-Smashing	**169**
Chapter 26	The Dying Trees in Kensington Gardens	**173**
Chapter 27	The Oleaginous Products of Thames Mud: Where They Come from and Where They Go	**177**
Chapter 28	Luminous Paint	**179**
Chapter 29	The Origin and Probable Duration of Petroleum	**183**
Chapter 30	The Origin of Soap	**189**
Chapter 31	Oiling the Waves	**193**
Chapter 32	On the So-Called "Crater Necks" and "Volcanic Bombs" of Ireland	**197**
Chapter 33	Travertine	**201**
Chapter 34	The Action of Frost in Water-Pipes and on Building Materials	**205**
Chapter 35	The Corrosion of Building Stones	**211**
Chapter 36	Fire-Clay and Anthracite	**215**
Chapter 37	Count Rumford's Cooking-Stoves	**221**
Chapter 38	The "Consumption of Smoke"	**225**
Chapter 39	The Air of Stove-Heated Rooms	**229**
Chapter 40	Ventilation by Open Fireplaces	**233**
Chapter 41	Domestic Ventilation	**237**
Chapter 42	Home Gardens for Smoky Towns	**243**
Chapter 43	Solids, Liquids, and Gases	**253**
Chapter 44	Murchison and Babbage	**265**
Chapter 45	Atmosphere versus Ether	**267**
Chapter 46	A Neglected Disinfectant	**269**
Chapter 47	Another Disinfectant	**271**
Chapter 48	Ensilage	**273**
Chapter 49	The Fracture of Comets	**275**
Chapter 50	The Origin of Comets	**277**
Index		**279**

PREFACE*

I am not aware that this reprint of some of my scattered notes and essays demands any apology.

The practice of making such collections and selections by the author himself has now become very general, and is much better done thus than by friends after his death.

Besides this, it supplies a growing want of these busy times, when so many of us are prevented by the struggles of business from sitting down to the consecutive systematic study of a formal treatise.

I have kept this demand steadily in view throughout, by selecting subjects which are likely to be interesting to all readers who are sufficiently intelligent to prefer sober fact to sensational fiction, but who, at the same time, do not profess to be scientific specialists.

In the writing of these papers my highest literary ambition has always been to combine clearness and simplicity with some attempt at philosophy.

W. M. W.
Willesden, *September, 1882.*

---

*This is an augmented, edited, and reformatted version of *Science in Short Chapters*, originally published in 1883 by John B. Alden Publisher.

Chapter 1

THE FUEL OF THE SUN

I offer the following sketch of the main argument which is worked out more fully in the essay I published in January, 1870, under the above title, hoping that many who hesitate to plunge into a presumptuous speculative work of more than 200 octavo pages may read this article, and reflect upon the subject.

The book has been handled in a most courteous and indulgent spirit by all the reviewers who have noticed it, but none have ventured to grapple with the argument it contains, although every possible opportunity and provocation for doing so is designedly afforded. It all rests upon the question which is discussed in the first three chapters, viz., whether the atmosphere which surrounds our earth is limited or unlimited in extent? If my reasoning upon this fundamental question is refuted, all that follows necessarily falls to the ground. If I am right, all our standard treatises on pneumatics and meteorology, which repeat the arguments contained in Dr. Wollaston's celebrated paper, must be remodeled. At the outset, I reprint that paper, and point out a very curious and monstrous fallacy which, for half a century, remained undetected, and had been continually repeated.

As the main point of issue between myself and Dr. Wollaston is merely a question of very simple arithmetic and geometry, nothing can be easier than to set me right if I am wrong; and, as the philosophical consequences depending upon this issue are of vast and fundamental importance, the question cannot be ignored by those who stand before the world as scientific authorities, without a practical abdication of their philosophical responsibilities. Any man who publishes an astronomical or meteorological treatise without discussing this question, which stands before him at the threshold of his subject, is unfit for the task he has undertaken, and unworthy of public confidence. This may appear a strong conclusion just now, but a few years will be sufficient to graft it firmly into the growth of scientific public opinion.[1]

"The Fuel of the Sun" is simply an attempt to trace some of the consequences which must of necessity result from the existence of an universal atmosphere, and it differs from other attempts to explain the great solar mystery, by making no demands whatever upon the imagination, *inventing* nothing,—no outside meteors, no new forces or materials. It supposes

---

[1] Up to the present date (1882) nobody, as far as I know, has questioned my figures or defended those of Wollaston. Sir William Grove has written to me, pointing out his own anticipations of my conclusions respecting the universality of atmospheric matter. Sir Charles Lyell, before his death, expressed very strong approval of my conclusions, and many other men of scientific eminence have done the same. To expect any immediate, unreserved adoption of such bold speculations would be unreasonable.

nothing whatever to exist but the known facts of the laboratory—the familiar materials of the earth and its atmosphere. It is shown that these materials and the forces residing within them must of necessity produce a sun, and manifest eternally all the observed solar phenomena, provided only they are aggregated in the quantities which our own central luminary presents, and are surrounded by attendant planets, such as his. Nothing is assumed or taken for granted beyond the simple fundamental hypothesis that the laws of nature are uniform throughout the universe. The argument thus conducted leads us step by step to a natural and connected explanation of the following important phenomena:—

1) The sources of solar and stellar heat and light.
2) The means by which the present amount of solar heat and light must be maintained so long as the solar system continues in existence.
3) The origin of the general and particular phenomena of the sun-spots.
4) The cause of the varying splendor of the photosphere, including such details as the "faculæ," "mottling," "granulations," etc., etc.
5) The forces which upheave the solar prominences.
6) The origin of the corona and zodiacal light.
7) The origin of the meteorites and the asteroids.
8) The meteorological phenomena of the planets.
9) The origin of the rings of Saturn.
10) The origin of the special structure of the nebulæ.
11) The source of terrestrial magnetism, and its connection with solar activity.

The first and second chapters are devoted to an examination of the limits of atmospheric expansibility. The experimental investigations of Dr. Andrews, Mr. Grove, Mr. Gassiot, and M. Geissler are cited to prove that the expansibility of the atmosphere is unlimited, and other cosmical evidence is adduced in support of this conclusion.

As this, which is really the foundation of the whole argument, is directly opposed to the views expressed by Dr. Wollaston, in his celebrated paper on "The Finite Extent of the Atmosphere," published in 1822, and generally accepted as established science, this paper is reprinted in the second chapter, and carefully examined.

Dr. Wollaston says "that air has been rarefied so as to sustain 1-100th of an inch of barometrical pressure," and further, that "beyond this limit we are left to conjectures founded on the supposed divisibility of matter; if this be infinite, so also must be the extent of our atmosphere."

I contend that our knowledge of the whole subject is fundamentally altered since these words were written. We are no longer "left to conjectures founded on the supposed divisibility of matter" to determine the possibility of further expansibility than that indicated by 1-100th of an inch of barometrical pressure, as we now have means of obtaining ten times, a hundred times, a thousand times, or even an infinitely greater rarefaction than Wollaston's supposed limit, an apparently absolute vacuum being now obtainable; and although the transmission of electricity affords a means of testing the existence of atmospheric matter with a degree of delicacy of which Wollaston had no conception, we are still unable to detect any indication of any limit to its expansibility.

The most remarkable part of Dr. Wollaston's paper is the *reductio ad absurdum* by which he seeks to finally demonstrate the finite extent of our atmosphere. He maintains, as I do, that if the elasticity of our atmosphere is unlimited, its extension must be commensurate with the universe, that every orb in space will, by gravitation, gather around itself an atmosphere proportionate to its gravitating power, and that, by taking the known quantity of the earth's atmosphere as our unit, we may calculate the amount of atmosphere possessed by any heavenly body of which the mass is known. On this basis Dr. Wollaston calculates the atmosphere of the sun, and concludes that its extent will be so great as to visibly affect the apparent motions of Mercury and Venus, when their declination makes its nearest approach to that of the sun. No such disturbance being actually observable, he concludes that such an atmosphere as he has calculated cannot exist. In like manner he calculates the atmosphere of Jupiter, and finds it to be so great, that its refraction would be sufficient "to render the fourth satellite visible to us when behind the centre of the planet, and consequently to make it appear on both (or all) sides at the same time."

On examining these calculations, I have discovered the very curious error above referred to. As this is a matter of figures that cannot be abridged, I must refer the reader to the original calculations. I will here merely state that Wollaston's method of calculating the solar gravitation atmosphere and that of Jupiter and the moon leads to the monstrous conclusion that, in ascending from the surface of the given orb, we always have the same limited amount of atmospheric matter above as that with which we started, although we are continually leaving a portion of it below.

Wollaston's mistake is based on the assumption that, under the circumstances supposed, the atmospheric pressure and density, at any given distance from the centre of the given orb, will vary inversely with the square of that distance. As the area of the base upon which such pressure is exerted varies *directly* with the square of the distance, the total atmosphere above every imaginable starting-distance would thus be ever the same. That this assumption, so utterly at variance with the known laws of atmospheric distribution, should have remained unchallenged for half a century, and that the conclusions based upon it should be accepted by the whole scientific world, and repeated in standard treatises, such as those of the "Encyclopedia Britannica," etc., etc., is, I think, one of the most remarkable curiosities presented by the history of science. If it were merely a little cobweb in some obscure corner of philosophy, there would be nothing surprising in its escape from the besom of scientific criticism; but this is so far from being the case, that it has hung, since 1822, like a dark veil obscuring another, a wider, and more interesting view of the universe which the idea of an universal atmosphere opens out. But I must now proceed to the next stage of the argument.

Starting from the conclusion reached in the previous chapters, that the atmosphere of our earth is but a portion of an universal elastic medium which it has attached to itself by its gravitation, and that all the other orbs of space must, in like manner, have obtained their proportion, I take the earth's mass, and its known quantity of atmospheric envelope as units, and calculating by the simple rule I have laid down in opposition to Wollaston's, I find that the total weight of the sun's atmosphere should be at least 117,681,623 times that of the earth's, and the pressure at its base equal, at least, to 15,233 atmospheres. What must be the results of such an atmospheric accumulation?

The experiment of compressing air in the condensing syringe, and thereby lighting a piece of German tinder, is familiar to all who have studied even the rudiments of physical science. Taking the formulæ of Leslie and Dalton, and applying them to the solar pressure of 15,233 atmospheres, we arrive according to Leslie, at the inconceivable temperature of 380,832°C., or 685,529°F., as that due to this amount of compression, or, according to

Dalton, at 761,665°F. What will be the effects of such a degree of heat upon materials similar to those of which our earth is composed?

Let us first take the case of water, which, for reasons I have stated, should be regarded as atmospheric, or universally diffused matter.

This brings us to a subject of the highest and widest philosophical and practical importance. I refer to the antagonism between the force of heat and that of chemical combination, to which the French chemists have given the name "dissociation." Having myself been unable to find any satisfactory English account of this subject at a time when it had already been well treated by French and German authors, in the form of published lectures and cyclopædia articles, I assume that others may have encountered a similar difficulty, and therefore dwell rather more fully upon this part of my present summary.

It appears that all chemical compounds may be decomposed by heat, and that, at a given pressure, there is a definite and special temperature at which the decomposition of each compound is effected. For the absolute and final establishment of the universality of this law further investigations are necessary, actual investigations having established it as far as they have gone, but these have not been exhaustive.

There appears to be a remarkable analogy between dissociation and evaporation. When a liquid is vaporized, a certain amount of heat is "rendered latent," and this quantity varies with the liquid and with the pressure, but is definite and invariable for each liquid at a given pressure. In like manner, when a compound is dissociated, a certain amount of heat is "rendered latent," or converted into dissociating force, and this varies with each compound and with the pressure, but is definite and invariable for each compound at a given pressure. Further, when condensation occurs, an amount of heat is evolved, as temperature, exactly equal to that which was rendered latent in the evaporation of the same substance under the same pressure; and, in like manner, when chemical re-combination of dissociated elements occurs, an amount of heat is evolved, as temperature, exactly equal to that which disappeared when the compound was dissociated by heat *alone* under the same pressure.

According to the recently adopted figures of M. Deville, the temperature at which the vapor of water becomes dissociated under ordinary atmospheric pressure is 2800°C., and the quantity of heat which disappears, as temperature, in the course of dissociation is 2153 *calorics*, i.e., sufficient to raise 2153 times its own weight of *liquid* water 1°C.; but, as the specific heat of aqueous vapor is to that of liquid water as 0·475 to 1, that latent heat expressed in the temperature it would have given to aqueous vapor is = 4532°C., or 8158°F.

In order to render the analogy between the ebullition and dissociation of water more evident and intelligible, I will state it as follows:

To commence the ebullition of water under ordinary pressure, a temperature of 100°C., or 212°F., must be attained.	To commence the dissociation of aqueous vapor under ordinary pressures, a temperature of 2800°C., or 5072°F., must be attained.
To complete the ebullition of a given quantity of water, an amount of heat must be applied, sufficient to have raised the water 537°C., or 968°F., above its boiling-point, had it not evaporated.	To complete the dissociation of a given quantity of aqueous vapor, an amount of heat must be applied sufficient to have raised the vapor 4532°C., or 8158°F., above its dissociation-point had it not decomposed.
In order that a given quantity of vapor of water shall condense, it must give off sufficient heat to raise its own weight of water 537°C., or 968°F.	In order that a given quantity of the elements of water may combine, they must give off sufficient heat to raise their own weight of aqueous vapor 4532°C., or 8158°F.

I have expressed these generalizations and analogies rather more definitely than they have been hitherto stated, but those who are acquainted with the researches of Deville, Cailletet, Bunsen, etc., will perceive that I am justified in doing so.[2]

With the general laws of the dissociation of water thus before us, we may follow out the necessary action of the above-stated pressure and consequent evolution of heat in the lower regions of the solar atmosphere upon the large proportion of aqueous vapor which I have shown that it should contain.

It is evident that the first result will be separation of this water into its elements, accompanied with a loss of temperature corresponding to the latent heat of dissociation. We may assume that in the lower regions of the solar atmosphere the free heat evolved by mechanical compression will be more than sufficient to dissociate the whole of the aqueous vapor, and thus the dissociated gases will be left at a higher temperature than was necessary to effect their dissociation. Their condition will thus be analogous to that of superheated steam: they will have to give off some heat before they can *begin* to combine.[3]

There will, however, be somewhere an elevation at which the heat evolved by the joint compression of the elementary and combined gases will be just sufficient to dissociate the latter, and here will be the meeting surface of the combined and the uncombined constituents of water. There will be a sphere containing combined oxygen and hydrogen surrounded by an atmospheric envelope containing large quantities of aqueous vapor, and the temperature at this limiting surface will be equal to that of the oxyhydrogen flame under a corresponding pressure.

What will occur under these conditions? Will the "detonating gases" behave as in the laboratory? Obviously not, as a glance at the third of the above parallel propositions will show. The dissociated gases cannot combine without giving off their 4532° of latent heat as actual temperature. This can only be effected by communication with matter which is cooler than itself.

If a bubble of steam is surrounded by water maintained at the boiling temperature, it will not condense at all, because any effort of condensation would be accompanied with an evolution of heat exactly sufficient to evaporate its own result. If, however, the surrounding water is slowly radiating, or otherwise losing its heat, the enclosed bubble of steam will condense proportionately, by giving off to its envelope an amount of its latent heat just sufficient to maintain the water at the boiling-point.

For further illustration, let us conceive the case of a certain quantity of the elements of water heated exactly to the temperature of dissociation, and confined in a vessel the sides of which are maintained externally at precisely the same temperature as the gases within, so that no heat can be added or taken away from them. No sensible amount of combination can take place, as the first infinitesimal effort of combustion, or combination, would set free just the

---

[2] Since the above was written these analogies have been generally accepted.
[3] Since the publication of "The Fuel of the Sun," Mr. Norman Lockyer has adopted this view of solar dissociation, and has gone so far as to suppose that it splits metals and other substances regarded by modern chemists as simple elements into more elementary and simple constituents. He assumes that the temperature of the solar atmosphere, growing higher at increasing depths, becomes somewhere capable of doing far greater dissociation work than that which separates the hydrogen of the prominences revealed by the spectroscope. In putting forth this "working hypothesis" he seems to have lost sight of the fact clearly proved by Deville's experiments, that the temperature of dissociation rises with the pressure to which the compound is subjected, and thus that within the bowels of the sun the metals will be far less dissociable than they are on the surface of our earth.

amount of heat required to decompose its own result. Let us now suppose a modification of these conditions, viz., that the vessel containing the dissociated gases, at the temperature of dissociation, shall be surrounded with bodies cooler than itself, i.e., capable of receiving more heat from it than they radiate towards it; there would then take place just so much combustion as would set free the amount of heat required to maintain the temperature of the vessel at the dissociation-point; or, in other words, combustion would go on to the extent of setting free just so much heat as the gaseous mass was capable of radiating, or otherwise transmitting to surrounding bodies; and this amount of combustion would continue till all the gases had combined.

We have only to give this hypothetical vessel a spherical form and an internal diameter of 853,380 miles—to construct its enveloping sides of a thick shell of aqueous vapor, etc., and then, by placing in the midst of the contained dissociated gases a nucleus of some kind, we are hypothetically supplied with, the main conditions which I suppose to exist in the sun.

A little reflection upon the application of the above-stated laws to these conditions will show that the stupendous ocean of explosive gases would constitute an enormous stock of fuel capable, by its combustion, of setting free exactly the same quantity of heat as had previously been converted into decomposing or separating force; the amount of combustion would always be limited by the possible amount of radiation, and the radiation would again be limited by the resisting envelope of aqueous vapor produced by this combustion.

If these conditions existed in a perfectly calm and undisturbed solar atmosphere, there would be a continually increasing external envelope of aqueous vapor, and a continually diminishing inner atmosphere of combustible gases; there would be a gradual diminution of the amount of solar radiation, and a slow and perpetually retarding progress towards solar extinction.

It should be noted that, according to this explanation, the *supply* of heat is originally derived from atmospheric condensation due to gravitation, that the *storage* of surplus heat is effected by dissociation, and its *evolution* mainly by recombination or combustion.

The great difficulty, that of the perpetual renewal of the solar fuel, still remains unsolved; the fact that during the millions of years of geological history we find no indications of any declining average of solar energy is so far still unexplained by this, as by every other, attempt to account for the origin of solar and stellar light and heat.

In his inaugural address to the British Association Meeting of 1866, Mr. Grove put the following very suggestive question:—"Our sun, our earth, and planets are constantly radiating heat into space; so, in all probability, are the other suns, the stars, and their attendant planets. What becomes of the heat thus radiated into space? If the universe has no limit—and it is difficult to conceive one—there is a constant evolution of heat and light; and yet more is given off than is received by each cosmical body, for otherwise night would be as light and as warm as day. What becomes of the enormous force thus apparently non-recurrent in the same form?"

This is a grand question, a philosophical thought worthy of the author of "The Correlation of Physical Forces." Most philosophical thinkers will, I believe, agree with me in concluding that a sound reply to it will solve the great mystery of the everlasting radiations of our sun and all the other suns of the universe. So long as we regard these suns as the *sources* of continually expended forces of light and heat, their everlasting and unabated renewal becomes a mystery utterly inscrutable to the human intellect, since the creation of new force, or any addition to the total forces of the universe, is as inconceivable to us as any addition to the total matter of the universe. The great solar question assumes a far more hopeful shape when we admit that all the forces of past radiations are somewhere diffused in space, and we ask whether a sun contains any mechanism by which it may collect and concentrate this diffused force, and thus perpetually gather from surrounding suns as much as it radiates towards them.

The next part of my work is an attempt to show that such a mechanism does exist in our solar system, and to explain its action.

We know that if atmospheric air is compressed it becomes heated, that if this heat is allowed to radiate and the air is again expanded to its original dimensions, it will be cooled below its original temperature to an extent precisely equal to the heat which it gave out when compressed. On this principle I endeavor to explain the everlasting maintenance of the solar and stellar radiations.

The sun is attended by his train of planets whose orbital motion he controls, but they in return react upon him as the moon does upon the earth. If this reaction were regular, like that of the moon upon the earth, a regular atmospheric tide would result; but the great irregularity of the dimensions, distances, and velocities of the planets produces a result equivalent to a number of clashing irregular tides in the solar atmosphere; or, otherwise stated, the centre of motion and centre of gravity of the whole system will be perpetually varying with the varying relative positions of the planets, and thus the solar nucleus and solar atmosphere will be subject to irregularities of motion, which, though very small relatively to the enormous magnitude of the sun, must be sufficient to produce mighty vortices, and thus effect a continual commingling between the outer and inner atmospheric strata.

It must be remembered that, according to the preceding, the inner or lower strata of the solar atmosphere should consist of our ordinary atmospheric mixture of oxygen and nitrogen, and the dissociated elements of water and carbonic acid, besides some of the more volatile elements of the solar nucleus. Outside of this there should be a boundary limit where the dissociated gases are combining as rapidly as their latent heat can be evolved by radiation; this will form a shell or sphere of flame,—the photosphere,—and above or beyond this will be the sphere of vapors resulting from this combustion, which, by their resistance to radiation, will limit the evolution of heat and consequent combustion.

Now the vortices above referred to will break through the shell of combustion, and drag down more or less of the outer vapor into the lower and hotter regions of dissociated gases.

As there can be no action without equal and contrary reaction, there can be no vortices, either in the solar atmosphere or a terrestrial stream, without corresponding upheavals. These upheavals will eject the lower dissociated gases more or less completely through the vaporous jacket which restrains their normal radiations, and, thus liberated, they will rush into combination with an explosive energy comparable to that which they display in our laboratories; not, however, with an instantaneous flash, but with a continuous rocket-like combustion, the rapidity of which will be determined by the possibility of radiation. The heat evolved by this combustion, acting simultaneously with the diminution of pressure, will effect a continually augmenting expansion of these upheaved gases, and as the rapidity of combustion will be accelerated in proportion to elevation above the restraining vapors, an outspreading far in excess of that which would be due to the original upheaving force, is to be expected.

The reader who is acquainted with the phenomena of the solar prominences will at once perceive how all these expectations are fulfilled by actual observations, especially by the more recent observations of Zöllner, Secchi, etc., which exhibit the typical solar prominence as a stem or jet rushing upwards through some restraining medium, and then expanding into a cloud-like or palm-tree form after escaping from this restraint. I need scarcely add that the clashing tide waves are the *faculæ*, and the vortices the sun-spots.

My present business, however, is to show how these vortices and eruptions—this down-rush in one part of the solar atmosphere and up-rush in another—contribute to the permanent maintenance of the solar light and heat. It must be understood that these outbursts are only visible to us as luminous prominences during the period of their explosive outburst, and while still subject to great expansive tension. Long after they have ceased to be visible to us their expansion must continue, until they finally and fully mingle with the medium into which they are flung, and attain a corresponding degree of rarefaction. This must occur at tens and hundreds of thousands of miles above the photosphere, according to the magnitude of the ejection. The spectroscopic researches of Frankland and Lockyer having shown that the atmospheric pressure at about the outer surface of the photosphere does not far exceed that of our atmosphere, I may safely regard all the upper portion of these solar ejections as having left the solar atmosphere proper, and become commingled with the general interstellar medium.

If the sun were stationary, or merely rotating, in the midst of this universal atmosphere, the same material that is ejected to-day would in the course of time return, and be whirled into the great sun-spot eddies; but such is not the case; the sun is driving through the ether with a velocity of about 450,000 miles per twenty four hours.

What must be the consequence of this motion? The sun will carry its own special atmospheric matter with it; but it cannot thus carry the whole of the interstellar medium. There must be a limit, graduated no doubt, but still a practical limit, at which its own atmosphere will leave behind, or pass through, the general atmospheric matter. There must be a heaping or condensation of this matter in the front, a rarefaction or wake in the rear, and a continuous bow of newly encountered atmosphere around the boundaries in the opposite direction to that of the sun's motion. The result of this must be that a great portion of the ejected atmospheric matter of the prominences will be swept permanently to the rear, and its place supplied by the material occupying the space into which the sun is advancing. We are thus presented with a mighty machinery of solar respiration; some of this newly arriving atmospheric matter must be stirred into the vortices, its quantity being exactly equivalent to that of the old material expired by the explosive eruptions, and left in the rear.

Now, the new atmospheric matter which is thus encountered and inspired, is the recipient of the everlasting radiations whose destination is the subject of Mr. Grove's inquiry; and these, when thus encountered and compressed, will of necessity evolve more or less of the heat which, through millions of millions of centuries they have been gradually absorbing; while, on the other hand, the expired or ejected matter of the gaseous eruptions will, like the artificially compressed air above referred to, have lost all the heat which during its solar existence it had by compression, dissociation, and re-combination contributed to the solar radiations. Therefore, when again fully expanded, it will be cooler than the general medium from which it was originally inspired by the advancing sun.

The daily supply of fresh atmospheric fuel will be a cylinder of ether of the same diameter as the sun, and 450,000 miles in length! I have calculated the weight of this cylinder of ether on the assumption (which of course is purely arbitrary) that the density of the interstellar medium is one ten-thousandth part of that of our atmosphere. It amounts to

14,313,915,000,000,000,000 tons, affording a supply of 165 millions of millions of tons per second; or, if we assume the interstellar medium to have a density of only one-millionth of that of our atmosphere, the supply would be rather more than one and a half millions of millions of tons per second. The proportion of this which is effective in the manner above stated is that which becomes stirred into the lower regions of the sun in exchange for the ejected matter of the prominences.

I will not here dwell upon the bombardment hypothesis, beyond observing that my explanation of solar phenomena supplies a continuous bombardment of the above-stated magnitude without adding anything to the magnitude of the sun.

So far, then, I answer Mr. Grove's question, by showing that the heat radiated into space by each of the solid orbs that people its profundities, is received by the universal atmospheric medium; is gathered again by the breathing of wandering suns, who inspire as they advance the breath of universal heat and light and life; then by impact, compression, and radiation, they concentrate and re-distribute its vitalizing power; and after its work is done, expire it in the broad wake of their retreat, leaving a track of cool exhausted ether—the ash-pits of the solar furnaces—to reabsorb the general radiations, and thus maintain the eternal round of life.

But ere this, a great difficulty has probably presented itself to the mind of the reader. He will refer to the calculations that have been made in order to determine the actual temperature of the solar surface and the intensity of its luminosity. Both of these are vastly in excess of those obtained in our laboratory experiments by the combustion of the elements of water. Even taking into consideration the dissociated carbonic acid whose elements should be burning in the photosphere with those of water, and adding to these the volatile metals of the solar nucleus whose dissociated vapors must, under the circumstances stated, be commingled with those of the solar atmosphere, and therefore contribute to the luminosity by their combustion, still by burning here on the earth a jet of such mixed gases and vapors we should not obtain any approach to either the luminosity or the temperature which is usually attributed to the sun.

I have made a very few simple experiments, the results of which remove these difficulties. They were conducted with the assistance of Mr. Jonathan Wilkinson, the official gas examiner to the Sheffield Corporation, using his photometric and gas-measuring apparatus. We first determined the amount of light radiated by a single fish-tail gas-burner consuming a measured quantity of gas per hour. We found when another was placed behind this, so that all the light of the second had to pass through the first, that the light of the two (measured by the illuminating intensity of their radiations upon a screen just as the solar luminosity has been measured) was just double that of one flame, three flames (still presenting to the photometric screen only the surface of one) gave it three times the amount of illumination, and so on with any number of flames we were able to test. Mr. Wilkinson has since arranged 100 flames on the same, principle, i.e., so that the 99 hinder flames shall all radiate through the one presented to the screen, thus affording the same surface as a single flame, but having 100 times its *thickness* or *depth*, and he finds that the law indicated by our first experiments is fully verified; that the 100 flames thus arranged illuminate the screen 100 times as intensely as the single flame. Other modifications of these experiments, described in Chapter vii. of "The Fuel of the Sun," establish the principle that a common hydrocarbon gas flame is transparent to its own radiations, or, in other words, that the amount of light radiated from such a flame, and its apparent intensity of luminosity, is proportionate to its thickness; therefore the luminosity of the sun may be produced by a photosphere having no greater intrinsic brilliancy than the flame of a tallow candle, provided the flame is of sufficient depth or thickness. I see good reasons for inferring that its intrinsic brilliancy is less than that of a candle—somewhere between that and a Bunsen's burner.

A similar series of experiments upon the radiation of the *heat* of flames through each other, indicated similar results; but my apparatus for these experiments was not so delicate and reliable as in the experiments on light, and, therefore, I cannot so decidedly affirm the absolute diathermancy of flame to its own radiations. Within the limits of error of these experiments, I found that with the same radiant surface presented to the thermometer, every addition to the thickness of the flame produced a proportionate increase of radiation.

This important law, though hitherto unnoticed by philosophers, is practically understood and acted upon by workmen who are engaged in furnace operations. Present space will not permit me to illustrate this by examples, but in passing I may mention the "mill furnaces," where armor-plates and other large masses of iron are raised to a welding temperature by radiant heat, and the ordinary puddling furnace, where iron is melted by radiant heat. In both of these special arrangements are made to obtain a "body" or thickness of radiant flame, while *intensity* of combustion is neglected and even carefully avoided.

According to this there are two factors engaged in producing the radiant effect from a given surface, *intensity* and *quantity*, i.e., *brilliancy* and *thickness* in the case of light, and *temperature* and *thickness* in the case of heat. In the Bude light, for example, consisting of concentric rings of coal-gas, we have small intensity with great quantity, in the lime-light we have a mere surface of great brilliancy but no thickness. If I am right, the surface of the moon maybe brighter than the luminous surface of the sun, the peculiarities of moonlight depending upon intensity, those of sunlight upon quantity of light.

The flame that roars from the mouth of a Bessemer converter has but small intrinsic brilliancy, far less than that of an ordinary gas flame, as may be seen by observing the thin waifs that sometimes project beyond the body of the flame. Nevertheless, its radiations are so effective that it is a painfully dazzling object even in the midst of sunny daylight; but then we have here not a hollow flame fed only by outside oxygen, but a solid body of flame several feet in thickness. Even the pallid carbonic acid flame which accompanies the pouring of the spiegeleisen has marvellous illuminating power.

The reader will now be able to understand my explanation of the sun-spots, of their nucleus, umbra, and penumbra. From what I have stated respecting the planetary disturbances or the solar rotation, the photosphere should present all the appearances due to the movements of a fiery ocean, raging and seething in the maddest conceivable fury of perpetual tempest. If the surface of a river flowing peacefully between its banks is perforated with conical eddies whenever it meets with a projecting rock or obstacle, or other agency which disturbs the regularity of its course, what must be the magnitude of the eddies in this ocean of flame and heated gases, when stirred to the lowest depths of its vast profundity by the irregular reeling of the solar nucleus within? Obviously, nothing less than the sunspots; those mighty maelströms into which a world might be dropped like a pea into an egg-cup.

When the photosphere or shell of combining gases is thus ripped open, the telescopic observer looks down the vortex, which, if deep enough, reveals to him the inner regions of dissociated gases and vapors. But these have the opposite property to that which I have shown to belong to flame; they are opaque to their own special radiations, while the flame is transparent to the light of the inner portions of itself. Thus, the dissociated interior of the solar envelope, though absolutely white-hot, will be comparatively dark (direct experiment has proved that the darkness of the spots is only relative).

The sides of the vortex funnel will consist of a mixture of dissociated gases, flaming gases, and combined gases, and will thus present various thicknesses of flame, and thereby display the various shades of the penumbra. Space will not permit me here to follow up the details of this subject, as I have done in the original work, where it is shown that if the telescope had not yet been invented, all the telescopic details of spot phenomena might have been described *à priori* as necessary consequences of the constitution I have above ascribed to the sun.

Not merely the great spot phenomena, but all the minor irregularities of the photosphere follow with similarly demonstrable necessity. Thus the many interfering solar tides must throw up great waves, literally mountainous in their magnitude, the summits and ridges of which, being raised into higher regions of the absorbing vaporous atmosphere that envelopes the photosphere, will radiate more freely, its dissociated matter will combine more abundantly, and will thicken the photosphere immediately below; this thicker flame will be more luminous than the normal surface, and thus produce the phenomena of the *faculæ*.

Besides these great ground-swells of the flaming ocean of the photosphere, there must be lesser billows, and ripples upon these, and mountain tongues of flame all over the surface. The crests of these waves, and the summits of these flame-alps, presenting to the terrestrial observer a greater depth of flaming matter, must be brighter than the hollows and valleys between; and their splendor must be further increased by the fact, that such upper ridges and summits are less deeply immersed in the outer ocean of absorbing vapors, which limits the radiation of the light as well as the heat of the photosphere. The effect of looking upon the surface of such a wild fury of troubled flame, with its confused intermingling of gradations of luminosity, must be very puzzling and difficult to describe; and hence the "willow leaves," "rice grains," "mottling," "granules," "things," "flocculi," "bits of white thread," "cumuli of cotton wool," "excessively minute fragments of porcelain," "untidy circular masses," "ridges," "waves," "hill knolls," etc., etc., to which the luminous irregularities have been compared.

At the time I wrote, the means of examination of the edge of the sun by the spectroscope was but newly discovered, and the results then published referred chiefly to the prominences proper. Since that, a new term has been introduced to solar technology, the "sierra," and the observations of the actual appearances of this sierra precisely correspond to my theoretical description of the limiting surface of the photosphere, which was written before I was acquainted with these observed facts. This will be seen by reference to Chapter x., the subject of which is, "The Varying Splendor of Different Portions of the Photosphere."[4]

But I must not linger any further upon this part of the subject, but proceed to another, where subsequent discoveries have strongly confirmed my speculations.

The mean specific gravity of the sun is not quite 1½ times that of water. The vapors of nickel, cobalt, copper, iron, chromium, manganese, titanium, zinc, cadmium, aluminium, magnesium, barium, strontium, calcium, and sodium, have been shown by the spectroscope to be floating on the outer regions of the sun. None of these could constitute the body of the sun in a solid or liquid state, and be subjected to the enormous pressure which such a mass must

---

[4] Still more recently (1882) the magnificent photographs of Jannsen have displayed further evidence of the flame-tongue character of the mottling.

exert upon itself without raising the mean specific gravity vastly above this; nor is there any other kind of matter with which we are acquainted which could exist within so large a mass in a liquid or solid state, and retain so low a density.

I must confess that my faith in the logical acumen of mathematicians has been rudely shaken by the manner in which eminent astronomers have described the umbra or nucleus of the sun-spots as the solid body of the sun seen through his luminous atmosphere, and the solid surface of Jupiter seen through his belts, and have discussed the habitability of Jupiter, Saturn, Uranus, and Neptune always on the assumption of their solidity, while the specific-gravity of all of these renders this surface solidity a demonstrable physical impossibility.

If the sun (or either of these planets) has a solid or liquid nucleus, it must be a mere kernel in the centre of a huge orb of gaseous matter, and though I have spoken rather definitely of the solar atmosphere in order to avoid complication, I must not, therefore, be understood to suppose that there exists in the sun any such definite boundary to the base of the atmospheric matter as we find here on the earth. The temperature, the density, and all we know of the chemistry of the sun justify the conclusion that in its outer regions, to a considerable depth below the photosphere, there must be a commingling of the atmospheric matter with the vapors of the metals whose existence the spectroscope has revealed. Some of these must be upheaved together with the dissociated elements of water. They are all combustible, and, with a few exceptions, the products of their combustion would solidify after they were projected beyond the photosphere. Much of the iron, nickel, cobalt, and copper might pass through the fiery ordeal of such projection, and solidify without oxidation, especially when more or less enveloped in uncombined hydrogen.

It is obvious that, under these circumstances, there must occur a series of precipitations analogous to those from the aqueous vapor of our atmosphere. These gaseous metals, or their oxides, must be condensed as clouds, rain, snow, and hail, according to their boiling and metal points, and the conditions of their ejection. We know that sudden and violent atmospheric disturbance, accompanied with fierce electrical discharges, especially favor the formation of hailstones in our terrestrial atmosphere. All such violence must be displayed on a hugely exaggerated scale in the solar outbursts, and therefore the hailstone formation should preponderate, especially as the metallic vapors condense more rapidly than those of water on account of the much smaller amount of their specific heat, and of the latent heat of their vapors.

What will become of these volleys of solid matter thus ejected with the furious and protracted explosions forming the solar prominences? In order to answer this question, we must remember that the spectroscope, as recently applied, merely displays the gaseous, chiefly the hydrogen, ejections; that these great gaseous flames bear a similar relation to the solid projectiles that the flash of a gun does to the grape-shot or cannon-ball. Mr. Lockyer says: "In one instance I saw a prominence 27,000 miles high change enormously in the space of ten minutes; and, lately, I have seen prominences much higher born and die in an hour." He has recently measured an actual velocity of 120 miles per second in the movements of this *gaseous* matter of the solar eruptions, the initial velocity of which must have been much greater.[5] If such is the velocity of the gaseous ejections, what must be that of the solid projectiles, and where must they go?

---

[5] Subsequent observations (1882) by Secchi, Young, and others have demonstrated velocities far exceeding this; quite sufficient to project the solid matter clearly beyond the sphere of solar attraction.

A cosmical cannonade is a necessary result of the conditions I have sketched, and as prominence-ejections are continually in progress, there must be a continual outpouring from the sun of solid fragments, which must be flung far beyond the limits of the gaseous prominences. As the luminosity of these glowing particles must be very small compared with that of the photosphere, they will be invisible in the glare of ordinary sunshine; but if our eyes be protected from this, they may then be rendered visible, both by their own glow and the solar light they are capable of reflecting. They should be seen during a total eclipse, and should exhibit radiant streams proceeding irregularly from different parts of the sun, but most abundantly from the neighborhood of the spot regions. As these spot regions occupy the intermediate latitudes between the poles and the equator of the sun, the greatest extensions of the outstreamings should be N.E. and S.W., and S.E. and N.W., while to the N., S., E., and W.—that is, opposite the poles and equator of the sun—there should be a lesser extension. The result of this must be an approximation to a quadrilateral figure, the diagonals of which should extend in a N.E. and S.W., and a S.E. and N.W. direction, or thereabouts. I say "thereabouts," because the zone of greatest activity is not exactly intermediate between the poles and the equator, but lies nearer to the solar equator.

Examined with the polariscope, these radiant streams should display a mixture of reflected light and self-luminosity. Examined with the spectroscope, a faint continuous spectrum due to such luminosity of solid particles should be exhibited, with possibly a few lines due to the small amount of vapor which, in their glowing condition, they might still give off. Besides this, there should appear the spectroscope indications of violent electrical discharges, which must occur as a necessary concomitant of the furious ejections of aqueous vapor and solid particles. All these metallic hailstones must be highly charged, like the particles of vesicular vapor ejected from the hydro-electric machine, or the vapors and projectiles of a terrestrial volcanic eruption.

I need scarcely add that this exactly describes the actually-observed results of the recent observations on the corona, and that all the phenomena of this great solar mystery are but necessary and predicable results of the constitution I ascribe to the sun.

There is a method of manufacturing hypotheses which has become rather prevalent of late, especially among mathematicians, who take observed phenomena, and then arbitrarily and purely from the raw material of their own imagination construct explanatory atoms, media, and actions, which are shaved and pared, scraped and patched, lengthened and shortened, thickened and narrowed, till they are made to fit the phenomena with mathematical accuracy. These laborious creations are then put forth as philosophical truths, and, *afterwards*, the accuracy of their fitting to the phenomena is quoted as evidence of the positive reality of the ethers, atoms, undulations, gyrations, collisions, or whatever else the mathematician may have thus skilfully created and fitted. It appears to me that such fitness only proves the ingenuity of the fitter—the skill of the mathematician—and that all such hypotheses belong to the poetry of science; they should be distinctly labelled as products of mathematical imagination, and nowise be confounded with objective natural truths. Such products of the imagination of the expert may assist the imagination of the student in comprehending some phenomena, just as "Jack Frost" and "Billy Wind" may represent certain natural forces to babies; but if Jack Frost, Billy Wind, electric and magnetic fluids, ultimate atoms, interatomic ethers, nervous fluids, etc., are allowed to invade the intellect, and are accepted as actual physical existences, they become very mischievous philosophical superstitions.

I make this digression in order to repudiate any participation in this kind of speculation. Though "The Fuel of the Sun" is avowedly a very bold attempt to unravel majestic mysteries, I have not sought *to elucidate the known by means of the unknown*, as do these inventors of imaginary agents, but have scrupulously followed the opposite principle. I have invented nothing, but have started from the experimental facts of the laboratory, the demonstrated laws of physical action, and have followed up step by step what I understand to be the necessary consequences of these. Many years ago I convinced myself that our atmosphere is but a portion of universal atmospheric matter; that Dr. Wollaston was wrong, and that the compression of this universal atmospheric matter is possibly the source of solar light and heat; but as this was long before M. Deville had investigated the subject of dissociation by heat,[6] I was unable to work out the problem at all satisfactorily. When I subsequently resumed the subject, I knew nothing about the corona, and had only read of the "red prominences" as possible lunar appendages, or solar clouds, or optical illusions. I had worked out the necessity of the gaseous eruptions, and their action in effecting an interchange of solar and general atmospheric matter, as the means of maintaining the solar light and heat, with no idea of proceeding further with the problem, when the announcement that the prominences were not merely unquestionable solar appendages, but were actually upheaved mountains of glowing hydrogen, suddenly and unexpectedly suggested their identity with my required atmospheric upheavals. It is true that their observed magnitude far exceeded my theoretical anticipations, and in this respect I have made some *à posteriori* adaptations, especially with the aid of a clearer understanding of the laws of dissociation which almost simultaneously became attainable.

In like manner, the necessity of the solid ejections presented themselves before I knew anything of the recently discovered details of the coronal phenomena—when I had merely read of a luminous halo which had been seen around the sun, and relying upon Mr. Lockyer, vaguely supposed it to be an effect of atmospheric illumination. I inferred that streams of solid particles must be pouring from the sun, and showering back again, but had no idea that such streams and showers were actually visible until I was rather startled on learning that the corona, instead of being, as I had loosely supposed, a mere uniform filmy halo, had been described by Mr. De la Rue, in his Bakerian Lecture on the Eclipse of 1860, as "softening off with very irregular outline, and sending off some *long streams*," etc. I was then living on the sides of a Welsh mountain far away from public libraries, and being no astronomer, my own books kept me better acquainted with the current progress of experimental than with astronomical science.

Even when "The Fuel of the Sun" was published I knew nothing of the American observations of the quadrangular figure of the corona, or should certainly have then quoted them, nor of the fact revealed by the Eclipse of December, 1870, that, "wherever on the solar disc a large group of prominences was seen on Mr. Seabroke's map, there a corresponding bulging out of the corona was chronicled on Professor Watson's drawing; and at the positions where no prominences presented themselves, there the bright portions of the corona extended to the smallest distances from the sun's limb;" and that Mr. Brothers's photographs *all* show the corona extending much further towards the west than towards the east, the west being "the region richest in solar prominences." I am sorry that the limits of this paper will not permit

---

[6] My first memorandum on this subject is dated April 23, 1840, in a *Register of Ideas*, then commenced in very early student days.

me to enter more fully into the bearings of the recent studies of the corona and the prominences upon my explanations of solar phenomena, especially as the differences between the inner and outer corona, which still appear to puzzle astronomers, are exactly what my explanation demands. I must make this the subject of a separate paper, and proceed at once to the next step of the general argument.

Assuming that such ejections of solid matter are poured from the prominences, to what distances may they travel? In attempting to answer this question, I avowedly ventured upon dangerous ground, for at the time of writing I only knew that the force of upheaval of the prominences must be enormous, *probably* sufficient to eject solid matter beyond the orbit of the earth and even beyond that of Mars. Actual measurements of the eruptive velocity of the solar prominences have since been made, and they are so great as to relieve me of my quantitative difficulty, and show that I was quite justified in the bold inference that these eruptions may account for the zodiacal light, the zones of meteors into which our earth is sometimes plunged, and even the outer zone of larger bodies, the asteroids.

But how, the reader will ask, can such solids, ejected from the sun, acquire orbital paths around him? "We have been taught that the parabola is the necessary path of such ejections." Mr. Proctor has evidently reasoned in this manner, for in last April number of "Fraser's Magazine" he says that some of my ideas are "opposed to any known laws, physical or dynamical," that "there is nothing absolutely incredible in the conception that masses of gaseous, liquid, or solid matter should be flung to a height exceeding manifold that of the loftiest of the colored prominences; whereas it is not only incredible, but impossible, that such matter should in any case come to circle in a closed orbit round the sun."

More careful reading would have shown Mr. Proctor that I have considered other conditions besides those of the textbooks, that the case is by no means one of simple radial projection from a fixed body into free space and undisturbed return. I distinctly stated that "the recent ejections may have any form of orbit within the boundaries of the conic sections," from a straight line returning upon itself, due to absolutely vertical projection, to a circular orbit produced by the tangential projection of such curving prominences *as the ram's horn*, etc. The outline of the zodiacal light would be formed by the termination or aphelion portion of these excursions, or of such a number of them as should be sufficient to produce a visible result.

Again, speaking of the asteroids, in Chapter xiv., I state that "I should have expected a still greater elongation and eccentricity in some of them, and such orbits may have existed; but an asteroid with an orbit of cometary eccentricity that would in the course of each revolution cross the paths of Mercury, Venus, the Earth, and Mars in nearly the same plane, and dive through the thickly scattered zodiacal cluster, both in going to the sun and returning from it, would be subject to disturbances which would continue until one of two things occurred. Its tangential force might become so far neutralized and its orbit so much elongated, that finally its perihelion distance should not exceed the solar radius, when it would finish its course by returning to the sun. On the other hand, its tangential velocity might be increased by heavy pulls from Jupiter, when slowly turning its aphelion path, and be similarly influenced by friendly jerks in crossing the orbits of the inferior planets; and thus its orbit might be widened, until it ceased periodically to cross the path of any of the planets by establishing itself in an orbit constantly intermediate between any two. Having once settled into such a path, it would remain there with comparative stability and permanency. If I am right in this view of the dynamical history of these older ejections, all the long elliptical paths of zodiacal particles, meteorites, or asteroids, would thus in the course of ages become eliminated, and the remaining orbits would be of planetary rather than cometary proportions."

A little reflection on the above-stated laws of dissociation will show that the maximum violence of hydrogen explosion will not occur at the birth of the ejections, but afterwards, when the dissociated gases have been already hurled beyond the sphere of restraining vapors. If my explanation is correct, the typical form of a solar prominence should be that of a spreading tree with a tall stem. At first the least resistance to radiation and consequent explosive combination must be in the vertical direction, as this will afford the shortest line that can be drawn through the thickness of the surrounding jacket of resisting vapor; but when raised above this envelope, the dissociated gases, cooled by their own expansion and comparatively free to radiate in all directions except downwards, will explode laterally as well as vertically, and thus spread out into a head. My theoretical prominence will be, in short, a monster rocket proceeding steadily upwards to a certain extent, and then gradually bursting and projecting its missiles in every direction from the vertical to the absolutely horizontal. Should the latter acquire a velocity of about 300 miles per second, not merely a closed but even an absolutely circular orbit would be possible. These and the multitude of weaker lateral ejections, reaching the sun by short parabolic paths, explain the mystery of the inner corona.

I need only refer Mr. Proctor to his own recently published book on the Sun, where he will find on plates 4, 5, and 6 a number of drawings from Zöllner and Respighi, which so thoroughly confirm my necessary theoretical deductions that they might be a series of fancy sketches of my own. When we consider that the base of a prominence is only visible when it happens to start exactly from the limb of the sun, while the vastly greater proportion of those which are observed, and have been drawn, have much of the stem cut off from view by the solar rotundity, the evidence afforded by such drawings in support of my theoretical deduction, that the typical form of the solar prominences is that of a palm-tree or bursting rocket, is greatly strengthened.[7]

In a paper by P. Secchi, dated Rome, March 20, 1871, and published in the "Comptes Rendus," March 27, this veteran solar observer speaks of the prominences as composed of jets, which, "upon reaching a certain elevation, stop and whirl upon themselves, giving birth to a brilliant cloud." This cloud is represented as spreading out on all sides from the summit of the combined jets. Again he says, "It is very common to see a little jet spot at a certain elevation above the chromosphere, and there spread itself out into a *wide hat* ("*un large chapeau*") of an absolutely nebulous constitution." This outspreading nebulosity is the flash of the incandescent vapors produced by the explosion which is theoretically demanded by my explanation to occur exactly in the manner and place described. These expanded incandescent gases will be rendered visible by the spectroscopic dilution of the continuous spectrum of the denser photosphere, while the solid projectiles that must proceed from them in every direction can only be seen during a solar eclipse.

The observations and drawings of Zöllner and Respighi were, for the most part, made while my book was in the press, and, like those of Secchi above quoted, were unknown to me when I wrote; I was then only able to quote, in support of my theoretical requirements, the

---

[7] Any reader of "The Fuel of the Sun" will perceive that the vaporous envelope which I have described as "an effectual jacket for limiting the amount of radiation," is a complete theoretical anticipation and explanation of the "solar crust" of Respighi and the "Trennungschicht" of Zöllner. We agree perfectly in our conclusions, though arriving at them by such very different paths, and so independently of each other.

evidences of actually observed tangential ejection afforded by Sir John Herschel's account of the great solar storm of September 1, 1859.

Besides this direct tangential projection there are other elements of motion contributing to the same result, such as the whirl of the prominences on themselves, their motion of translation on the sun's disk, and the rotation of the sun itself.

I must now bring this sketch to a close by stating that, in order to submit the fundamental question of an universal atmosphere to an *experimentum crucis* analogous to that by which Pascal tested the atmospheric theory of Torricelli, I have calculated the theoretical density of the atmosphere of the moon and of each of the planets, and compared the results as severely as I could with the observed facts. As Jupiter is 27,100 times heavier than the moon, and between these wide extremes there are six planets presenting great variations of mass, the probabilities of accidental coincidence are overwhelmingly against me, and a close concurrence of observed telescopic refraction and other phenomena with the theoretical atmospheric density must afford the strongest possible confirmation of the soundness of the basis of my whole argument. Such a concurrence exists, and some new and very curious light is unexpectedly thrown upon the meteorology of Mars and the constitution of the larger planets. The latter, if I am right, must be miniature suns, *permanently* red or white-hot, must be something like a photosphere, surrounded by a sphere of vapor (the outside of which we see), must have mimic spot vortices and prominences, and in the case of Saturn must eject volleys of meteoric matter, some of which should finally settle down into orbital paths, and thus produce the rings.

These are startling conclusions, and when I reached them they were utterly at variance with general astronomical opinion, but I find since their publication that some astronomers have already shown considerable readiness to adopt them. In my case this view of the solar constitution of the larger planets is not a matter of mere opinion, or guessing, or probability, but it follows of necessity, and as stated on page 200, "the great mystery of Saturn's rings is resolved into a simple consequence, a demonstrable and necessary result of the operation of the familiar forces, whose laws of action have been demonstrated here upon the earth by experimental investigation in our laboratories. No strained hypotheses of imaginary forces are required, no ethers or other materials are demanded, beyond those which are beneath our feet and around our heads here upon our own planet; all that is necessary is to grant that the well-known elements and compounds of the chemist, and the demonstrated forces of the experimental physicist, exist and operate in the places, and have the quantities and modes of distribution described by the astronomer; this simple postulate admitted, these wondrous appendages spring into rational existence, and like the eternal fires of the sun, the barren surface of the moon, the dry valleys of Mercury, the hazy equivocations of Venus, the seas and continents and polar glaciers of Mars, and the cloud-covered face of Jupiter, follow as necessary consequences of an universal atmosphere."

If I am right in ascribing a gaseous condition to the sun and the larger planets, and tracing the maintenance of this condition to the disturbing gravitation of the attendant planets or satellites, a solution of the riddle of the nebulæ at once presents itself. We have only to suppose a star cluster or group composed of orbs of solar or great planetary dimensions, and that these act mutually upon each other as the planets on our sun, or the satellites upon Saturn, but in a far more violent degree owing to the far greater relative masses of the reacting elements, and we obtain the conditions under which great gaseous orbs would be not merely pitted on their surface, but riven to their very centres, moulded and shaped throughout by the whirling hurricane of their whole substance. When thus in the centre of a tornado of opposing gravitations the tortured orb would be twisted bodily into a huge vorticose crater, into the bowels of which the aqueous vapor would be dragged and dissociated, and then, entangled with the inner matter of the riven sphere, would be hurled upwards, again to burst forth in an explosion of such magnitude that the original body would be measurably presented as a mere appendage, the rocket case of the flood of fire it had vomited forth.

The reader must complete the picture. If he will take a little trouble in doing so he will find that it becomes a portrait of one or the other of the nebulæ, according to the kind of intergravitating star-cluster from which he starts. I have endeavored to work out some of the details of the nebular conditions in Chapter xx. In Chapter xxi. I have concluded by showing the analogy between a sun and the hydro-electric machine, the sun being the cylinder and the prominences the steam jets. If issuing jets of high-pressure steam have the same properties at a distance of 93 millions of miles from the earth as upon its surface, the body of the sun and the issuing steam must be in opposite electrical conditions, and furious electrical excitation must result; and if the laws of electrical induction are constant throughout the universe, the earth must be as necessarily subject to solar electrical influence as to his thermal radiations. Thus the same reasoning which explains the origin and maintenance of the solar heat and light, the sun-spots, the photosphere, the chromosphere, the sierra, the prominences, the zodiacal light, the aerolites and asteroids; the meteorology of the planets and the rings of Saturn, also shows how the electrical disturbances which produce the aurora borealis and direct the needle may originate.

Electrical theories of the corona and zodiacal light, and their connection of some kind with the aurora borealis, have been put forth in many shapes, but so far as I have learned none afford any explanation of the *origin* of the electrical disturbance. Without this they are like the vortices of Descartes, which explained the movements of the planets by supposing another kind of motion still more incomprehensible.

Explanations which are more difficult to explain than the phenomena they propose to elucidate only obscure the light of true science, and stand as impedimente to the progress of sound philosophy.

Chapter 2

DR. SIEMENS' THEORY OF THE SUN

A paper was read on March 2, 1882, by Dr. C. W. Siemens at the Royal Society, and he published an article on "A New Theory of the Sun" in the April number of the *Nineteenth Century*. All who have read my essay on "The Fuel of the Sun" are surprised at the statement with which the magazine article opens, viz.: that this "may be termed a first attempt to open for the sun a debtor and creditor account, inasmuch as he has hitherto been regarded only as a great almoner pouring forth incessantly his boundless wealth of heat, without receiving any of it back."

Some of my friends suppose that Dr. Siemens has wilfully ignored the most important element of my theory, and have suggested indignation and protest on my part. I am quite satisfied, however, that they are mistaken. I see plainly enough that although Dr. Siemens quotes my book, he had not read it when he did so; that in stating that "Grove, Humboldt, Zoellner, and Mattieu Williams have boldly asserted the existence of a space filled with matter," he derived this information from the paper of Dr. Sterry Hunt which he afterward quotes. This inference has been confirmed by subsequent correspondence with Dr. Siemens, who tells me that he saw the book some years since but *had not read it*. My contributions to the philosophy of solar physics would have been far more widely known and better appreciated had I followed the usual course of announcing firstly "a working hypothesis," to warn others off the ground, then reading a preliminary paper, then another and another, and so on during ten or a dozen years, instead of publishing all at once an octavo volume of 240 pages, which has proved too formidable even to many of those who are specially interested in the subject.

I am compelled to infer that this is the reason why so many of the speculations, which were physical heresies when expounded therein, have since become so generally adopted, without corresponding acknowledgment. This is not the place for specifying the particulars of such adoptions, but I may mention that in due time "An Appendix to the Fuel of the Sun," including the whole history of the subject, will be published. The materials are all in hand, and only await arrangement. In the meantime I will briefly state some of the points of agreement and difference between Dr. Siemens and myself.

In the first place, we both take as our fundamental basis of speculation the idea of an universal extension of atmospheric matter, and we both regard this as the recipient of the diffused solar radiations, which are afterwards recovered and recondensed, or concentrated.

Thus our "fuel of the sun" is primarily the same, but, as will presently be seen, our machinery for feeding the solar furnace is essentially different.

Certain desiccated pedants have sneered at my title, "The Fuel of the Sun," as "sensational," and have refused to read the book on this account; but Dr. Sterry Hunt has provided me with ample revenge. He has disentombed an interesting paper by Sir Isaac Newton, dated 1675, in which the same sensationalism is perpetrated with very small modification, Sir Isaac Newton's title being "Solary Fuel." Besides this, his speculations are curiously similar to my own, his fundamental idea being evidently the same, but the chemistry of his time was too vague and obscure to render its development possible. This paper was neglected and set aside, was not printed in the Transactions of the Royal Society, and remained generally unknown till a few months ago, when the energetic American philosopher brought it forth, and discussed its remarkable anticipations.

Dr. Siemens supposes that the rotation of the sun effects a sort of "fan action," by throwing off heated atmospheric matter from his equatorial regions, which atmospheric matter is afterwards reclaimed and passed over to the polar regions of the sun. This interchange he describes as effected by the differences of pressure on the fluid envelope of the sun; the portion over the polar regions being held down by the whole force of solar gravitation, while the equatorial atmosphere is subject to this pressure, or attraction, minus the centrifugal impulse due to solar rotation. He maintains that this "centrifugal action, however small in amount as compared with the enormous attraction of the sun, would destroy the balance, and determine a motion towards the sun as regards the mass opposite the polar surface, and into space as regards the equatorial mass." He adds that "the equatorial current so produced, owing to its mighty proportions, would flow outwards into space, to a practically unlimited distance."

I will not here discuss the dynamics of this hypothesis; whether the reclaiming action of the superior polar attraction would occur at the vast distances from the sun supposed by Dr. Siemens, or much nearer home, and produce an effect like the recurving of the flame of his own regenerative gas-burner; or, whether he is right in comparing the centrifugal force at the solar equator with that of the earth, by simply measuring the relative velocity of translation irrespective of angular velocity. I will merely suggest that in discussing these, it is necessary, in order to do justice to Dr. Siemens, to always keep in mind the assumed condition of an universal and continuous atmospheric medium, and not to reason, as some have done already, upon the basis of a limited solar atmosphere with a definite boundary, from beyond which particles of atmospheric matter are to be flung away into vacuous space, without the intervention of all-pervading fluid pressure.

It is evident that *if* such fan action can bring back *all* the material that has received the solar radiations, and which holds them either as temperature or otherwise, the restoration and perpetuation of solar energy will be complete, for even the heat received by our earth and its brother and sister planets would still remain in the family, as they would radiate it into the interplanetary atmospheric matter supposed to be reclaimed by the sun.

But, as Mr. Proctor has clearly shown, the rays of the sun cannot do all the work thus required for his own restoration without becoming extinguished as regards the outside universe; and if the other suns—i.e., the stars—do the same they could not be visible to us.

Thus Dr. Siemens' theory removes our sun from his place among the stars, and renders the great problem of stellar radiation more inscrutable than ever by thus putting the evidence of our great luminary altogether out of court.

My theory, on the contrary, demands only a gradual absorption of solar and stellar rays, such as actual observation of their varying splendor indicates.

If space were absolutely transparent, and its infinite depths peopled throughout, the firmament would present to our view one continuous blazing dome, as all the spaces between the nearer stars would be filled by the infinity of radiations from the more distant.

Chapter 3

ANOTHER WORLD DOWN HERE

What a horrible place must this world appear when regarded according to our ideas from an insect's point of view! The air infested with huge flying hungry dragons, whose gaping and snapping mouths are ever intent upon swallowing the innocent creatures for whom, according to the insect, if he were like us, a properly constructed world ought to be exclusively adapted. The solid earth continually shaken by the approaching tread of hideous giants—moving mountains—that crush out precious lives at every footstep, an occasional draught of the blood of these monsters, stolen at life-risk, affording but poor compensation for such fatal persecution.

Let us hope that the little victims are less like ourselves than the doings of ants and bees might lead us to suppose; that their mental anxieties are not proportionate to the optical vigilance indicated by the four thousand eye-lenses of the common house-fly, the seventeen thousand of the cabbage butterfly and the wide-awake dragon-fly, or the twenty-five thousand possessed by certain species of still more vigilant beetles.

Each of these little eyes has its own cornea, its lens, and a curious six-sided, transparent prism, at the back of which is a special retina spreading out from a branch of the main optic nerve, which, in the cockchafer and some other creatures, is half as large as the brain. If each of these lenses forms a separate picture of each object rather than a single mosaic picture, as some anatomists suppose, what an awful army of cruel giants must the cockchafer behold when he is captured by a schoolboy!

The insect must see a whole world of wonders of which we know little or nothing. True, we have microscopes, with which we can see one thing at a time if carefully laid upon the stage; but what is the finest instrument that Ross can produce compared to that with twenty-five thousand object-glasses, all of them probably achromatic, and each one a living instrument, with its own nerve-branch supplying a separate sensation? To creatures thus endowed with microscopic vision, a cloud of sandy dust must appear like an avalanche of massive rock-fragments, and everything else proportionally monstrous.

One of the many delusions engendered by our human self-conceit and habit of considering the world as only such as we know it from our human point of view, is that of supposing human intelligence to be the only kind of intelligence in existence. The fact is, that what we call the lower animals have special intelligence of their own as far transcending our intelligence as our peculiar reasoning intelligence exceeds theirs. We are as incapable of following the track of a friend by the smell of his footsteps as a dog is of writing a metaphysical treatise.

So with insects. They are probably acquainted with a whole world of physical facts of which we are utterly ignorant. Our auditory apparatus supplies us with a knowledge of sounds. What are these sounds? They are vibrations of matter which are capable of producing corresponding or sympathetic vibrations of the drums of our ears or the bones of our skull. When we carefully examine the subject, and count the number of vibrations that produce our world of sounds of varying pitch, we find that the human ear can only respond to a limited range of such vibrations. If they exceed three thousand per second, the sound becomes too shrill for average people to hear it, though some exceptional ears can take up pulsations or waves that succeed each other more rapidly than this.

Reasoning from the analogy of stretched strings and membranes, and of air vibrating in tubes, etc., we are justified in concluding that the smaller the drum or the tube the higher will be the note it produces when agitated, and the smaller and the more rapid the aerial wave to which it will respond. The drums of insect ears, and the tubes, etc., connected with them, are so minute that their world of sounds probably begins where ours ceases; that the sound which appears to us as continuous is to them a series of separated blows, just as vibrations of ten to twelve per second appear to us. We begin to hear such vibrations as continuous sounds when they amount to about thirty per second. The insect's continuous sound probably begins beyond three thousand. The blue-bottle may thus enjoy a whole world of exquisite music of which we know nothing.

There is another very suggestive peculiarity in the auditory apparatus of insects. Its structure and position are something between those of an ear and of an eye. Careful examination of the head, of one of our domestic companions—the common cockroach or black-beetle—will reveal two round white points, somewhat higher than the base of the long outer antennæ, and a little nearer to the middle line of the head. These white projecting spots are formed by the outer transparent membrane of a bag or ball filled with fluid, which ball or bag rests inside another cavity in the head. It resembles our own eye in having this external transparent tough membrane, which corresponds to the cornea or transparent membrane forming the glass of our eye-window; which, like the cornea, is backed by the fluid in an ear-ball corresponding to our eye-ball, and the back of this ear-ball appears to receive the outspreadings of a nerve, just as the back of our eye is lined with that outspread of the optic nerve forming the retina. There does not appear to be in this or other insects a tightly stretched membrane which, like the membrane of our ear-drum, is fitted to take up bodily air-waves and vibrate responsively to them. But it is evidently adapted to receive and concentrate some kind of vibration, or motion, or tremor.

What kind of motion can this be? What kind of perception does this curious organ supply? To answer these questions we must travel beyond the strict limits of scientific induction and enter the fairyland of scientific imagination. We may wander here in safety, provided we always remember where we are, and keep a true course guided by the compass-needle of demonstrable facts.

I have said that the cornea-like membrane of the insect's ear-bag does not appear capable of responding to *bodily* air-waves. This adjective is important, because there are vibratory movements of matter that are not bodily but molecular. An analogy may help to render this distinction intelligible. I may take a long string of beads and shake it into wavelike movements, the waves being formed by the movements of the whole string. We may now conceive another kind of movement or vibration by supposing one bead to receive a blow pushing it forward, this push to be communicated to the next, then to the third, and so on, producing a minute running tremor passing from end to end. This kind of action may be rendered visible by laying a number of billiard balls or marbles in line and bowling an outside ball against the end one of the row. The impulse will be rapidly and invisibly transmitted all along the line, and the outer ball will respond by starting forward.

Heat, light, and electricity are mysterious internal movements of what we call matter (some say "ether," which is but a name for imaginary matter). These internal movements are as invisible as those of the intermediate billiard balls; but if there be a line of molecules acting thus, and the terminal one strikes an organ of sense fitted to receive its motion, some sort of perception may follow. When such movements of certain frequency and amplitude strike our organs of vision, the sensation of light is produced. When others of greater amplitude and smaller frequency strike the terminal outspread of our common sensory nerves, the sensation of heat results. The difference between the frequency and amplitude of the heat waves and the light waves is but small, or, strictly speaking, there is no actual line of separation lying between them; they run directly into each other. When a piece of metal is gradually heated, it is first "black-hot;" this is while the waves or molecular tremblings are of a certain amplitude and frequency; as the frequency increases and amplitude diminishes (or, to borrow from musical terms, as the pitch rises), the metal becomes dull red-hot; greater rapidity, cherry red; greater still, bright red; then yellow-hot and white-hot: the luminosity growing as the rapidity of molecular vibration increases.

There is no such gradation between the most rapid undulations or tremblings that produce our sensation of sound and the slowest of those which give rise to our sensations of gentlest warmth. There is a huge gap between them, wide enough to include another world or several other worlds of motion, all lying between our world of sounds and our world of heat and light, and there is no good reason whatever for supposing that matter is incapable of such intermediate activity, or that such activity may not give rise to intermediate sensations, provided there are organs for taking up and sensifying (if I may coin a desirable word) these movements.

As already stated, the limit of audible tremors is three to four thousand per second, but the smallest number of tremors that we can perceive as heat is between three and four millions of millions per second. The number of waves producing red light is estimated at four hundred and seventy-four millions of millions per second; and for the production of violet light, six hundred and ninety-nine millions of millions. These are the received conclusions of our best mathematicians, which I repeat on their authority. Allowing, however, a very large margin of possible error, the world of possible sensations lying between those produced by a few thousands of waves and any number of millions is of enormous width.

In such a world of intermediate activities the insect probably lives, with a sense of vision revealing to him more than our microscopes show to us, and with his minute eye-like ear-bag sensifying material movements that lie between our world of sounds and our other far-distant worlds of heat and light.

There is yet another indication of some sort of intermediate sensation possessed by insects. Many of them are not only endowed with the thousands of lenses of their compound eyes, but have in addition several curious organs that have been designated "ocelli" and "stemmata." These are generally placed at the top of the head, the thousand-fold eyes being at the sides. They are very much like the auditory organs above described—so much so that in consulting different authorities for special information on the subject I have fallen into some confusion, from which I can only escape by supposing that the organ which one anatomist describes as the ocelli of certain insects is regarded as the auditory apparatus when examined in another insect by another anatomist. All this indicates a sort of continuity of sensation connecting the sounds of the insect world with the objects of their vision.

But these ocular ears or auditory eyes of the insect are not his only advantage over us. He has another sensory organ to which, with all our boasted intellect, we can claim nothing that is comparable, unless it be our olfactory nerve. The possibility of this I will presently discuss.

I refer to the *antennæ*, which are the most characteristic of insect organs, and wonderfully developed in some, as may be seen by examining the plumes of the crested gnat. Everybody who has carefully watched the doings of insects must have observed the curiously investigative movements of the antennæ, which are ever on the alert, peering and prying to right and left and upwards and downwards. Huber, who devoted his life to the study of bees and ants, concluded that these insects converse with each other by movements of the antennæ, and he has given to the signs thus produced the name of "antennal language." They certainly do communicate information or give orders by some means; and when the insects stop for that purpose, they face each other and execute peculiar wavings of these organs that are highly suggestive of the movements of the old semaphore telegraph arms.

The most generally received opinion is that these antennæ are very delicate organs of touch, but some recent experiments made by Gustav Hansen indicate that they are organs of smelling or of some similar power of distinguishing objects at a distance. Flies deprived of their antennæ ceased to display any interest in tainted meat that had previously proved very attractive. Other insects similarly treated appear to become indifferent to odors generally. He shows that the development of the antennæ in different species corresponds to the power of smelling which they seem to possess.

I am sorely tempted to add another argument to those brought forward by Hansen, viz.: that our own olfactory nerves, and those of all our near mammalian relations, are curiously like a pair of antennæ.

There are two elements in a nervous structure—the gray and the white; the gray, or ganglionic portion, is supposed to be the centre or seat of nervous power, and the white medullary or fibrous portion merely the conductor of nervous energy.

The nerves of the other senses have their ganglia seated internally, and bundles of tubular white threads spread outwards therefrom; but not so with the olfactory nervous apparatus. These present two horn-like projections that are thrust forward from the base of the brain, and have white or medullary stems that terminate outwardly or anteriorly in ganglionic bulbs resting upon what I may call the roof of the nose; these bulbs throw out fibres that are composed, rather paradoxically, of more gray matter than white. In some quadrupeds with great power of smell, the olfactory nerves extend so far forward as to protrude beyond the front of the hemispheres of the brain, with bulbous terminations relatively very much larger than those of man.

They thus appear like veritable antennæ. In some of our best works on anatomy of the brain (Solly, for example) a series of comparative pictures of the brains of different animals is shown, extending from man to the cod-fish. As we proceed downwards, the horn-like projection of the olfactory nerves beyond the central hemispheres goes on extending more and more, and the relative magnitude of the terminal ganglia or olfactory lobes increases in similar order.

We have only to omit the nasal bones and nostrils, to continue this forward extrusion of the olfactory nerves and their bulbs and branches, to coat them with suitable sheaths provided with muscles for mobility, and we have the antennæ of insects. I submit this view of the comparative anatomy of these organs as my own speculation, to be taken for what it is worth.

There is no doubt that the antennæ of these creatures are connected by nerve-stalks with the anterior part of their supra-œsophageal ganglia, i.e., the nervous centres corresponding to our brain.

But what kind and degree of power must such olfactory organs possess? The dog has, relatively to the rest of his brain, a much greater development of the olfactory nerves and ganglia than man has. His powers of smell are so much greater than ours that we find it difficult to conceive the possibility of what we actually see him do. As an example, I may describe an experiment I made upon a bloodhound of the famous Cuban breed. He belonged to a friend whose house is situated on an eminence commanding an extensive view. I started from the garden and wandered about a mile away, crossed several fields by sinuous courses, climbing over stiles, and jumping ditches, always keeping the house in view; I then returned by quite a different track. The bloodhound was set upon the beginning of my track. I watched him from a window galloping rapidly, and following all its windings without the least halting or hesitation. It was as clear to his nose as a gravelled path or a luminous streak would be to our eyes. On his return I went down to him, and without approaching nearer than five or six yards, he recognized me as the object of his search, proving this by circling round me, baying deeply and savagely though harmlessly, as he always kept at about the same distance.[8]

If the difference of development between the human and canine internal antennæ produces all this difference of function, what a gulf may there be between our powers of perceiving material emanations and those possessed by insects! If my anatomical hypothesis is correct, some insects have protruding nasal organs or out-thrust olfactory nerves as long as all the rest of their bodies. The power of movement of these in all directions affords the means of sensory communication over a corresponding range, instead of being limited merely to the direction of the nostril openings. In some insects, such as the plumed gnat, the antennæ do not appear to be thus moveable, but this want of mobility is more than compensated by the multitude of branchings of these wonderful organs, whereby they are simultaneously exposed in every direction. This structure is analogous to the fixed but multiplied eyes of insects, which, by seeing all round at once, compensate for the want of that mobility possessed by others that have but a single eyeball mounted on a flexible and mobile stalk; that of the spider, for example.

Such an extension of such a sensory function is equivalent to living in another world of which we have no knowledge and can form no definite conception. We, by our senses of touch and vision, know the shapes and colors of objects, and by our very rudimentary olfactory organs form crude ideas of their chemistry or composition, through the medium of their material emanations; but the huge exaggeration of this power in the insect should supply him with instinctive perceptive powers of chemical analysis, a direct acquaintance with the

---

[8] What did he smell? Was it an emanation from the soles of my feet? If so, how did this aura get through the soles of my boots, which were thick? It could scarcely have been the odor of the boot soles themselves that he followed, as he recognized me afterwards at some distance. This suggests an interesting experiment, that anybody owning one of these dogs may easily try. Make a similar track to mine, but when on the way, take off the boots you wore on starting and change them for some one else's boots, or a new pair, and watch the result from the window.

inner molecular constitution of matter far clearer and deeper than we are able to obtain by all the refinements of laboratory analyses or the hypothetical formulating of molecular mathematicians. Add this to the other world of sensations producible by the vibratory movements of matter lying between those perceptible by our organs of hearing and vision, then strain your imagination to its cracking point, and you will still fail to picture the wonderland in which the smallest of our fellow-creatures may be living, moving, and having their being.

Chapter 4

THE ORIGIN OF LUNAR VOLCANOES

Many theoretical efforts, some of considerable violence, have been made to reconcile the supposed physical contradiction presented by the great magnitude and area of former volcanic activity of the Moon, and the present absence of water on its surface. So long as we accept the generally received belief that water is a necessary agent in the evolution of volcanic forces, the difficulties presented by the lunar surface are rather increased than diminished by further examination and speculation.

We know that the lava, scoriæ, dust and other products of volcanic action on this earth are mainly composed of mixed silicates—those of alumina and lime preponderating. When we consider that the solid crust of the Earth is chiefly composed of silicic acid, and of basic oxides and carbonates which combine with silicic acid when heated, a natural necessity for such a composition of volcanic products becomes evident.

If the Moon is composed of similar materials to those of the Earth, the fusion of its crust must produce similar compounds, as they are formed independently of any atmospheric or aqueous agency.

This being the case, the phenomena presented by the cooling of fused masses of mixed silicates in the absence of water become very interesting. Opportunities of studying such phenomena are offered at our great iron-works, where fused masses of iron cinder, composed mainly of mixed silicates, are continually to be seen in the process of cooling under a variety of circumstances.

I have watched the cooling of such masses very frequently, and have seen abundant displays of miniature volcanic phenomena, especially marked where the cooling has occurred under conditions most nearly resembling those of a gradually cooling planet or satellite; that is, when the fused cinder has been enclosed by a solid resisting and contracting crust.

The most remarkable that I have seen are those presented by the cooling of the "tap cinder" from puddling furnaces. This, as it flows from the furnace, is received in stout iron boxes ("cinder-bogies") of circular or rectangular horizontal section. The following phenomena are usually observable on the cooling of the fused cinder in a circular bogie.

First a thin solid crust forms on the red-hot surface. This speedily cools sufficiently to blacken. If pierced by a slight thrust from an iron rod, the red-hot matter within is seen to be in a state of seething activity, and a considerable quantity exudes from the opening. If a bogie filled with fused cinder is left undisturbed, a veritable spontaneous volcanic eruption takes place through some portion, generally near the centre, of the solid crust. In some cases, this eruption is sufficiently violent to eject small spurts of molten cinder to a height equal to four or five diameters of the whole mass.

The crust once broken, a regular crater is rapidly formed, and miniature streams of lava continue to pour from it; sometimes slowly and regularly, occasionally with jerks and spurts due to the bursting of bubbles of gas. The accumulation of these lava-streams forms a regular cone, the height of which goes on increasing. I have seen a bogie about 10 or 12 inches in diameter, and 9 or 10 inches deep, thus surmounted by a cone above 5 inches high, with a base equal to the whole diameter of the bogie. These cones and craters could be but little improved by a modeler desiring to represent a typical volcano in miniature.

Similar craters and cones are formed on the surface of cinder which is not confined by the sides of the bogie. I have seen them well displayed on the "running-out beds" of refinery furnaces. These, when filled, form a small lake of molten iron covered with a layer of cinder. This cinder first skins over, as in the bogies, then small crevasses form in this crust, and through these the fused cinder oozes from below. The outflow from this chasm soon becomes localized, so as to form a single crater, or a small chain of craters; these gradually develop into cones by the accumulation of outflowing lava, so that when the whole mass has solidified, it is covered more or less thickly with a number of such hillocks. These, however, are much smaller than in the former case, reaching to only one or two inches in height, with a proportionate base. It is evident that the dimensions of these miniature volcanoes are determined mainly by the depth of the molten matter from which they are formed. In the case of the bogies, they are exaggerated by the overpowering resistance of the solid iron bottom and sides, which force all the exudation in the one direction of least resistance, viz., towards the centre of the thin upper crust, and thus a single crater and a single cone of the large relative dimensions above described are commonly formed.

The magnitude and perfection of these miniature volcanoes vary considerably with the quality of the pig-iron and the treatment it has received, and the difference appears to depend upon the evolution of gases, such as carbonic oxide, volatile chlorides, fluorides, etc. I mention the fluorides particularly, having been recently engaged in making some experiments on Mr. Henderson's process for refining pig-iron, by exposing it when fused to the action of a mixture of fluoride of calcium and oxides of iron, alumina, manganese, etc. The cinder separated from this iron displayed the phenomena above described very remarkably, and jets of yellowish flame were thrown up from the craters while the lava was flowing. The flame was succeeded by dense white vapors as the temperature of the cinder lowered, and a deposit of snow-like, flocculent crystals was left upon and around the mouth or crater of each cone. The miniature representation of cosmical eruptions was thus rendered still more striking, even to the white deposit of the haloid salts which Palmieri has described as remaining after the recent eruption of Vesuvius.

The gases thus evolved have not yet been analytically examined, and the details of the powerful reactions displayed in this process still demand further study; but there can be no doubt that the combination of silicic acid with the base of the fluor spar is the fundamental reaction to which the evolution of the volatile fluorides, etc., is mainly due.

A corresponding evolution of gases takes place in cosmical volcanic action, whenever silicic acid is fused in contact with limestone or other carbonate, and a still closer analogy is presented by the fusion of silicates in contact with chlorides and oxides, in the absence of water. If the composition of the Moon is similar to that of the Earth, chlorides of sodium, etc., must form an important part of its solid crust; they should correspond in quantity to the great deposit of such salts that would be left behind if the ocean of the Earth were evaporated to dryness. The only assumptions demanded in applying these facts to the explanation of the surface configuration of the Moon are, 1st, that our satellite resembles its primary in chemical composition; 2d, that it has cooled down from a state of fusion; and 3d, that the magnitude of the eruptions, due to such fusion and cooling, must bear some relation to the quantity of matter in action.

The first and second are so commonly made and understood, that I need not here repeat the well-known arguments upon which they are supported, but may remark that the facts above described afford new and weighty evidence in their favor.

If the correspondence between the form of a freely suspended and rotating drop of liquid and that of a planet or satellite is accepted as evidence of the exertion of the same forces of cohesion, etc., on both, the correspondence between the configuration of the lunar surface, and that of small quantities of fused and freely cooled earth-crust matter, should at least afford material support to the otherwise indicated inference, that the materials of the Moon's crust are similar to those of the Earth's, and that they have been cooled from a state of fusion.

I think I may safely generalize to the extent of saying, that no considerable mass of fused earthy silicates can cool down under circumstances of free radiation without first forming a heated solid crust, which, by further radiation, cooling, and contraction, will assume a surface configuration resembling more or less closely that of the Moon. Evidence of this is afforded by a survey of the spoil-banks of blast furnaces, where thousands of blocks of cinder are heaped together, all of which will be found to have their upper surfaces (that were freely exposed when cooling) corrugated with radiating miniature lava streams, that have flowed from one or more craters or openings that have been formed in the manner above described.

The third assumption will, I think, be at once admitted, inasmuch as it is but the expression of a physical necessity.

According to this, the Earth, if it has cooled as the Moon is supposed to have done, should have displayed corresponding irregularities, and generally, the magnitude of mountains of solidified planets and satellites should be on a scale proportionate to their whole mass. In comparing the mountains of the Moon and *Mercury* with those of the Earth, a large error is commonly made by taking the customary measurements of terrestrial mountain-heights from the sea-level. As those portions of the Earth which rise above the waters are but its upper mountain slopes, and the ocean bottom forms its lower plains and valleys, we must add the greatest ocean depths to our customary measurements, in order to state the full height of what remains of the original mountains of the Earth. As all the stratified rocks have been formed by the wearing down of the original upper slopes and summits, we cannot expect to be able to recognize the original skeleton form of our water-washed globe.

If my calculation of the atmosphere of *Mercury* is correct, viz., that its pressure is equal to about one seventh of the Earth's, or 4¼ inches of mercury, there can be no liquid water on that planet, excepting perhaps over a small amount of circumpolar area, and during the extremes of its aphelion winter. Thus the irregularities of the terminator, indicating mountain elevations calculated to reach to 1/253 of the diameter of the planet, are quite in accordance with the above-stated theoretical consideration.

There is one peculiar feature presented by the cones of the cooling cinder which is especially interesting. The flow of fused cinder from the little crater is at first copious and continuous; then it diminishes and becomes alternating, by a rising and falling of the fused mass within the cone. Ultimately the flow ceases, and then the inner liquid sinks, more or less, below the level of the orifice. In some cases, where much gas is evolved, this sinking is so considerable as to leave the cone as a mere hollow shell; the inner liquid having settled down and solidified with a flat or slightly rounded surface, at about the level of the base of the cone, or even lower. These hollow cones were remarkably displayed in some of the cinder of the Henderson iron, and their formation was obviously promoted by the abundant evolution of gas.

If such hollow cones were formed by the cooling of a mass like that of the Moon, they would ultimately and gradually subside by their own weight. But how would they yield? Obviously by a gradual hinge-like bending at the base towards the axis of the cone. This would occur with or without fracture, according to the degree of viscosity of the crust, and the amount of inclination. But the sides of the hollow-cone shell, in falling towards the axis, would be crushing into smaller circumferences. What would result from this? I think it must be the formation of fissures, extending, for the most part, radially from the crater towards the base, and a crumpling up of the shell of the cone by foldings in the same direction. Am I venturing too far in suggesting that in this manner may have been formed the mysterious rays and rills that extend so abundantly from several of the lunar craters?

The upturned edges or walls of the broken crust, and the chasms necessarily gaping between them, appear to satisfy the peculiar phenomena of reflection which these rays present. These edges of the fractured crust would lean towards each other, and form angular chasms; while the foldings of the crust itself would form long concave troughs, extending radially from the crater.

These, when illuminated by rays falling upon them in the direction of the line of vision, must reflect more light towards the spectator than does the general convex lunar surface, and thus they become especially visible at the full Moon.

Such foldings and fractures would occur after the subsidence and solidification of the lava-forming liquid—that is, when the formation of new craters had ceased in any given region; hence they would extend across the minor lateral craters formed by outbursts from the sides of the main cone, in the manner actually observed.

The fact that the bottoms of the great walled craters of the Moon are generally lower than the surrounding plains must not be forgotten in connection with this explanation.

I will not venture further with the speculations suggested by the above-described resemblances, as my knowledge of the details of the telescopic appearances of the Moon is but second-hand. I have little doubt, however, that observers who have the privilege of direct familiarity with such details, will find that the phenomena presented by the cooling of iron cinder, or other fused silicates, are worthy of further and more careful study.

Chapter 5

NOTE ON THE DIRECT EFFECT OF SUN-SPOTS ON TERRESTRIAL CLIMATES

Professor Langley determines quantitatively the effects respectively produced by the radiations from the solar spots, penumbra, and photosphere upon the face of a thermopile, and infers that these effects measure their relative influence on terrestrial climate.

In thus assuming that the heat communicated to the thermopile measures the solar contribution to terrestrial climate, Professor Langley omits an important factor, viz., the amount of heat absorbed in traversing the earth's atmosphere; and in measuring the relative efficiency of the spots, penumbra, and photosphere, he has not taken into account the variations of diathermancy of the intervening atmospheric matter, which are due to the variations in the source of heat.

Speaking generally, it may be affirmed that the radiations of obscure heat are more largely absorbed by the gases and vapors of our atmosphere than those of luminous heat, and the great differences in the mere luminosity of the spots, penumbra, and photosphere justify the assumption that the radiations of a sun-spot will (to use the expressive simile of Tyndall) lose far more by atmospheric sifting than will those from the photosphere.

But the spot areas will be none the less effective on terrestrial climate on that account. A given amount of heat arrested by the earth's atmosphere will have even greater climatic efficiency than if received upon its solid surface, inasmuch as the gases are worse radiators than the rocks, and will therefore, *cæteris paribus*, retain a larger proportion of the heat they receive.

I have long ago endeavored to show[9] that the depth of the photosphere, from the solar surface inwards, is limited by dissociation; that the materials of the Sun within the photosphere exist in a dissociated, elementary condition; that at the photosphere they are, for the most part, combined. This view has since been adopted by many eminent solar physicists, and if correct, demands a much higher temperature within the depths revealed by that withdrawal of the photospheric veil which constitutes a sun-spot.

If I am right in this, and also in supposing the spot-radiations to be so much more abundantly absorbed than those of the photosphere, and if in spite of this higher temperature of the spots, the *surface* of the earth receives from them the lower degree of heat measured by Professor Langley, another interesting consequence must follow. The excess of spot-heat

---

[9] "The Fuel of the Sun," Chapters iv. to x.

directly absorbed by the atmosphere, and mainly by the water dissolved or suspended in its upper regions, must be especially effective in dissipating clouds and checking or modifying their formation. The meteorological results of this may be important, and are worthy of careful study.

In thus venturing to question some of Professor Langley's inferences I am far from underrating the interest and importance of his researches. On the contrary, I regard the quantitative results he has obtained as especially valuable and opportune, in affording means of testing the above-named and other speculations in solar physics. Similar observations repeated at different elevations would decide, so far as the lower regions are concerned, whether or not there is any difference in the quantity of heat imparted by the bright and obscure portions of the Sun to our atmosphere. If the differences already observed by Professor Langley vary in ascending, a new means will be afforded of studying the constitution of the interior of the Sun and its relations to the photosphere. Direct evidence of selective absorption by our atmosphere may thus be obtained, which would go far towards solving one of the crucial solar problems, viz., whether the darker regions are hotter or cooler than the photosphere.

The obscure radiations from the moon must be absorbed by our atmosphere like those from the sun-spot, and may be sufficiently effective to account for the alleged dissipation of clouds by the full moon.

In both cases the climatic influence is greatly heightened by the fact that all the heat thus absorbed is directly effective in raising the temperature of the air. The action of the absorbed heat in reference to cloud-formation is directly opposite to that of the transmitted solar heat, as this reaching the surface of the earth evaporates the superficial water, and thereby produces the material of clouds. On the other hand, the heat which is absorbed by the air increases its vapor-holding capacity, and thus prevents the formation of clouds, or even effects the dissolution of clouds already formed.

Chapter 6

THE PHILOSOPHY OF THE RADIOMETER AND ITS COSMICAL REVELATIONS

So much speculation, and not a little extravagant speculation, has been devoted to the dynamics of the radiometer, that I feel some compunction in adding another stone to the heap, my only apology and justification for so doing being that I propose to regard the subject from a very unsophisticated point of view, and with somewhat heretical directness of vision—i.e., quite irrespective of atoms, molecules, or ether, or any other specific preconceptions concerning the essential kinetics of radiant forces, beyond that of regarding such forces as affections or conditions of matter which are transmitted radially in constant quantity, and therefore obey the necessary law of radial diffusion or inverse squares.

The primary difficulty which appears to have generally been suggested by the movements of the radiometer, is the case which it seems to present of mechanical action without any visible basis of corresponding reaction: a visible tangible object pushed forward, without any visible pushing agent or resisting fulcrum against which the moving body reacts.

This difficulty has been met by the invocation of obedient and vivacious molecules of residual atmospheric matter, which have been called upon to bound and rebound between the vanes and the inner surfaces of the glass envelope of the instrument.

How is it that the advocates of these activities have not sought to verify their speculations by modifying the shape and dimensions of the exhausted glass bulb or receiver?[10] If the motion of the radiometer is due to such excursions and collisions, the length of excursion and the angles of collision must modify its motions; and such modification under given conditions would form a fine subject for the exercise of the ingenuity of molecular mathematicians. If their hypothetical data are sound, they should be able to predict the relative velocities or torsion-force of a series of radiometers of similar construction in all other respects, but with variable shapes and diameters of enclosing vessels.

If we divest our minds of all visions of hypothetical atoms, molecules, ethers, etc., and simply look at the facts of radiation with the same humility of intellect as we usually regard gravitation, this primary difficulty of the radiometer at once vanishes. The force of gravitation is a radiant force acting somehow between, or upon, or by distant bodies; and these bodies, however far apart, act and react upon each other with mutual forces, precisely equal and exactly contrary. We conceive the sun pulling the earth in a certain direction, and receiving

---

[10] Since this was written some such modifications have been made with equivocal results.

from the earth an equal pull in a precisely contrary direction, and we have hitherto demanded no ethereal or molecular link for the transmission of these mutually attractive forces. Why, then, should we not regard radiant repulsive energy in the same simple manner?

If we do this there is no difficulty in finding the ultimate reaction fulcrum of the radiometer vanes. It is simply the radiating body, the match, the candle, the lamp, the sun, or whatever else may be the source of the impelling radiations. According to this view, the radiant source must be repelled with precisely the same energy as the arms or pendulum of the radiometer; and it would move backward or in opposite direction if equally free to move. If, by any means, we cause the glass envelope of the radiometer to become the radiant source, it should be repelled, and may even rotate in opposite direction to the vanes, or *vice versâ*. This has been done with floating radiometers.

Viewed thus as simple matter of fact, irrespective of any preconceived kinetics of intervening media, the net result of Mr. Crookes's researches become nothing less than the discovery of a new law of nature of great magnitude and the broadest possible generality, viz., that the sun and all other radiant bodies—i.e., all the materials of the universe—exert a mechanical repulsive force, in addition to the calorific, luminous, actinic, and electrical forces with which they have hitherto been credited. He has shown that this force is refrangible and dispersible, that it is outspread with the spectrum, but is most concentrated, or active, in the region of the ultra-red rays, and progressively feeblest in the violet; or, otherwise stated, it exists in closer companionship with heat than with light, and closer with light than with actinism.

According to the doctrine of exchanges, which has now passed from the domain of theory to that of demonstrated law, all bodies, whatever be their temperature, are perpetually radiating heat-force, the amount of which varies, *cæteris paribus*, with their temperature. If we now add to this generalization that all bodies are similarly radiating mechanical force and suffering corresponding mechanical reaction, the theoretical difficulties of the radiometer vanish. What must follow in the case of a freely suspended body unequally heated on opposite sides?

It must be repelled in a direction perpendicular to the surface of its hottest side. If two rockets were affixed to opposite sides of a pendant body, and were to exert unequal ejective forces, the reaction of the stronger rocket would repel the body in the opposite direction to its preponderating ejection. This represents the radiometer vane with one side blackened and the other side bright. When exposed to luminous rays the black side becomes warmer than the bright side by its active absorption and conversion of light into heat, and thus the blackened face radiates in excess and recedes.

We may regard it thus as acting by its own radiations, or otherwise as acted upon by the more powerful radiant whose rays are differentially received by the black and bright sides. These different modes of regarding the action are perfectly consistent with each other, and analogous to the two different modes of regarding gravitation, when we describe the sun as attracting the earth, or, otherwise, the earth as gravitating to the sun. Strictly speaking, neither of these descriptions is correct, as the gravitation is mutual, and the total quantity exerted between the sun and the earth is equal to the sum of their energies, but it is sometimes convenient to regard the action from a solar standpoint, and at others from a terrestrial. So with the radiometer and the strictly mutual repulsions between it and the predominating radiant.

It appears to me that this unsophisticated conception of radiant mechanical repulsive force, and its necessary mechanical reaction on the radiant body, meets all the facts at present revealed by the experiments of Mr. Crookes and others.

The attraction which occurs when the disc of the radiometer is surrounded with a considerable quantity of atmospheric matter is probably due to inequality of atmospheric pressure. The absorbing face of the disc becomes heated above the temperature of the opposite face, the film of air in contact with the warmer face rises, leaving a relatively vacuous space in front. This produces a rush of air from back to front which carries the radiometer vane with it. When the exhaustion of the radiometer is carried so far that the residual air is only just sufficiently dense to neutralize the direct repulsion of radiation, the neutral point is reached. When exhaustion is carried beyond this, repulsion predominates.

Taking Mr. Crookes's estimate of the mechanical energy of solar radiation at 32 grains per square foot, 2 cwts. per acre, 57 tons per square mile, etc., and accepting these as they are offered, i.e., merely as provisional and approximate estimates, we are led to a cosmical inference of the highest importance, one that must materially modify our interpretations of some of the grandest phenomena of the universe. Although the estimated sunlight pressure upon the earth, the three thousand millions of tons, is too small a fraction of the earth's total weight to effect an easily measurable increase of the length of our year, the case is quite otherwise with the asteroids and the zones of meteoric matter revolving around the sun.

The mechanical repulsion of radiation is a superficial action, and must, therefore, vary with the amount of surface exposed, while that of gravitation varies with the mass. Thus the ratio of radiant repulsion to the attraction of gravitation goes on increasing with the subdivison of masses, and becomes an important fraction in the case of the smaller bodies of the solar system. A zone of meteorites traveling around the sun would be broken up, sifted, and sorted into different orbits, according to their diameters, if this superficial repulsion operated against gravitation without any compensating agency. Gravitation would be opposed in various degrees, neutralized, and, in the case of cosmic dust, even reversed. Comets presenting so large a surface in proportion to their mass would either be driven away altogether or forced to move in orbits utterly disobedient to present calculations. This would occur if the inter-planetary spaces were as nearly vacuous as the torsion instrument with which Mr. Crookes made his measurements.

Regarding the properties of our atmosphere only in the light of experimental data, irrespective of imaginary molecules, and their supposed gyrations or oscillations, we see at once that an inter-planetary or inter-stellar vacuum must act like a Sprengel pump upon our atmosphere, upon the atmosphere of other planets, and upon those of the sun and the stars, and would continue such action until an equilibrium between the repulsive energy of the gas and the gravitation of the solid orbs had been established. Atmospheric matter would thus be universally diffused, with special accumulations around solid orbs, varying in quantity with their respective gravitating energy. Such a universal atmosphere would accelerate orbital motion, and this acceleration would vary with the surface of bodies. Its action being thus exactly opposed to that of radiant repulsion, it must, at a certain density, exactly neutralize it. That it does this is evident from the obedience of all the elements of the solar system to the calculated action of gravitation; and thus Mr. Crookes's researches not only confirm the idea of universal atmospheric diffusion, but they afford a means by which we may ultimately measure the actual density of the universal atmosphere. If, as I have endeavored to show in my essay on "The Fuel of the Sun," the initial radiant energy of every star depends upon its mass, and its consequent condensation of atmospheric matter, the density of inter-planetary atmosphere sufficient to neutralize the radiant mechanical energy of our sun may be the same as is demanded to perform the same function for all the stars of the universe, and all their attendant worlds, comets, and meteors.

In order to prevent misunderstanding of the above, I must add that I have therein studiously assumed a negative position in reference to all hypothetical conceptions of the nature of heat, light, etc., and their modes of transmission, simply because I feel satisfied that the subject has hitherto been obscured and complicated by overstrained efforts to fit the phenomena to the excessively definite hypotheses of modern molecular mathematicians. The atoms invented by Dalton for the purpose of explaining the demonstrated laws of chemical combination performed this function admirably, and had great educational value, so long as their purely imaginary origin was kept in view; but when such atoms are treated as facts, and physical dogmas are based upon the assumption of their actual existence, they become dangerous physical superstitions. Regarding matter as continuous, i.e., supposing it to be simply as it appears to be, and co-extensive with the universe, in accordance with the experimental evidences of the unlimited expansibility of gaseous matter, we need only assume that our sensations of heat, light, etc., are produced by active conditions of such matter analogous to those which are proved to produce our sensations of sound. On this basis there is no difficulty in conceiving the rationale of the reaction which produces the repulsion of the radiometer. I may even go further, and affirm that it is impossible to rationally conceive radiation producing any mechanical effects without mechanical reaction. If heat be motion, and actual motion of actual matter, mechanical force must be exerted to produce it, and a body which is warmer on one side than the other, i.e., which is exerting more outward motion-producing force on one side than on the other, must be subject to proportionally unequal reaction, and, therefore, if free to move, must retreat in a direction contrary to that of its greater activity. Regarded thus, the residual air of the radiometer does act, not by collisions of particles between the vane and inside of the glass vessel, but by the direct reaction of the radiant energy which would operate irrespective of vessels, i.e., upon naked radiometer vanes if carried halfway to the moon, or otherwise freed from excess of atmospheric embarrassment.

The recent experiments of Mr. Crookes, showing retardation of the radiometer with extreme exhaustion, seem to indicate that heat-rays, like the electric discharge, demand a certain amount of atmospheric matter as their carrier.

I cannot conclude these hasty and imperfect notes, written merely with suggestive intent, without quoting a passage from the preface to the "Correlation of Physical Forces," which, though written so long ago, appears to me worthy of the profoundest present consideration.

"It appears to me that heat and light maybe considered as affections; or, according to the undulatory theory, vibrations of matter itself, and not of a distinct ethereal fluid permeating it: these vibrations would be propagated just as sound is propagated by vibrations of wood or as waves by water. To my mind all the consequences of the undulatory theory flow as easily from this as from the hypothesis of a specific ether; to suppose which, namely, to suppose a fluid *sui generis* and of extreme tenuity penetrating solid bodies, we must assume, first, the existence of the fluid itself; secondly, that bodies are without exception porous; thirdly, that these pores communicate; fourthly, that matter is limited in expansibility. None of these difficulties apply to the modification of this theory which I venture to propose: and no other difficulty applies to it which does not equally apply to the received hypothesis."

Chapter 7

ON THE SOCIAL BENEFITS OF PARAFFIN

To the inhabitants of Jupiter, who have always one, two, or three of their four moons in active and efficient radiation, or of Saturn displaying the broad luminous oceans of his mighty rings in addition to the minor lamps of his eight ever-changeful satellites, the relative merits of rushlights, candles, lamps, and gaslights may be a question of indifference; but to us, the residents of a planet which has but one small moon that only displays her nearly full face during a few nights of each month, the subject of artificial light is only second in importance to those of food and artificial heat, and every step that is made in the improvement of our supplies of this primary necessary must have a momentous influence on the physical comfort, and also upon the intellectual and moral progress, of this world's human inhabitants.

If a cockney Rip Van Winkle were to revisit his old haunts, the changes produced by the introduction of gas would probably surprise him the most of all he would see. He would be astonished to find respectable people, and even unprotected females, going alone, unarmed and without fear, at night, up the by-streets which in his days were deemed so dangerous, and he would soon perceive that the bright gaslights had done more than all the laws, the magistrates, and the police, to drive out those crimes which can only flourish in darkness. The intimate connection between physical light and moral and intellectual light and progress is a subject well worthy of an exhaustive treatise.

We must, however, drop the general subject and come down to our particular paraffin lamp. In the first place, this is the cheapest light that has ever been invented—cheaper than any kind of oil lamp—cheaper than the cheapest and nastiest of candles, and, for domestic purposes, cheaper than gas. For large warehouses, shops, streets, public buildings, etc., it is not so cheap as gas should be, but is considerably cheaper than gas actually is at the price extorted by the despotism of commercial monopoly.

The reason why it is especially cheaper for domestic purposes is, first, because the small consumer of gas pays a higher price than the large consumer; and secondly, because a lamp can be placed on a table or wherever else its light is required, and therefore a small lamp flame will do the work of a much larger gas flame. We must remember that the intensity of light varies inversely with the square of the distance from the source of light; thus the amount of light received by this page from a light at one foot distance is four times as great as if it were two feet distant, nine times as great as at three feet, sixteen times as great as at four feet, one hundred times as great as at ten feet, and so on. Hence the necessity of two or three great flames in a gas chandelier suspended from the ceiling of a moderate-sized room.

In a sitting-room lighted thus with gas, we are obliged, in order to read comfortably by the distant source of light, to burn so much gas that the atmosphere of the room is seriously polluted by the products of this extravagant combustion. A lamp at a moderate distance—say eighteen inches or two feet, or thereabouts—will enable us to read or work with one-tenth to one-twentieth the amount of combustion, and therefore with so much less vitiation of the atmosphere, and, if we use a paraffin lamp, at much less expense.

But the chief value of the paraffin lamp is felt where gas is not obtainable—in the country mansion or villa, the farmhouse, and, most of all, in the poor man's cottage. We have Bible Societies for providing cheap Bibles; we have cheap standard works, cheap magazines, cheap newspapers, etc.; but all these are unavailable to the poor man until he can get a good and cheap light wherewith to read them at the only time he has for reading, viz., in the evenings, when his work is done. One shilling's worth of cheap literature will require two shillings' worth of dear candles to supply the light necessary for reading it. Therefore, the cheapening of light has quite as much to do with the poor man's intellectual progress as the cheapening of books and periodicals.

For a man to read comfortably, and his wife to do her needlework, they must have a candle for each, if dependent on tallow dips. They may, and do, struggle on with one such candle, but the inconvenience soon sickens them of their occupation; the man lolls out for an idle stroll, soon encounters a far more bright and cheerful room than the gloomy one he has just left, and, moth-like, he is attracted by the light, and finishes up his evening in the public-house.

We may preach, we may lecture, we may coax, wheedle, or anathematize, but no amount of words of any kind will render a gloomy ill-lighted cottage so attractive as the bright bar and tap-room; and human nature, irrespective of conventional distinctions of rank and class, always seeks cheerfulness after a day of monotonous toil. Fifty years ago the middle classes were accustomed to spend their evenings in taverns, but now they prefer their homes, simply because they have learned to make their homes more comfortable and attractive.

We have not yet learned how to supply the working millions with suburban villas, but if their small rooms can be made bright and cheerful during the long evenings, a most important step is made towards that general improvement of social habits which necessarily results from a greater love of home. We may safely venture to predict that the paraffin lamp will have as much influence in elevating the domestic character of the poorer classes as the street lamps have had in purging the streets of our cities from the crimes of darkness that once infested them.

A great deal has been said about the poisonous character of paraffin works. I admit that they have much to answer for in reference to trout—that the clumsy and wasteful management of certain ill-conducted works has interfered with the sport of the anglers of one or two of the trout streams of the United Kingdom—but all the assertions that have been made relative to injury to human health are quite contrary to truth.

The fact is that the manufacture of mineral oils from cannel and shale is an unusually healthful occupation. The men certainly have dirty faces, but are curiously exempt from those diseases which are most fatal among the poor. I allude to typhus fever, and all that terrible catalogue of ills usually classed under the head of zymotic diseases. This has been strikingly illustrated in the Flintshire district. The very sudden development of the oil trade in the neighborhood of Leeswood caused that little village and the scattered cottages around to be crowded to an extent that created the utmost alarm among all who are familiar with the results of such overcrowding in poor, ill-drained, and ill-ventilated cottages. Rooms were commonly filled with lodgers who economized the apartments on the Box and Cox principle, the night workers sleeping during the day, and the day workers during the night, in the same beds. The extent to which this overcrowding was carried in many instances is hardly credible.

Mr. R. Platt, who is surgeon to most of the collieries and oilworks of this district, reports that Leeswood has enjoyed a singular immunity from typhus and fever—that, during a period when it was prevalent as a serious epidemic among the agricultural population living on the slopes of the surrounding mountains, no single case occurred among the oil-making population of Leeswood, though its position and overcrowding seemed so directly to court its visitation. If space permitted I might give further illustrations in reference to allied diseases.

There is no difficulty in accounting for this. Carbolic acid, one of the most powerful of our disinfectants, is abundantly produced in the oilworks, and this is carried by the clothes of the men, and with the fumes of the oil, into the dwellings of the workmen and through all the atmosphere of the neighborhood, and has thereby counteracted some of the most deadly agencies of organic poisons. Besides this, the paraffin oil itself is a good disinfectant.

Even the mischief done to the trout is more than counterbalanced by the destruction of those mysterious fungoid growths which result from the admixture of sewage matter with the water of our rivers, and are so destructive to human health and life. The carbolic acid and paraffin oil, in destroying these as well as the trout, are really acting as great purifiers of the river, so that, after all, the only interest that has suffered is the sporting interest. This same interest has otherwise suffered. The old haunts of the snipe and woodcock, of partridges, hares, and pheasants, are being ruthlessly and barbarously destroyed, and—horrible to relate—hundreds of cottages, inhabited by vulgar, hard-handed, thick-booted human beings, are taking their place. Churches are being extended, school-houses and chapels built; penny readings, lectures, concerts, etc., are in active operation, and even drinking fountains are in course of construction; but the trout have suffered and the woodcocks are gone.

We may thus measure the good against the evil as it stands here in the headquarters of oil-making, and should add to one side the advantages which the cheap and brilliant light affords—advantages which we might continue to enumerate, but they are so obvious that it is unnecessary to go further.

There is one important and curious matter which must not be omitted. This, like the moral and intellectual advantages of the cheap paraffin light, has hitherto remained unnoticed, viz., that the introduction of mineral oils and solid paraffin for purposes of illumination and lubrication has largely increased the world's supply of food.

This may not be generally obvious at first sight; but to him, who, like the writer, has had many a supper at an Italian osteria with peasants and carbonari, it is obvious enough. He will remember how often he has seen the lamp that has lighted himself and companions to their supper filled from the same flask as supplied the salad which formed so important a part of the supper itself. Throughout the South of Europe salads are most important elements of national food, and when thus abundantly eaten the oil is quite necessary, the oil is also used for many of the cookery operations where butter is used here, and this same olive oil has hitherto been the chief, and in some places the sole, illuminating agent. The poor peasant of the South looks jealously at his lamp, and feeds its stingily, for it consumes his richest and choicest food, and, if well supplied, would eat as much as a fair-sized baby.

The Russian peasant and other Northern people have a similar struggle in the matter of tallow. It is their choicest dainty, and yet, to their bitter grief, they have been compelled to burn it. Hundreds and thousands of tons of this and of olive oil have been annually consumed for the lubrication of our steam engines and other machines. A better time is approaching now that paraffin lamps are so rapidly becoming the chief illuminators of the whole civilized world, superseding the crude tallow candle and the antique olive-oil lamp, while, at the same time, the tallow candle is gradually being replaced by the beautiful sperm-like paraffin candle; and, in addition to this, the greedy engines that have consumed so much of the olive oil and the tallow are learning to be satisfied with lubricators made from minerals kindred to themselves.

The peasants of the sunny South will feed upon salads made doubly unctuous and nutritious by the abundant oil; their fried meats, their pastry, omelettes, and sauces will be so much richer and better than heretofore, and the Russian will enjoy more freely his well-beloved and necessary tallow, when the candle is made and the engine lubricated with the fat extracted from coals and stones which no human stomach can envy. I might travel on to China and tell of the work that paraffin and paraffin oils have yet to do among the many millions there and in other countries of the East. The great wave of mineral light has not yet fairly broken upon their shores; but when it has once burst through the outer barriers, it will, without doubt, advance with great rapidity, and with an influence whose beneficence can scarcely be exaggerated.

(The above was written in the early days of paraffin lamps, and while the writer was engaged in the distillation of paraffin oils, etc., from the Leeswood cannel. These are now practically superseded by American petroleum of similar composition, but distilled in Nature's oilworks. The anticipations that appeared Utopian at the time of writing have since been fully realized, or even exceeded, as the wholesale price of mineral oil has fallen from two shillings per gallon to an average of about eightpence, and lamps have been greatly improved. At this price the cost of maintaining a light of given power in an ordinary lamp is about equal to that of ordinary London gas, if it were supplied at one shilling per thousand cubic feet. The mineral oil, being a fine hydrocarbon, does far less mischief than gas by its combustion, as may be proved by warming a conservatory with a paraffin stove and another with a stove. In the latter all the delicate plants will be killed; in the first they scarcely suffer at all. If these facts were generally understood we should be in a better position for battle with the gas monopolies. The importation of petroleum to the United Kingdom during the first five months of 1882 amounted to 26,297,346 gallons).

Chapter 8

THE SOLIDITY OF THE EARTH

In his opening address to the Mathematical and Physical Section of the British Association, Sir William Thomson affirmed, "with almost perfect certainty, that, whatever may be the relative densities of rock, solid and melted, or at about the temperature of liquefaction, it is, I think, quite certain that cold solid rock is denser than hot melted rock; and no possible degree of rigidity in the crust could prevent it from breaking in pieces and sinking wholly below the liquid lava," and that "this process must go on until the sunk portions of the crust build up from the bottom a sufficiently close-ribbed skeleton or frame, to allow fresh incrustations to remain bridged across the now small areas of lava-pools or lakes."[11]

This would doubtless be the case if the material of the earth were chemically homogeneous or of equal specific gravity throughout, and if it were chemically inert in reference to its superficial or atmospheric surroundings. But such is not the case. All we know of the earth shows that it is composed of materials of varying specific gravities, and that the range of this variation exceeds that which is due to the difference between the theoretical internal heat of the earth and its actual surface temperature.

We know by direct experiment that these materials, when fused together, arrange themselves according to their specific gravities, with the slight modification due to their mutual diffusibilities. If we take a mixture of the solid elements of which the earth, so far as we know it, is composed, fused them, and leave them exposed to atmospheric action, what will occur?

The heavy metals will sink, the heaviest to the bottom, the lighter metals (i.e., those that we call the metals of the earths, because they form the basis of the earth's superficial crust) will rise along with the silicon, etc., to the surface; these and the silicon will oxidize and combine, forming silicates, and with a sufficient supply of carbonic acid, some of them, such as calcium, magnesium, etc., will form carbonates when the temperature sinks below that of the dissociation of such compounds.

The scoria thus formed will float upon the heavy metals below and protect them from cooling by resisting their radiation; but if in the course of contraction of this crust some fissures are formed reaching to the melted metals below, the pressure of the floating solid will inject the fluid metal upwards into these fissures to a height corresponding to the flotation

---

[11] *Nature*, vol. xiv. p. 429.

depth of the solid, and thus form metallic veins permeating the lower strata of the crust. I need scarcely add that this would rudely but fairly represent what we know of the earth.

But it may be objected that I only describe an imaginary experiment. This is true as regards the whole of the materials united in a single fusion. Nobody has yet produced a complete model with platinum and gold in the centre, and all the other metals arranged in theoretical order with the oxidized, silicated, and carbonated crust outside; but with a limited number of elements this has been done, is being done daily, on a scale of sufficient magnitude to amply refute Sir William Thomson's description of a fused earth solidifying from the centre outwards.

This refutation is to be seen in our blast furnaces, refining furnaces, puddling furnaces, Bessemer ladles, steel melting-pots, cupels, foundry crucibles; in fact, in almost every metallurgical operation down to the simple fusion of lead or solder in a plumber's ladle, with its familiar floating crust of dross or oxide.

As an example I will, on account of its simplicity, take the open hearth finery and the refining of pig-iron. Here a metallic mixture of iron, silicon, carbon, sulphur, etc., is simply fused and exposed to the superficial action of atmospheric air. What is the result?

Oxidation of the more oxidizable constituents takes place, and these oxides at once arrange themselves according to their specific gravities. The oxidized carbon forms atmospheric matter and rises above all as carbonic acid, then the oxidized silicon, being lighter than the iron, floats above that, and combines with aluminium or calcium that may have been in the pig and with some of the iron; thus forming a silicious crust closely resembling the predominating material of the earth's crust.

When the oxidation in the finery is carried far enough, the melted material is tapped out into a rectangular basin or mould, usually about 10 feet long and about 3 feet wide, where it settles and cools. During this cooling the silica and silicates—i.e., the rock matter—separate from the metallic matter and solidify on the surface as a thin crust, which behaves in a very interesting and instructive manner. At first a mere skin is formed. This gradually thickens, and as it thickens and cools becomes corrugated into mountain chains and valleys much higher and deeper, in proportion to the whole mass, than the mountain chains and valleys of our planet. After this crust has thickened to a certain extent volcanic action commences. Rifts, dykes, and faults are formed by the shrinkage of the metal below, and streams of lava are ejected. Here and there these lava streams accumulate around their vent and form insulated conical volcanic mountains with decided craters, from which the eruption continues for some time. These volcanoes are relatively far higher than Chimborazo. The magnitude of these actions varies with the quality of the pig-iron.

The open hearth finery is now but little used, but probably some are to be seen at work occasionally in the neighborhood of Glasgow, and I am sure that Sir William Thomson will find a visit to one of them very interesting. Failing this, he may easily make an experiment by tapping into a good-sized "cinder bogie" some melted pig-iron from a pudding furnace (taking it just before the iron "comes to nature"), and leaving the melted mixture to cool slowly and undisturbed.

The cinder of the blast furnace, which in like manner floats on the top of the melted pig-iron, resembles still more closely the prevailing rock-matter of the earth, on account of the larger proportion, and the varied compounds, of earth-metals it contains.

For the volcanic phenomena alone he need simply watch what occurs when in the ordinary course of puddling the cinder is run into a large bogie, and the bogie is left to cool standing upright. I need scarcely add that these phenomena strikingly illustrate and confirm Mr. Mallett's theory of earthquakes, volcanoes, and mountain-formation.

In merely passing through an iron-making district one may see the results of what I have called the volcanic action, by simply observing the form of those oyster-shaped or cubical blocks of cinder that are heaped in the vicinity of every blast furnace that has been at work for some time. Radial ridges or consolidated miniature lava-streams are visible on the exposed face of nearly, if not quite all of these. They were ejected or squeezed up from below while the mass was cooling, when the outer crust had consolidated but the inner portion still remained liquid. Many of these are large enough, and sufficiently well-marked, to be visible from a railway carriage passing a cinder heap near the road.[12]

---

[12] See Chapter on "The Origin of Lunar Volcanoes."

Chapter 9

A Contribution to the History of Electric Lighting

As the subject of lighting by electricity is occupying so much public attention, and the merits of various inventors and inventions are so keenly discussed, the following facts may have some historical interest in connection with it.

In October, 1845, I was consulted by some American gentlemen concerning the construction of a large voltaic battery for experimenting upon an invention, afterwards described and published in the specification of "King's Patent Electric Light" (Letters Patent granted for Scotland, November 26, 1845; enrolled March 25, 1846; English Patent sealed November 4, 1845).

Mr. King was not the inventor, but he and Mr. Dorr supplied capital, and Mr. Snyder also held a share, which was afterwards transferred to myself. The inventor was Mr. Starr, a young man about twenty-five years of age, and one of the ablest experimental investigators with whom I have ever had the privilege of near acquaintance.

He had been working for some years on the subject, commencing with the ordinary arc between charcoal points. His first efforts were directed to maintaining constancy, and he showed me, in January of 1846, an arrangement by which he succeeded in effecting an automatic renewal of contact by means of an electro-magnet, the armature of which received the electric flow, when the arc was broken, and which thus magnetized brought the carbons together and then allowed them to be withdrawn to their required separation, when the flow returned. This device was almost identical with that subsequently re-invented and patented by Mr. Staite (quite independently, I believe), and which, with modifications, has since been rather extensively used.

Although successful so far, he was not satisfied. He reasoned out the subject, and concluded that the electric spark between metals, the electric arc between the carbons, and other luminous electric phenomena are secondary effects due to the heating and illumination of electric carriers; that the electric spark of the conductors of ordinary electrical machines is simply a transfer of incandescent particles of metal, which effect a kind of electric convection, known as the disruptive discharge; and that the more brilliant arc between the carbon points is simply due to the use of a substance which breaks up more readily, and gives a longer, broader, and more continuous stream of incandescent convection particles.

This is now readily accepted, but at that time was only dawning upon the understanding of electricians. I am satisfied that Mr. Starr worked out the principle quite originally. He therefore concluded that, the light being due to solid particles heated by electric disturbance, it would be more advantageous—as regards steadiness, economy, and simplicity—to place in the current a continuous solid barrier, which should present sufficient resistance to its passage to become bodily incandescent without disruption.

This was the essence of the invention specified in King's Patent as "a communication from abroad," which claims the use of continuous metallic and carbon conductors, intensely heated by the passage of a current of electricity, for the purposes of illumination.

The metal selected was platinum, which, as the specification states, "though not so infusible as iridium, has but little affinity for oxygen, and offers a great resistance to the passage of the current." The form of thin sheets known by the name of leaf-platinum is described as preferable. These to be rolled between sheets of copper in order to secure uniformity, and to be carefully cut in strips of equal width, and with a clean edge, in order that one part may not be fused before the other parts have obtained a sufficiently high temperature to produce a brilliant light. This strip to be suspended between forceps.

I need not describe the arrangement for regulating the distance between the forceps, for directing the current, etc., as we soon learned that this part of the invention was of no practical value, on account of the narrow margin between efficient incandescence and the fusion of the platinum. The experiments with the large battery that I made—consisting of 100 Daniell cells, with two square feet of working surface of each element in each cell, and the copper-plates about three-quarters of an inch distant from the zinc—satisfied all concerned that neither platinum nor any available alloy of platinum and iridium could be relied upon; especially when the grand idea of subdividing the light by interposing several platinum strips in the same circuit, and working with a proportionally high power, was carried out.

This drove Mr. Starr to rely upon the second part of the specification, viz., that of using a small stick of carbon made incandescent in a Torricellian vacuum. He commenced with plumbago, and, after trying many other forms of carbon, found that which lines gas-retorts that have been long in use to be the best.

The carbon stick of square section, about one tenth of an inch thick and half an inch working length, was held vertically, by metallic forceps at each end, in a barometer tube, the upper part of which, containing the carbon, was enlarged to a sort of oblong bulb. A thick platinum wire from the upper forceps was sealed into the top of the tube and projected beyond; a similar wire passed downwards from the lower forceps, and dipped into the mercury of the tube, which was so long that when arranged as a barometer the enlarged end containing the carbon was vacuous.

Considerable difficulty was at first encountered in supporting this fragile stick. Metallic supports were not available, on account of their expansion; and, finally, little cylinders of porcelain were used, one on each side of the carbon stick, and about three eighths of an inch distant.

By connecting the mercury cup with one terminal of the battery, and the upper platinum-wire with the other, a brilliant and perfectly steady light was produced, not so intense as the ordinary disruption arc between carbons, but equally if not more effective, on account of the magnitude of brilliant radiating surface.

Some curious phenomena accompanied this illumination of the carbon. The mercury column fell to about half its barometric height, and presently the glass opposite the carbon stick became slightly dimmed by the deposition of a thin film of sooty deposit.

At first the depression of the mercury was attributed to the formation of mercurial vapor, and is described accordingly in the specification; but further observation refuted this theory, for no return of the mercury took place when the tube was cooled. The depression was permanent. The formation of vaporous carbon was suggested by one of the capitalists; but neither Mr. Starr nor myself was satisfied with this, nor with any other surmise we were able to make during Mr. Starr's lifetime, nor up to the period of final abandonment of the enterprise.

When this occurred the remaining apparatus was assigned to me, and I retained possession of the finally arranged tube and carbon for many years, and have shown it in action worked by a small Grove's battery in the Town Hall of Birmingham, and many times to my pupils at the Birmingham and Midland Institute.

These exhibitions suggested an explanation of the mysterious gaseous matter, which I believe to be the correct one, and also of the carbon deposit. It is this:—That the carbon contains occluded oxygen; that when the carbon is heated some of this oxygen combines with the carbon, forming carbonic oxide and carbonic acid, and a little smoke. I proved the presence of carbonic acid by the usual tests, but did not quantitatively determine its proportion of the total atmosphere.

If I were fitting up another tube on this principle I should wash it with a strong solution of caustic potash before filling with mercury, and allow some of the potash solution to float on the mercury surface, by filling the tube while the glass remained moistened with the solution. My object would be to get rid of the carbonic acid as soon as formed, as the observations I have made lead me to believe that—when the carbon stick is incandescent in an atmosphere of carbonic acid or carbonic oxide—a certain degree of dissociation and re-combination is continually occurring, which weakens and would ultimately break up the carbon stick, and increases the sooty deposit.

The large battery was arranged for intensity, but even then it was found that the quantity (I use the old-fashioned terms) of electricity was excessive, and that it worked more advantageously when the cells were but partially filled with acid and sulphate. A larger stick of carbon might have been used with the whole surface in full action.

After working the battery in various ways, and duly considering the merits of the other forms of battery then in use, Mr. Starr was driven to the conclusion that for the purposes of practical illumination the voltaic battery is a hopeless source of power, and that magneto-electric machinery driven by steam-power must be used. I fully concurred with him in this conclusion, so did Mr. King, Mr. Dorr, and all concerned.

Mr. Starr then set to work to devise a suitable dynamo-electric machine, and, following his usual course of starting from first principles, concluded that all the armatures hitherto constructed were defective in one fundamental element of their arrangement. The thick copper wire surrounding the soft iron core necessarily follows a spiral course, like that of a coarse screw-thread; but the electric current or lines of force, which it is designed to pick up and carry, circulate at right angles to the axis of the core, and extend to some distance beyond its surface. The problem thus presented is to wind around the soft iron a conductor that shall be broad enough to grasp a large proportion of this outspread force, and yet shall follow its course as nearly as possible by standing out at right angles to the axis of the armature. This he endeavored to effect by using a core of square section, and winding round it a broad ribbon of sheet copper, insulated on both sides by cementing on its surfaces a layer of silk ribbon. This armature was laid with one edge against one side of the core, and carried on thus to the angle; then turned over so that its opposite edge should be presented to the next side of the core; this side to be followed in like manner, the ribbon similarly turned again at the next corner, and so on till the core became fully enclosed or armed with the continuous ribbon, which thus encircled the core with its edges outwards, and nearly at right angles to the axis, in spite of its width, which might be increased to any extent found by experiment to be desirable.

At this stage my direct co-operation and confidential communication with Mr. Starr ceased, as I remained in London while he went to Birmingham in order to get his machinery constructed, and to apply it at the works of Messrs. Elkington, who had then recently introduced the principle of dynamo-electric motive-power for electro-plating, etc., and were, I believe, using Woolrich's apparatus, the patent for which was dated August 1, 1842, and enrolled February 1, 1843.

I am unable to state the results of his efforts in Birmingham. I only heard the murmurs of the capitalists, who loudly complained of expenditure without results. They had dreamed the same dream that Mr. Edison has recently re-dreamed, and has told the world so loudly. They supposed that the mechanically excited current might be carried along great lengths of wire, and the carbons interposed wherever required, and that the same electricity would flow on and do the duty of illumination over and over again as a river may fall over a succession of weirs and turn water-wheels at each. Mr. Starr knew better; his scepticism was misinterpreted; he was taunted with failure and non-fulfilment of the anticipations he had raised, and with the fruitless expenditure of large sums of other people's money. He was a high-minded, honorable, and very sensitive man, suffering already from overworked brain before he went to Birmingham. There he worked again still harder, with further vexation and disappointment, until one morning he was found dead in his bed. Having, during my short acquaintance with him, enjoyed his full confidence in reference to all his investigations, I have no hesitation in affirming that his early death cut short the career of one who otherwise would have largely contributed to the progress of experimental science, and have done honor to his country.

His martyrdom, for such it was, taught me a useful lesson I then much needed, viz., to abstain from entering upon a costly series of physical investigations without being well assured of the means of completing them, and, above all, of being able to afford to fail.

There are many others who sorely need to be impressed with the same lesson, especially at this moment and in connection with this subject.

The warning is the most applicable to those who are now misled by a plausible but false analogy. They look at the progress made in other things, the mighty achievements of modern Science, and therefore infer that the electric light—even though unsuccessful hitherto—may be improved up to practical success, as other things have been. A great fallacy is hidden here. As a matter of fact the progress made in electric lighting since Mr. Starr's death, in 1846, has been very small indeed. As regards the lamp itself no progress whatever has been made. I am satisfied that Starr's continuous carbon stick, properly managed in a true vacuum, or an atmosphere free from oxygen, carbonic oxide, carbonic acid, or other oxygen compound, is the best that has yet been placed before the public for all purposes where exceptionally intense illumination (as in lighthouses) is not demanded.[13]

Comparing electric with gas-lighting, the hopeful believers in progressive improvement appear to forget that gas-making and gas-lighting are as susceptible of further improvement as electric lighting, and that, as a matter of fact, its practical progress during the last forty years is incomparably greater than that of the electric light. I refer more particularly to the practical and crucial question of economy. The bi-products, the ammoniacal salts, the liquid hydrocarbons, and their derivatives, have been developed into so many useful forms by the achievements of modern chemistry, that these, with the coke, are of sufficient value to cover the whole cost of manufacture, and leave the gas itself as a volatile residuum that costs nothing. It would actually and practically cost nothing, and might be profitably delivered to the burners of gas consumers (of far better quality than now supplied in London) at one shilling per thousand cubic feet, if gas-making were conducted on sound commercial principles,—that is, if it were not a corporate monopoly, and were subject to the wholesome stimulating influence of free competition and private enterprise. As it is, our gas and the price we pay for it are absurdities; and all calculations respecting the comparative cost of new methods of illumination should be based not on what we *do* pay per candle-power of gas-light, but what we *ought* to pay and *should* pay if the gas companies were subjected to desirable competition, or visited with the national confiscation I consider they deserve.

Having had considerable practical experience in the commercial distillation of coal for the sake of its liquid and solid hydrocarbons, I speak thus plainly and with full confidence.

There is yet another consideration, and one of vital importance, to be taken into account, viz., that—whether we use the electric light derived from a dynamo-electric source, or coal-gas—our primary source of illuminating power is coal, or rather the chemical energy derivable from the combination of its hydrogen and carbon with oxygen. Now this chemical energy is a limited quantity, and the progress of Science can no more increase this quantity than it can make a ton of coal weigh 21 cwts. by increasing the quantity of its gravitating energy.

The demonstrable limit of scientific possibilities is the economical application of this limited store of energy, by converting it into the demanded form of force without waste. The more indirect and roundabout the method of application, the greater must be the loss of power in the course of its transfer and conversion. In heating the boiler that sets the dynamo-electric machine to work, about one-half the energy of the coal is wasted, even with the best constructed furnaces. This merely as regards the quantity of water evaporated. In converting the heat-force into mechanical power—raising the piston, etc., of the steam-engine—this working half is again seriously reduced. In further converting this residuum of mechanical power into electrical energy, another and considerable loss is suffered in originating and sustaining the motion of the dynamo-electric machine, in the dissipation of the electric energy that the armature cannot pick up, and in overcoming the electrical resistances to its transfer.

I am unable to state the amount of this loss in trustworthy figures, but should be very much surprised to learn that, with the best arrangements now known, more than one-tenth of the original energy of the coal is made practically available. This small illuminating residuum

---

[13] The burnt card, burnt bamboo, and other flimsy incandescent threads now (1882) in vogue, merely represent Starr's preliminary failures prior to his adoption of the hard adamantine stick of retort-carbon, which I suppose will be duly re-invented, patented again, and form the basis of new Limited Companies, when the present have collapsed.

may, and doubtless will, be increased by the progress of practical improvement; but from the necessary nature of the problem, the power available for illumination at the end of the series must always be but a small portion of that employed at the beginning.

In burning the gas derived from coal we obtain its illuminating power *directly*, and if we burn it properly we obtain nearly all. The coke residuum is also directly used as a source of heat. The chief waste of the original energy in the gas-works is represented by that portion of the coke that is burned under the retorts, and in obtaining the relatively small amount of steam-power demanded in the works. These are far more than paid for by the value of the liquid hydrocarbons and the ammonia salts, when they are properly utilized.

In concluding my narrative, I may add that after Mr. Starr's death the patentees offered to engage me on certain terms to carry on his work. I declined this, simply because I had seen enough to convince me of the impossibility of any success at all corresponding to their anticipations. During the intervening thirty years I have abstained from further meddling with the electric light, because all that I had seen then, and have heard of since, has convinced me that—although as a scientific achievement the electric light is a splendid success—its practical application to all purposes where cost is a matter of serious consideration is hopeless, and must of necessity continue to be so.

Whoever can afford to pay some shillings per hour for a single splendid light of solar completeness can have it without difficulty, but not so where the cost in pence per hour per burner has to be counted.

I should add that before the publication of King's specification, Mr. (now Sir William) Grove proposed the use of a helix or coil of platinum, made incandescent by electricity, as a light to be used for certain purposes. This was shown at the Royal Society on or about December 1, 1845.

Since the publication of the above in 1879, I have learned, from a paper in the "Quarterly Journal of Science," by Professor Ayrton, that in 1841 an English patent was granted to De Moylens for electric lighting by incandescence.

Chapter 10

THE FORMATION OF COAL

In the course of a pedestrian excursion made in the summer of 1855 I came upon the Aachensee, one of the lakes of North Tyrol, rarely visited by tourists. It is situated about 30 miles N.E. of Innispruck, and fills the basin of a deep valley, the upper slopes of which are steep and richly wooded. The water of this lake is remarkably transparent and colorless. With one exception, that of the Fountain of Cyane—a deep pool forming the source of the little Syracusan river—it is the most transparent body of water I remember to have seen. This transparency revealed a very remarkable sub-aqueous landscape. The bottom of the lake is strewn with branches and trunks of trees, which in some parts are in almost forest-like profusion. As I was alone in a rather solitary region, and carrying only a satchel of luggage, my only means of further exploration were those afforded by swimming and diving. Being an expert in these, and the July summer day very calm and hot, I remained a long time in the water, and, by swimming very carefully to avoid ripples, was able to survey a considerable area of the interesting scene below.

The fact which struck me the most forcibly, and at first appeared surprising, was the upright position of many of the large trunks, which are of various lengths—some altogether stripped of branches, others with only a few of the larger branches remaining. The roots of all these are more or less buried, and they present the appearance of having grown where they stand. Other trunks were leaning at various angles and partly buried, some trunks and many branches lying down.

On diving I found the bottom to consist of a loamy powder of gray color, speckled with black particles of vegetable matter—thin scaly fragments of bark and leaves. I brought up several twigs and small branches, and with considerable difficulty, after a succession of immersions, succeeded in raising a branch about as thick as my arm and about eight feet long, above three-fourths of which was buried, and only the end above ground in the water. My object was to examine the condition of the buried and immersed wood, and I selected this as the oldest piece I could reach.

I found the wood very dark, the bark entirely gone, and the annual layers curiously loosened and separable from each other, like successive rings of bark. This continued till I had stripped the stick to about half of its original thickness, when it became too compact to yield to further stripping.

This structure apparently results from the easy decomposition of the remains of the original cambium of each year, and may explain the curious fact that so many specimens of fossilized wood exhibit the original structure of the stem, although all the vegetable matter has been displaced by mineral substances. If this stem had been immersed in water capable of precipitating or depositing mineral matter in very small interstices, the deposit would have filled up the vacant spaces between these rings of wood as the slow decomposition of the vegetable matter proceeded. At a later period, as the more compact wood became decomposed, it would be substituted by a further deposit, and thus concentric strata would be formed, presenting a mimic counterpart of the vegetable structure.

The stick examined appeared to be a branch of oak, and was so fully saturated with water that it sank rapidly upon being released.

On looking around the origin of this sub-aqueous forest was obvious enough. Here and there the steep wooded slopes above the lake were broken by long alleys or downward strips of denuded ground, where storm torrents, or some such agency, had cleared away the trees and swept most of them into the lake. A few uprooted trees lying at the sides of these bare alleys told the story plainly enough. Most of these had a considerable quantity of earth and stones adhering to their roots: this explains the upright position of the trees in the lake.

Such trees falling into water of sufficient depth to enable them to turn over must sink root downwards, or float in an upright position, according to the quantity of adhering soil. The difference of depth would tend to a more rapid penetration of water in the lower parts, where the pressure would be greatest, and thus the upright or oblique position of many of the floating trunks would be maintained till they absorbed sufficient water to sink altogether.

It is generally assumed that fossil trees which are found in an upright position have grown on the spot where they are found. The facts I have stated show that this inference is by no means necessary, not even when the roots are attached and some soil is found among them. In order to account for the other surroundings of these fossil trees a very violent hypothesis is commonly made, viz., that the soil on which they grew sank down some hundreds of feet without disturbing them. This demands a great strain upon the scientific imagination, even in reference to the few cases where the trees stand perpendicular. As the majority slope considerably the difficulty is still greater. I shall presently show how trees like those immersed in Aachensee may have become, and are now becoming, imbedded in rocks similar to those of the Coal Measures.

In the course of subsequent excursions on the fjords of Norway I was reminded of the sub-aqueous forest of the Aachensee, and of the paper which I read at the British Association meeting of 1865, of which the above is an abstract—not by again seeing such a deposit under water, for none of the fjords approach the singular transparency of the lake, but by a repetition on a far larger scale of the downward strips of denuded forest ground. Here, in Norway, their magnitude justifies me in describing them as vegetable avalanches. They may be seen on the Sognefjord, and especially on those terminal branches of this great estuary, of which the steep slopes are well wooded. But the most remarkable display that I have seen was in the course of the magnificent, and now easily made, journey up the Storfjord and its extension and branches, the Slyngsfjord, Sunelvsfjord, Nordalsfjord, and Geirangerfjord. Here these avalanches of trees, with their accompaniment of fragments of rock, are of such frequent occurrence that sites of the farm-houses are commonly selected with reference to possible shelter from their ravages. In spite of this they do not always escape. In the October previous to my last visit a boat-house and boat were swept away; and one of the most recent among the tracks that I saw reached within twenty yards of some farm-buildings.

What has become of the millions of trees that are thus falling, and have fallen, into the Norwegian fjords during the whole of the present geological era? In considering this question we must remember that the mountain slopes forming the banks of these fjords continue downwards under the waters of the fjords which reach to depths that in some parts are to be counted in thousands of feet.

It is evident that the loose stony and earthy matter that accompanies the trees will speedily sink to the bottom and rest at the foot of the slope somewhat like an ordinary sub-aerial talus, but not so the trees. The impetus of their fall must launch them afloat and impel them towards the middle of the estuary, where they will be spread about and continue floating, until by saturation they become dense enough to sink. They will thus be pretty evenly distributed over the bottom. At the middle part of the estuary they will form an almost purely vegetable deposit, mingled only with the very small portion of mineral matter that is held in suspension in the apparently clear water. This mineral matter must be distributed among the vegetable matter in the form of impalpable particles having a chemical composition similar to that of the rocks around. Near the shores a compound deposit must be formed consisting of trees and fragments of leaves, twigs, and other vegetable matter mixed with larger proportions of the mineral *débris*.

If we look a little further at what is taking place in the fjords of Norway we shall see how this vegetable deposit will ultimately become succeeded by an overlying mineral deposit which must ultimately constitute a stratified rock.

All these fjords branch up into inland valleys down which pours a brawling torrent or a river of some magnitude. These are more or less turbid with glacier mud or other detritus, and great deposits of this material have already accumulated in such quantity as to constitute characteristic modern geological formations bearing the specific Norsk name of *ören*, as *Laerdalsören*, *Sundalsören*, etc., describing the small delta plains at the mouth of a river where it enters the termination of the fjord, and which, from their exceptional fertility, constitute small agricultural settlements bearing these names, which signify the river sands of *Laerdal*, *Sundal*, etc. These deposits stretch out into the fjord, forming extensive shallows that are steadily growing and advancing further and further into the fjord. One of the most remarkable examples of such deposits is that brought by the Storelv (or Justedals Elv), which flows down the Justedal, receiving the outpour from its glaciers, and terminates at Marifjören. When bathing here I found an extensive sub-aqueous plain stretching fairly across that branch of the Lyster fjord into which the Storelv flows. The waters of the fjord are whitened to a distance of two or three miles beyond the mouth of the river. These deposits must, if the present conditions last long enough, finally extend to the body, and even to the mouth, of the fjords, and thus cover the whole of the bottom vegetable bed with a stratified rock in which will be entombed, and well preserved, isolated specimens of the trees and other vegetable forms corresponding to those accumulated in a thick bed below, but which have been lying so long in the clear waters that they have become soddened into homogeneous vegetable pulp or mud, only requiring the pressure of solid superstratum to convert them into coal.

The specimens of trees in the upper rock, I need scarcely add, would be derived from the same drifting as that which produced the lower pulp; but these coming into the water at the period of its turbidity and of the rapid deposition of mineral matter, would be sealed up one by one as the mineral particles surrounding it subsided. Fossils of estuarine animals would, of course, accompany these, or of fresh-water animals where, instead of a fjord, the scene of these proceedings is an inland lake. In reference to this I may state that at the inner extremities of the larger Norwegian fjords the salinity of the water is so slight that it is imperceptible to taste. I have freely quenched my thirst with the water of the Sörfjord, the great inner branch of the Hardanger, where pallid specimens of bladder wrack were growing on its banks.

In the foregoing matter-of-fact picture of what is proceeding on a small scale in the Aachensee, and on a larger in Norway, we have, I think, a natural history of the formation, not only of coal seams, but also of the Coal Measures around and above them.

The theory which attributed our coal seams to such vegetable accumulations as the rafts of the Mississippi is now generally abandoned. It fails to account for the state of preservation and the position of many of the vegetable remains associated with coal.

There is another serious objection to this theory that I have not seen expressed. It is this: rivers bringing down to their mouths such vegetable deltas as are supposed, would also bring considerable quantities of earthy matter in suspension, and this would be deposited with the trees. Instead of the 2 or 3 per cent of incombustible ash commonly found in coal, we should thus have a quantity more nearly like that found in bituminous shales which may thus be formed, viz., from 20 to 80 per cent.

The alternative hypothesis now more commonly accepted—that the vegetation of our coal-fields actually grew where we find it—is also refuted by the composition of coal-ash. If the coal consisted simply of the vegetable matter of buried forests its composition should correspond to that of the ashes of plants; and the refuse from our furnaces and fireplaces would be a most valuable manure. This we know is not the case. Ordinary coal-ash, as Bischof has shown, nearly corresponds to that of the rocks with which it is associated; and he says that "the conversion of vegetable substances into coal has been effected by the agency of water;" and also that coal has been formed, not from dwarfish mosses, sedges, and other plants which now contribute to the growth of our peat-bogs, but from the stems and trunks of the forest trees of the Carboniferous Period, such as *Sigillariæ*, *Lepdodendra*, and *Coniferæ*.[14] All we know of these plants teaches us that they could not grow in a merely vegetable soil containing but 2 or 3 per cent of mineral matter. Such must have been their soil for hundreds of generations in order to give a depth sufficient for the formation of the South Staffordshire ten-yard seam.

All these and other difficulties that have stood so long in the way of a satisfactory explanation of the origin of coal appear to me to be removed if we suppose that during the Carboniferous Period Britain and other coal-bearing countries had a configuration similar to that which now exists in Norway, viz., inland valleys terminating in marine estuaries, together with inland lake basins. If to this we superadd the warm and humid climate usually attributed to the Carboniferous Period, on the testimony of its vegetable fossils, all the conditions requisite for producing the characteristic deposits of the Coal Measures are fulfilled.

We have first the under-clay due to the beginning of this state of things, during which the hill slopes were slowly acquiring the first germs of subsequent forest life, and were nursing them in their scanty youth. This deposit would be a mineral mud with a few fossils and that fragmentary or fine deposit of vegetable matter that darkens the carboniferous shales and strips the sandstones. Such a bed of dark consolidated mud, or fine clay, is found under every seam of coal, and constitutes the "floor" of the coal pit. The characteristic striped rocks—the

---

[14] Hull, "On the Coal-fields of Great Britain."

"linstey" or "linsey" of the Welsh colliers—is just such as I found in the course of formation in the Aachensee near the shore, as described above.

The prevalence of estuarine and lacustrine fossils in the Coal Measures is also in accordance with this: the constitution of coal-ash is perfectly so. Its extreme softness and fineness of structure; its chemical resemblance to the rocks around, and above, and below; and oblong basin form common to our coal seams; the apparent contradiction of such total destruction of vegetable structure common to the true coal seams, while immediately above and below them are delicate structures well preserved, is explained by the more rapid deposition of the latter, and the slow soddening of the former as above described.

I do not, however, offer this as an explanation of the formation of *every kind of coal*. On the contrary, I am satisfied that cannel coal, and the black shales usually associated with it, have a different origin from that of the ordinary varieties of bituminous coal. The fact that the products of distillation of cannel and these shales form different series of hydrocarbons from those of common coal, and that they are nearly identical with those obtained by the distillation of peat, is suggestive of origin in peat-bogs, or something analogous to them.

To the above I may add the concluding sentences of the chapter on Coal in Lyell's "Elements of Geology." Speaking of fossils in the Coal Measures, he says: "The rarity of air-breathers is a very remarkable fact when we reflect that our opportunities of examining strata *in close connection with ancient land* exceed in this case all that we enjoy in regard to any other formations, whether primary, secondary, or tertiary. We have ransacked hundreds of soils replete with the fossil roots of trees, have dug out hundreds of erect trunks and stumps which stood in the position in which they grew, have broken up myriads of cubic feet of fuel still retaining its vegetable structure, and, after all, we continue almost as much in the dark respecting the invertebrate air-breathers of this epoch, *as if the coal had been thrown down in mid-ocean*. The early date of the carboniferous strata cannot explain the enigma, because we know that while the land supported a luxuriant vegetation, the contemporaneous seas swarmed with life—with Articulata, Mollusca, Radiata, and Fishes. We must, therefore, collect more facts if we expect to solve a problem which, in the present state of science, cannot but excite our wonder; and we must remember how much the conditions of this problem have varied within the last twenty years. We must be content to impute the scantiness of our data and our present perplexity partly to our want of diligence as collectors, and partly to our want of skill as interpreters. We must also confess that our ignorance is great of the laws which govern the fossilization of land animals, whether of low or high degree."

The explanation of the origin of coal which I have given in the foregoing meets all these difficulties. It shows how vast accumulations of vegetable matter may have been formed "in close connection with the ancient land," and yet "as if the coal had been thrown down in mid-ocean" as far as the remains of terrestrial animals are concerned. It explains the nearly total absence of land shells, and of the remains of other animals that must have lived in the forests producing the coal, and which would have been buried there with the coal had it been formed on land as usually supposed. It also meets the cases of the rare and curious exceptions, seeing that occasionally a land animal would here and there be drowned in such fjords under circumstances favorable to its fossilization.

Chapter 11

THE SOLAR ECLIPSE OF 1871

THE FIRST TELEGRAMS

This time we may fairly expect some approach to a solution of the riddle of the corona, as the one essential which neither scientific skill nor Government liberality could secure to the eclipse observers, has been afforded, viz., fine weather. The telegraph has already informed us of this, and also that good use has been made of the good weather. From one station we are told: "Thin mist; spectroscope satisfactory; reversion of lines entirely confirmed; six good photographs." From another: "Weather fine; telescopic and camera photographs successful; ditto polarization; good sketches; many bright lines in spectrum."

This is very different from the gloomy accounts of the expedition of last year; when we consider that the different observers are far apart, and that if all or some of them are similarly favored we shall have in the photographs a series of successive pictures taken at intervals of time sufficiently distant to reveal any progressive changes that may have occurred in the corona while the moon's shadow was passing from one station to the other. I anticipate some curious revelations from these progressive photographs, that may possibly reconcile the wide differences in the descriptions that competent observers have given of the corona of former eclipses, which they had seen at stations distant from each other.

Barely two years have elapsed since I suggested, in "The Fuel of the Sun," that the great solar prominences and the corona are due to violent explosions of the dissociated elements of water; that the prominences are the gaseous flashes, and the corona the ejected scoria, or solidified metallic matter belched forth by the furious cannonade continually in progress over the greater portion of the solar surface.

This explanation at first appeared extravagant, especially as it was carried so far as to suggest that not merely the corona, but the zodiacal light, the zone of meteors which occasionally drop showers of solid matter upon the earth, and even the "pocket-planets" or asteroids so irregularly scattered between the orbits of Mars and Jupiter, consist of solid matter thus ejected by the great solar eruptions. Even up to the spring of the present year, when Mr. Lockyer and other leaders of the last year's expeditions reported their imperfect results, and compared them with various theories, this one was not thought worthy of their attention.

Since that time—during the past six or eight months—a change has taken place which strikingly illustrates the rapid progress of solar discovery. Observations and calculations of the force and velocity of particular solar eruptions have been made, and the results have proved that they are amply sufficient to eject solid missiles even further than I supposed them to be carried.

Mr. Proctor, basing his calculations upon the observations of Respighi, Zöllner, and Professor Young, has concluded that it is even possible that meteoric matter may be ejected far beyond the limits of our solar system into the domain of the gravitation of other stars, and that other stars may in like manner bombard the sun.

This appears rather startling; but, as I have already said, the imagination of the poet and the novelist is beggared by the facts revealed by the microscope, so I may now repeat the assertion, and state it still more strongly, in reference to the revelations of the telescope and the spectroscope.

As a sample of these, I take the observations of Professor Young, made on September 7th last, and described fully in "Nature" on October 19.

He first observed a number of the usual flame-prominences having the typical form which has been compared to a "banyan grove." One of these banyans was greater than the rest. This monarch of the solar flame-forest measured *fifty-four thousand miles in height*, and its outspreading measured in one direction about *one hundred thousand miles*. It was a large eruption-flame, but others much larger have been observed, and Professor Young would probably have merely noted it among the rest, had not something further occurred. He was called away for twenty-five minutes, and when he returned "the whole thing had been literally blown to shreds by some inconceivable uprush from beneath." The space around "was filled with flying *débris*—a mass of detached vertical fusiform filaments, each from 10 sec. to 30 sec. long by 2 sec. or 3 sec. wide, brighter and closer together where the pillars had formerly stood, *and rapidly ascending*." Professor Young goes on to say, that "When I first looked, some of them had already reached a height of 100,000 miles, and while I watched they rose, with a motion almost perceptible to the eye, until in ten minutes the uppermost were 200,000 miles above the solar surface. This was ascertained by careful measurement."

Here, then, we have an observed velocity of 10,000 miles per minute, and this is the gaseous matter, merely the flash of the gun by which the particles of solidified solar matter are supposed to be projected.

The reader must pause and reflect, in order to form an adequate conception of the magnitudes here treated—100,000 miles long and 54,000 miles high! What does this mean? Twelve and a half of our worlds placed side by side to measure the length, and six and three quarters, piled upon each other, to measure the height! A few hundred worlds as large as ours would be required to fill up the whole cubic contents of this flame-cloud. The spectroscope has shown that these prominences are incandescent hydrogen. Most of my readers have probably seen a soap-bubble or a bladder filled with the separated elements of water, and then exploded, and have felt the ringing in their ears that has followed the violent detonation.

Let them struggle with the conception of such a bubble or bladder magnified to the dimensions of only one such a world as ours, and then exploded; let them strain their power of imagination even to the splitting point, and still they must fail most pitifully to picture the magnitude of this solar explosion observed on September 7th last, which flashed out to a magnitude of more than five hundred worlds, and then expanded to the size of more than five thousand worlds, even while Professor Young was watching it. Professor Young concludes his description by stating that "it seems far from impossible that the mysterious coronal streamers, if they turn out to be truly solar, as now seems likely, may find their origin and explanation in such events."

This, and a number of similar admissions, suggestions, and conclusions from the leading astronomers, indicate that the eruption theory of the corona will not be passed over in silence by the observers of this eclipse, and it is to this that I have referred in the above remarks respecting the interest attaching to a series of photographs showing successive states of this outspreading enigma.

Father Secchi's spectroscopic observations on the uneclipsed sun led him to assert the existence of a stratum of glowing metallic vapors immediately below the envelope connected with the hydrogen of the eruptions. This is just what is required by my eruption theory to supply the solid materials of the ejections forming the corona.

Professor Young's announcement of the reversal of the spectroscopic lines at the moment when the stratum was seen independently of the general solar glare, startled Mr. Lockyer and others who had disputed the accuracy of the observations of the great Italian observer, as it confirmed them so completely. Scepticism still prevailed, and Young's observation was questioned; but now even our slender telegraphic communication from Colonel Tenant to Dr. Huggins indicates that the question must be no longer contested. "Reversion of lines entirely confirmed" is a message so important that if the expeditions had done no more than this, all their cost in money and scientific labor would be amply repaid in the estimation of those who understand the value of pure truth.

A few more fragments of intelligence respecting the Eclipse Expedition have reached us, the last Indian mail having started just after the eclipse occurred. They fully confirm the first telegraphic announcement, rather strengthening than otherwise the expectations of important results, especially in reference to the photographs of the corona.

I have read in the Ceylon newspapers some full descriptions by amateur observers, in which the general magnificence of the phenomena is described. From these it is evident that the corona must have been displayed in its full grandeur; but as the writers do not attempt to describe those features which have at the present moment a special scientific interest, I shall not dwell upon them, but await the publication of the official report of the chief, and of the more important collateral observing expeditions.

The unsophisticated reader may say "Are not one man's eyes as good as another's, and why should the observations of the learned men of the expeditions be so much better than those of any other clear-sighted persons?" This is a perfectly fair question, and admits of a ready answer. All that can be known by mere unprepared naked-eye observation is tolerably well known already; the questions which await solution can only be answered by putting the sun to torture by means of instruments specially devised for that purpose; and by a skillful organization, and division of labor among the observers.

There is so much to be seen during the few seconds of total obscuration that no one human being, however well trained in the art of observing, could possibly see all. Therefore it is necessary to pre-arrange each observer's part, to have careful rehearsals of what is to be done by each during the precious seconds; and each man must exercise a vast amount of self-control in order to confine his attention to his own particular bit of observation, while he is surrounded with such marvellous phenomena as a total eclipse presents.

The grandeur of the gloomy landscape, the sudden starting out of the greater stars, the seeming falling of the vault of heaven, the silence of the animal world, the closing of the flowers, and all that the ordinary observer would regard with so much awe and wondering delight, must be sacrificed by the philosopher, whose business is to confine his gaze to a narrow slit between two strips of metal, and to watch nothing else but the exact position and appearance of a few bright or dark lines across what appears but a strip of colored riband. He must resist the temptation to look aside and around with the stubbornness of self-denial of another St. Antonio. Besides this, he must thoroughly understand exactly what to look for, and how to find it. By combining the results of his observations with those of the others, who in like manner have undertaken to work with another instrument, or upon another part of the phenomena, we get a scientific result comparable to that which in a manufactory we obtain by the division of labor of many skilled workmen, each doing only that which by his training he has learned to do the best and the most expeditiously.

Further Details by Post

Although the formal official reports of the Eclipse Expedition are not yet published, and may not be for some weeks or months, we are able from the letters of Lockyer, Jannsen, Respighi, Maclear, etc., to form some idea of the general results. We may already regard two or three important questions as fairly answered. The reversal of the dark solar lines of the spectrum which was first announced by the great Roman observer, Father Secchi, and seen by him without an eclipse, may now be considered as established. It is true that all the observers of 1871 did not witness this. Some were doubtful, but others observed it positively and distinctly.

In such a case negative results do not refute the positive observations of qualified men, especially when several of such observations have been made independently; the phenomenon is but instantaneous, a mere flash of bright stripes in place of dark lines across the colored riband of the spectroscope, which happens just at the moment before and after totality, and is presented only when the instrument is accurately directed to the delicate curved vanishing thread of light which is the last visible fragment of the solar outline, and that which makes the first flash of his re-appearance.

A little explanation is necessary to render the significance of this "reversal" intelligible to those who have not specially studied the subject.

1st. When the spectroscope is directed to a luminous solid a simple rainbow-band or "continuous spectrum" is seen. When, on the other hand, the object is a luminous gas or vapor of moderate density, the spectrum is not a continuous band with its colors actually blending; it consists only of certain luminous stripes with blank spaces between them, each particular gas or vapor showing its own particular set of stripes of certain colors, and always appearing at exactly the same place, so invariably and certainly, that, by means of such luminous stripes, the composition of the gas or vapor may be determined. If, however, the gas be much compressed, the stripes widen as the condensation proceeds; they may even spread out sufficiently to meet and form a continuous spectrum like that from a solid. Liquids also produce continuous spectra.

2d. When a luminous solid or liquid, or very dense gas, capable of producing a continuous spectrum, is viewed through an intervening body of other gas or vapor of moderate or small density, fine *dark lines* cross the spectrum in precisely the same places as the bright stripes would appear if this intervening gas or vapor were luminous and seen by itself.

When the spectroscope is directed to the face of the sun under ordinary circumstances, it presents a brilliant continuous spectrum, striped with a multitude of the dark lines. From this it has been inferred that the luminous face of the sun is that of an incandescent solid or liquid, and that it is surrounded by the gases and vapors whose bright stripes, when artificially produced, occupy precisely the same places as the dark lines of the solar spectrum. This was the theory of Kirchoff and others in the early days of spectrum analysis, when it was only known that solids and liquids were capable of producing a continuous spectrum. The important discovery that gases and vapors, if sufficiently condensed, will also produce a continuous spectrum, opened another speculation, far more consistent with the other known facts concerning the constitution of the sun, viz., that the sun may be a great gaseous orb, blazing at its surface and gradually increasing in density from the surface towards the centre.

According to this, the metals sodium, calcium, barium, magnesium, iron, chromium, nickel, copper, zinc, strontium, cobalt, manganese, aluminium, and titanium, whose vapors, with those of some few other substances, give the dark lines that cross the solar spectrum, should exist neither as solids nor liquids on the solar surface, but as blazing gases. But such blazing gases, according to what I have stated above, should give us bright stripes instead of dark lines. Why, then, are not such bright stripes seen under ordinary circumstances?

This is easily answered. These blazing gases must, as we proceed from the surface of the sun downwards, become so condensed by the pressure of their own superincumbent strata, as to produce a continuous spectrum of great brilliancy. With such a background the bright stripes would be confounded and lost to sight. Besides this, the outer film of cooler vapor through which our vision must necessarily penetrate before reaching the luminous solar surface, will produce the dark lines exactly where the bright stripes should be, and thus effectually obliterate them; or, in other words, the intervening non-luminous vapors are opaque to the particular rays of light which the bright vapors of the same substance emits.

Therefore, according to this theory, if we could sweep away these outside darkening vapors, and screen off the inner layers of denser blazing matter which produces the continuous background, we should have a spectrum displaying a multitude of bright stripes exactly where the black lines of the ordinary solar spectrum appear.

Secchi announced that these bright lines were to be seen under favorable circumstances, when, by skillful management, the rays from the edge of the sun were so caught by the slit of the spectroscope as to exhibit only the spectrum of the superficial layer of the sun's bright surface. This was disputed at the time by Mr. Lockyer, who, I suspect, omitted to consider the atmospheric difficulties under which English astronomers work, and the fact that the atmosphere of Italy is exceptionally favorable for delicate astronomical observation.

If he had fairly considered this I think he would agree with me in concluding that an observation of this kind, avowedly made with great difficulty and questionable distinctness by so skillful a spectroscopic observer as Father Secchi, could not possibly be seen by any human eyes through a London atmosphere.

Subsequently Professor Young startled the astronomical world by the announcement that, at the moment when the thinnest perceptible thread of the sun's edge was alone displayed during the eclipse which he observed, the whole of the dark lines of the solar spectrum flashed out as bright stripes in a most unmistakable manner. This observation is now fully confirmed. The first telegrams from Mr. Pogson, the Government astronomer of Madras, and from Colonel Tennant, both announce this most positively, Colonel Tennant's words being, "the reversion of the lines fully confirmed." A similar result was obtained by some, but not by all, of the Ceylon observers.

To understand this clearly, we must consider the fact that what appears to us as the outline of a flat disc is really that part of the sun which we see by looking horizontally athwart his rotundity, just as we look at the ocean surface of our own earth when we stand upon the shore and see its horizon outline. When the moon obscures all but the last film of this solar edge, we see only the surface of the supposed gaseous orb, just that portion of the blazing gases which are not greatly compressed by those above them, and which accordingly should, if they consist of the vapors or the gases above named, display a bright-striped spectrum, provided the intervening non-luminous vapors of the same metals are not sufficiently abundant to obscure them—at this particular moment, when only the absolute horizon-line is seen, and the body of the moon cuts off all the intervening solar surface, and the lower or denser portion of the intervening super-solar vapors, though, of course, these are not so entirely cut off as the continuous background.

The reversion of the dark lines therefore reveals to us the stupendous fact that the surface of the mighty sun, which is as big as a million and a quarter of our worlds, consists of a flaming ocean of hydrogen and of the metals above-named in a gaseous condition, similar to that of the hydrogen itself.

This fact, coupled with the other revelations of the spectroscope, which, without the help of an eclipse, reveals the surface outline of the sun, the "sierra" and the "prominences" tell us that this flaming ocean is in a state of perpetual tempest, heaving up its billows and flame-Alps hundreds and thousands of miles in height, and belching forth above all these still taller pillars of fire that even reach an elevation of more than a hundred thousand miles, and then burst out into mighty clouds of flame and vapor, bigger than five hundred worlds.

What does the last eclipse teach us in reference to the corona? Firstly and clearly, that Lockyer's explanation which attributed it to an illumination of the upper regions of the earth's atmosphere must be now forever abandoned. This theory has died hard, but, in spite of Mr. Lockyer's proclamation of "victory all along the line," it is now past galvanizing. There can be no further hesitation in pronouncing that the corona actually belongs to the sun itself, that it is a marvelous solar appendage extending from the sun in all directions, but by no means regularly.

The immensity of this appendage will be best understood by the fact that the space included within the outer limits of the visible corona is at least twenty times as great as the bulk of the sun itself, that above twenty-five millions of our worlds would be required to fill it.

Jannsen says: "I believe the question whether the corona is due to the terrestrial atmosphere is settled, and we have before us the prospect of the study of the extra-solar regions, which will be very interesting and fertile."

The spectroscope, the polariscope, and ordinary vision all concur in supporting the explanation that the corona is composed of solid particles and gaseous matter intermingled. It fulfils exactly all the requirements of the hypothesis which attributes it to the same materials as those which in a gaseous state cause the reversion of the dark lines above described, but which have been ejected with the great eruptions forming the solar prominences, and have become condensed into glowing metallic hailstones as their distance from the central heat has increased. These must necessarily be accompanied by the vapors of the more volatile materials, and should give out some of the lighter gases, such as hydrogen, which, under greater pressure, would be occluded within them, just as the hydrogen gas occluded within the substance of the Lenarto meteor (a mass of iron which fell from the sky upon the earth) was extracted by the late Master of the Mint by means of his mercurial air-pump.

The rifts or gaps between the radial streamers, which have been so often described and figured, but were regarded by some as optical illusions, are now established as unquestionable facts. Mr. Lockyer, the last to be convinced, is now compelled to admit this, which overthrows the supposition that this solar appendage is a luminous solar atmosphere of any kind. If it were gaseous or true vapor, it must obey the law of gaseous diffusion, and could not present the phenomena of bright radial streamers, with dark spaces between them, unless it were in the course of very rapid radial motion either to or from the sun.

The photographs have not yet been published. When they have all arrived, and can be compared, we shall learn something that I anticipate will be extremely interesting respecting the changes of the corona, as they have been taken at the different stations at different times. I alluded to this subject before, when it was only a matter of possibility that such a succession of pictures might have been taken. We now have the assurance that such pictures have been obtained. There can be no question about optical illusion in these; they are original affidavits made by the corona itself, signed, sealed, and delivered as its own act and deed.

Chapter 12

METEORIC ASTRONOMY

The number of the *Quarterly Journal of Science* for May, 1872, contains some articles of considerable interest. The first is by the indefatigable Mr. Proctor, on "Meteoric Astronomy," in which he embodies a clear and popular summary of the researches which have earned for Signor Schiaparelli this year's gold medal of the Astronomical Society. Like all who venture upon a broad, bold effort of scientific thought, extending at all into the regions of philosophical theory, Schiaparelli has had to wait for recognition. A simple and merely mechanical observation of a bare fact, barely and mechanically recorded without the exercise of any other of the intellectual faculties than the external senses and observing powers, is at once received and duly honored by the scientific world; but any higher effort is received at first indifferently, or sceptically, and is only accepted after a period of probation, directly proportionate to its philosophical magnitude and importance, and inversely proportionate to the scientific status of the daring theorist.

At first sight this appears unjust, it looks like honoring the laborers who merely make the bricks, and despising the architect who constructs the edifice of philosophy from the materials they provide. Many a disappointed dreamer, finding that his theory of the universe has not been accepted, and that the expected honors have not been showered upon him, has violently attacked the whole scientific community as a contemptible gang of low-minded mechanical plodders, void of imagination, blind to all poetic aspirations, and incapable of any grand and comprehensive flight of intellect.

Had these impulsive gentlemen been previously subjected to the strict discipline of inductive scientific training, their position and opinions would have been very different. Their great theories would either have had no existence, or have been much smaller, and they would understand that philosophic caution is one of the characteristic results of scientific training.

Simple facts, which can be immediately proved by simple experiments and simple observations, are at once accepted, and their discoverers duly honored, without any hesitation or delay, but the grander efforts of generalization require careful thought and laborious scrutiny for their verification, and therefore the acknowledgment of their merits is necessarily delayed; but when it does arrive full justice is usually done.

Thus Grove's "Correlation of the Physical Forces," the greatest philosophical work on purely physical science of this generation, was commenced in 1842, when its author occupied but a humble position at the London Institution. The book was but little noticed for many years, and, had Mr. Grove (now Sir William Grove) not been duly educated by the discipline above referred to, he might have become a noisy cantankerous martyr, one of those "ill-used men" who have been made familiar to so many audiences by Mr. George Dawson.

Instead of this, he patiently waited, and, as we have lately seen, the well-deserved honors have now been liberally awarded.

In a very few years hence we shall be able to say the same of the once diabolical Darwin, and eight or nine other theorists, who must all be content to take their trial and patiently await the verdict; the time of waiting being of necessity proportionate to the magnitude of the issue.

The theories of Schiaparelli, which, as Mr. Proctor says, "after the usual term of doubt have so recently received the sanction of the highest astronomical tribunal of Great Britain," are not of so purely speculative a character as to demand a very long "term of doubt." They are directly based on observations and mathematical calculations which bring them under the domain of the recognized logic of mathematical probability. Those who are specially interested in the modern progress of astronomy should read this article in the *Quarterly Journal of Science*, which is illustrated with the diagrams necessary for the comprehension of the researches and reasoning of Schiaparelli and others who have worked on the same ground.

I can only state the general results, which are that the meteors which we see every year, more or less abundantly, on the nights of the 10th and 11th of August, and which always appear to come from the same point in the heavens, are then and thus visible because they form part of an eccentric elliptical zone of meteoric bodies which girdle the domain of the sun; and that our earth, in the course of its annual journey around the sun, crosses and plunges more or less deeply into this ellipse of small attendant bodies, which are supposed to be moving in regular orbits around the sun.

Schiaparelli has compared the position, the direction, and the velocity of motion of the August meteors with the orbit of the great comet of 1862, and infers that there is a close connection between them, so close that the meteors may be regarded as a sort of trail which the comet has left behind. He does not exactly say that they are detached vertebræ of the comet's tail, but suggests the possibility of their original connection with its head.

Similar observations have been made upon the November meteoric showers, which by similar reasoning, are associated with another comet; and further yet, it is assumed upon analogy that other recognized meteor systems, amounting to nearly two hundred in number, are in like manner associated with other comets.

If these theories are sound, our diagrams and mental pictures of the solar system must be materially modified. Besides the central sun, the eight planets and the asteroids moving in their nearly circular orbits, and some eccentric comets traveling in long ellipses, we must add a countless multitude of small bodies clustered in elliptical rings, all traveling together in the path marked by their containing girdle, and following the lead of a streaming vaporous monster, their parent comet.

We must count such comets, and such rings filled with attendant fragments, not merely by tens or hundreds, but by thousands and tens of thousands, even by millions; the path of the earth being but a thread in space, and yet a hundred or two are strung upon it.

In this article Mr. Proctor seems strongly disposed to return to the theory which attributes solar heat and light to a bombardment of meteors from without, and the solar corona and zodiacal light as visible presentments of these meteors. Still, however, he clings to the more recent explanation which regards the corona, the zodiacal light, and the meteors as matter ejected from the sun by the same forces as those producing the solar prominences. For my own part I shall not be at all surprised if we find that, ere long, these two apparently conflicting hypotheses are fully reconciled.

The progress of solar discovery has been so great since January, 1870, when my ejection theory was published, that I may now carry it out much further than I then dared, or was justified in daring to venture. Actual measurement of the projectile forces displayed in some of the larger prominences renders it not merely possible, but even very probable, that some of the exceptionally great eruptive efforts of the sun may be sufficiently powerful to eject solar material beyond the reclaiming reach of his own gravitating power.

In such a case the banished matter must go on wandering through the boundless profundity of space until it reaches the domain of some other sun, which will clutch the fragment with its gravitating energies, and turn its straight and ever onward course into the curved orbit. Thus the truant morsel from our sun will become the subject of another sun—a portion of another solar system.

What one sun may do, another and every other may do likewise, and, if so, there must be a mutual bombardment, a ceaseless interchange of matter between the countless suns of the universe. This is a startling view of our cosmical relations, but we are driving rapidly towards a general recognition of it.

The November star showers have perpetrated some irregularities this year. They have been very unpunctual, and have not come from their right place. We have heard something from Italy, but not the tidings of the Leonides that were expected. Instead of the great display of the month occurring on the 13th and 14th, it was seen on the 27th. We have accounts from different parts of England, Ireland, Scotland, and Wales, also from Italy, Greece, Egypt, etc.

Mr. Slinto, in a letter to the *Times*, estimates the number seen at Suez as reaching at least 30,000, while in Italy and Athens about 200 per minute were observed. They were not, however, the Leonides, that is, they did not radiate from a point in the constellation Leo, but from the region of Andromeda. Therefore they were distinct from that system of small wanderers usually designated the "November meteors," were not connected with Tempel's comet (comet 1, 1866), but belong to quite another set.

The question now discussed by astronomers is whether they are connected with any other comet, and, if so, with which comet?

In the "Monthly Notices" of the Royal Astronomical Society, published October 24th last, is a very interesting paper by Professor Herschel, on "Observations of Meteor Showers," supposed to be connected with "Biela's comet," in which he recommends that "a watch should be kept during the last week in November and the first week in December," in order to verify "the ingenious suggestions of Dr. Weiss," which, popularly stated, amount to this, viz., that a meteoric cloud is revolving in the same orbit as Biela's comet, and that in 1772 the earth dashed through this meteoric orbit on December 10th. In 1826 it did the same, on December 4th; in 1852 the earth passed through the node on November 28th, and there are reasons for expecting a repetition at about the same date in 1872.

The magnificent display of the 27th has afforded an important verification of these anticipations, which become especially interesting in connection with the curious history of Biela's comet, which receives its name from M. Biela, of Josephstadt, who observed it in 1826, calculated its orbit, and considered it identical with the comets of 1772, 1805, etc. It travels in a long eccentric ellipse, and completes its orbit in 2410 days—about 6¾ years. It appeared again, as predicted, in 1832 and 1846.

Its orbit very nearly intersects that of the earth, and thus affords a remote possibility of that sort of collision which has excited so much terror in the minds of many people, but which an enthusiastic astronomer of the present generation would anticipate with something like the sensational interest which stirs the soul of a London street-boy when he is madly struggling to keep pace with a fire-engine.

The calculations for 1832 showed that this comet should cross the earth's orbit a little before the time of the earth's arrival at the same place; but as such a comet, traveling in such an orbit, is liable to possible retardations, the calculations could only be approximately accurate, and thus the sensational astronomer was not altogether without hope. This time, however, he was disappointed; the comet was punctual, and crossed the critical node about a month before the earth reached it.

As though to compensate for this disappointment, the comet at its next appearance exhibited some entirely new phenomena. It split itself into two comets, in such a manner that the performance was visible to the telescopic observer. Both of these comets had nuclei and short tails, and they alternately varied in brightness, sometimes one, then the other, having the advantage. They traveled on at a distance of about 156,000 miles from each other, with parallel tails, and with a sort of friendly communication in the form of a faint arc of light, which extended as a kind of bridge, from one to the other. Besides this, the one which was first the brighter, then the fainter, and finally the brighter again, threw out two additional tails, one of which extended lovingly towards its companion.

The time of return in 1852 was of course anxiously expected by astronomers, and careful watch was kept for the wanderers. They came again at the calculated time, still separated as before.

They were again due in 1859, in 1866, and, finally, at about the end of last November, or the beginning of the present month. Though eagerly looked for by astronomers in all parts of the civilized world, they have been seen no more since 1852.

What, then, has become of them? Have they further subdivided? Have they crumbled into meteoric dust? Have they blazed or boiled into thin air? or have they been dragged by some interfering gravitation into another orbit? The last supposition is the most improbable, as none of the visible inhabitants of space have come near enough to disturb them.

The possibility of a dissolution into smaller fragments is suggested by the fact that, instead of the original single comet, or the two fragments, meteoric showers have fallen towards the earth at the time when it has crossed the orbit of the original comet, and these showers have radiated from that part of the heavens in which the comet should have appeared. Such was the case with the magnificent display of November 27th, and astronomers are inclining more and more to the idea that comets and meteors have a common origin—the meteors are little comets, or comets are big meteors.

In the latest of the "Monthly Notices," of the Royal Astronomical Society, published last week, is a paper by Mr. Proctor, in which he expands the theory expounded three years ago by an author whom your correspondent's modesty prevents him from naming, viz., that the larger planets—Jupiter, Saturn, Uranus, and Neptune—are minor suns, ejecting meteoric matter from them by the operation of forces similar to those producing the solar prominences.

Mr. Proctor subjects this bold hypothesis to mathematical examination, and finds that the orbit of Tempel's comet and its companion meteors correspond to that which would result from such an eruption occurring on the planet Uranus. An eruptive force effecting a velocity of about thirteen miles per second, which is vastly smaller than the actually measured velocity of the matter of the solar eruptions, would be sufficient to thrust such meteoric or cometary matter beyond the reclaiming reach of the gravitation of Uranus, and hand it over to the sun, to make just such an orbit as that of Tempel's comet and the Leonides meteors.

He shows that other comets and meteoric zones are similarly allied to other planets, and thus it may be that the falling stars and comets are fragments of Jupiter, Saturn, Uranus, or Neptune. Verily, if an astronomer of the last generation were to start up among us now, he would be astounded at modern presumption.

The star shower of November 27th, and its connection with Biela's broken and lost comet, referred to in my last letter, are still subjects of research and speculation. On November 30th Professor Klinkerfues sent to Mr. Pogson, of the Madras Observatory, the following startling telegram: "Biela touched earth on 27th. Search near Theta Centauri."

Mr. Pogson searched accordingly from comet-rise to sunrise on the two following mornings, but in vain; for even in India they have had cloudy weather of late. On the third day, however, he had "better luck," saw something like a comet through an opening between clouds, and on the following days was enabled to deliberately verify this observation and determine the position and some elements of the motion of the comet, which displayed a bright nucleus, and faint but distinct tail.

This discovery is rather remarkable in connection with the theoretical anticipation of Professor Klinkerfues; but the conclusion directly suggested is by no means admitted by astronomers. Some, have supposed that it is not the primary Biela, but the secondary comet, or offshoot, which grazed the earth, and was seen by Mr. Pogson; others that it was neither the body, the envelope, nor the tail of either of the comets which formed the star shower, but that the meteors of November 27th were merely a trail which the comet left behind.

A multitude of letters were read at the last and previous meeting of the Astronomical Society, in which the writers described the details of their own observations. As these letters came from nearly all parts of the world, the data have an unusual degree of completeness, and show very strikingly the value of the work of amateur astronomical observers.

By the collation and comparison of these, important inductions are obtainable. Thus Professor A. S. Herschel concludes that the earth passed through seven strata of meteoric bodies, having each a thickness of about 50,000 miles—in all about 350,000 miles. As the diameter of the visible nebulosity of Biela's comet was but 40,000 miles when nearest the earth in 1832, the great thickness of these strata indicates something beyond the comet itself.

Besides this, Mr. Hind's calculation for the return of the primary comet shows that on November 27th it was 250 millions of miles from the earth.

Those, however, who are determined to enjoy the sensation of supposing that they really have been brushed by the tail of a comet, still have the secondary comet to fall back upon. This, as already described, was broken off the original, from which it was seen gradually to diverge, but was still linked to it by an arch of nebulous matter.

If this divergence has continued, it must now be far distant—sufficiently far to afford me an opportunity of safely adding another to the numerous speculations, viz., that we may, on November 27th, have plunged obliquely through this connecting arm of nebulous matter, which was seen stretching between the parent comet and its offshoot. The actual position of the meteoric strata above referred to is quite consistent with the hypothesis.

Chapter 13

THE "GREAT ICE AGE" AND THE ORIGIN OF THE "TILL"

The growth of science is becoming so overwhelming that the old subdivisions of human knowledge are no longer sufficient for the purpose of dividing the labor of experts. It is scarcely possible now for any man to become a naturalist, a chemist, or a physicist in the full sense of either term; he must, if he aims at thoroughness, be satisfied with a general knowledge of the great body of science, and a special and a full acquaintance with only one or two of its minor subdivisions. Thus geology, though but a branch of natural history, and the youngest of its branches, has now become so extensive that its ablest votaries are compelled to devote their best efforts to the study of sections which but a few years ago were scarcely definable.

Glaciation is one of these, which now demands its own elementary text-books over and above the monographs of original investigators. This demand has been well supplied by Mr. James Geikie in the "The Great Ice Age,"[15] of which a second edition has just been issued. Every student of glacial phenomena owes to Mr. Geikie a heavy debt of gratitude for the invaluable collection of facts and philosophy which this work presents. It may now be fairly described as a standard treatise on the subject which it treats.

One leading feature of the work offers a very aggressive invitation to criticism. Scotchmen are commonly accused of looking upon the whole universe through Scotch spectacles, and here we have a Scotchman treating a subject which affects nearly the whole of the globe, and devoting about half of his book to the details of Scottish glacial deposits; while England has but one-third of the space allowed to Scotland, Ireland but a thirtieth, Scandinavia less than a tenth, North America a sixth, and so on with the rest of the world. Disproportionate as this may appear at first glance, further acquaintance with the work justifies the pre-eminence which Mr. Geikie gives to the Scotch glacial deposits. Excepting Norway, there is no country in Europe which affords so fine a field for the study of the vestiges of extinct glaciers as Scotland, and Scotland has an advantage even over Norway in being much better known in geological detail. Besides this, we must always permit the expounder of any subject to select his own typical illustrations, and welcome his ability to find them in a region which he himself has directly explored.

---

[15] "The Great Ice Age, and its Relation to the Antiquity of Man." By James Geikie, F.R.S., etc. Second edition, revised, 1877. Daldy and Isbister.

Mr. Geikie's connection with the geological survey of Scotland has afforded him special facilities for making good use of Scottish typical material, and he has turned these opportunities to such excellent account that no student after reading "The Great Ice Age" will find fault with its decided nationality.

The leading feature—the basis, in fact—of this work deserves especial notice, as it gives it a peculiar and timely value of its own. This feature is that the subject—as compared with its usual treatment by other leading writers—is turned round and presented, so to speak, bottom upwards. De Saussure, Charpentier, Agassiz, Humboldt, Forbes, Hopkins, Whewell, Stark, Tyndall, etc., have studied the living glaciers, and upon the data thus obtained have identified the work of extinct glaciers. Chronologically speaking, they have proceeded backwards, a method absolutely necessary in the early stages of the inquiry, and which has yielded admirable results. Geikie, in the work before us, proceeds exactly in the opposite order. Availing himself of the means of identifying glacial deposits which the retrogressive method affords, he plunges at once to the lowest and oldest of these deposits, which he presents the most prominently, and then works upwards and onwards to recent glaciation.

The best illustration I can offer of the timely advantage of this reversed treatment is (with due apology for necessary egotism) to state my own case. In 1841, when the "glacial hypothesis," as it was then called, was in its infancy, Professor Jamieson, although very old and nearly at the end of his career, took up the subject with great enthusiasm, and devoted to it a rather disproportionate number of lectures during his course on Natural History. Like many of his pupils, I became infected by his enthusiasm, and went from Edinburgh to Switzerland, where I had the good fortune to find Agassiz and his merry men at the "Hotel des Neufchatelois"—two tents raised upon a magnificent boulder floating on the upper part of the Aar glacier. After a short but very active sojourn there I "did," not without physical danger, many other glaciers in Switzerland and the Tyrol, and afterwards practically studied the subject in Norway, North Wales, and wherever else an opportunity offered, reading in the meantime much of its special literature; but, like many others, confining my reading chiefly to authors who start with living glaciers and describe their doings most prominently. When, however, I read the first edition of Mr. Geikie's "Great Ice Age," immediately after its publication, his mode of presenting the phenomena, bottom upwards, suggested a number of reflections that had never occurred before, leading to other than the usual explanations of many glacial phenomena, and correcting some errors into which I had fallen in searching for the vestiges of ancient glaciers. As these suggestions and corrections may be interesting to others, as they have been to myself, I will here state them in outline.

The most prominent and puzzling reflection or conclusion suggested by reading Mr. Geikie's description of the glacial deposits of Scotland was, that the great bulk of them are quite different from the deposits of existing glaciers. This reminded me of a previous puzzle and disappointment that I had met in Norway, where I had observed such abundance of striation, such universality of polished rocks and rounded mountains, and so many striking examples of perched blocks, with scarcely any decent vestiges of moraines. This was especially the case in Arctic Norway. Coasting from Trondhjem to Hammerfest, winding round glaciated islands, in and out of fjords banked with glaciated rock-slopes, along more than a thousand miles of shore line, displaying the outlets of a thousand ancient glacier valleys, scanning eagerly throughout from sea to summit, landing at several stations, and climbing the most commanding hills, I *saw only one ancient moraine*—that at the Oxfjord station described in "Through Norway with Ladies."[16]

But this negative anomaly is not all. The ancient glacial deposits are not only remarkable on account of the absence of the most characteristic of modern glacial deposits, but in consisting mainly of something which is quite different from any of the deposits actually formed by any of the modern glaciers of Switzerland or any other country within the temperate zones.

I have seen nothing either at the foot or the sides of any living Alpine or Scandinavian glacier that even approximately represents the "till" or "boulder clay," nor any description of such a formation by any other observer; and have met with no note of this very suggestive anomaly by any writer on glaciers. Yet the till and boulder clay form vast deposits, covering thousands of square miles even of the limited area of the British Isles, and constitute the main evidence upon which we base all our theories respecting the existence and the vast extent and influence of the "Great Ice Age."

Although so different from anything at present produced by the Alpine or Scandinavian glaciers, this great deposit is unquestionably of glacial origin. The evidences upon which this general conclusion rests are fully stated by Mr. Geikie, and may safely be accepted as incontrovertible. Whence, then, the great difference?

One of the suggestions to which I have already alluded as afforded by reading Mr. Geikie's book was a hypothetical solution of this difficulty, but the verification of the hypothesis demanded a re-visit to Norway. An opportunity for this was afforded in the summer of 1874, during which I traveled round the coast from Stavanger to the Arctic frontier of Russia, and through an interesting inland district. The observations there made and strengthened by subsequent reflections, have so far confirmed my original speculative hypothesis that I now venture to state it briefly as follows:

That the period appropriately designated by Mr. Geikie as the "Great Ice Age" includes at least two distinct periods or epochs—the first of very great intensity or magnitude, during which the Arctic regions of our globe were as completely glaciated as the Antarctic now are, and the British islands and a large portion of Northern Europe were glaciated as completely, and nearly in the same manner, as Greenland is at the present time; that long after this, and immediately preceding the present geological epoch, there was a minor glacial period, when only the now existing valleys, favorably shaped and situated for glacial accumulations, were

---

[16] The terminal moraine at the Oxfjord station, which I have already mentioned as the only ancient example of an ordinary moraine that I have seen in Arctic Norway, was, of course, a special object of interest to me. Further observation showed that it does not merely consist of the heap of stones I noticed in 1856, which appears like a disturbed talus cut through and heaped up at its lower part, but that there is another moraine adjoining it, or in continuation with it, which is covered with vegetation, and stretches quite across the mouth of the valley. The Duke of Roxburgh, who is well acquainted with this neighborhood, having spent sixteen summers in Arctic Norway, was one of our fellow-passengers, and told me that this moraine forms a barrier that dams up the waters of a considerable lake, abounding with remarkably fine char. I learned this just as the packet was starting, too late to go on shore even for a few minutes, and obtain a view of this lake and the valley beyond. This I regret, as it might have revealed some explanation of the exceptional nature of this moraine. It would be interesting to learn whether it belongs to the greater ice age, or to that period of minor glaciation that fashioned the farm patches already described. The formation of the lake is easily understood in the latter case. It is only required that such a minor reglaciated valley as one of these should be of larger magnitude and of very gentle inclination at its lower part, so that the secondary glacier should die out before reaching the present seashore. It would then deposit its moraine across the mouth of the valley, and this moraine would dam up the waters which such a valley must necessarily receive from the drainage of its hilly sides. Llyn Idwal, in North Wales, is a lake thus formed.

partially or wholly filled with ice. There may have been many intermediate fluctuations of climate and glaciation, and probably were such, but as these do not affect my present argument they need not be here considered.

So far I agree with the general conclusions of Mr. Geikie as I understand them, and with the generally received hypotheses, but in what follows I have ventured to diverge materially.

It appears to me that the existing Antarctic glaciers and some of the glaciers of Greenland are essentially different in their conformation from the present glaciers of the Alps, and from those now occupying some of the fjelds and valleys of Norway; and that the glaciers of the earlier or greater glacial epoch were similar to those now forming the Antarctic barrier, while the glaciers of the later or mino glacial epoch resembled those now existing in temperate climates, or were intermediate between these and the Antarctic glaciers. The nature of the difference which I suppose to exist between the two classes of glaciers is this: The glaciers (properly so called) of temperate climates are the overflow of the *nevé* (the great reservoir of ice and snow above the snow line). They are composed of ice which is protruded below the snow-line into the region where the summer thaw exceeds the winter snow-fall. This ice is necessarily subject to continual thinning or wasting from its *upper* or exposed surface, and thus finally becomes liquefied, and is terminated by direct solar action.

Many of the characteristic phenomena of Alpine glaciers depend upon this; among the more prominent of which are the superficial extrusion of boulders or rock fragments that have been buried in the *nevé* or have fallen into the crevasses of the upper part of the true glacier, and the final deposit of these same boulders of fragments at the foot of the glaciers forming ordinary moraines.

But this is not all. The thawing which extrudes, and finally deposits the larger fragments of rock, sifts from them the smaller particles, the aggregate bulk of which usually exceeds very largely that of the larger fragments. This fine silt or sand thus washed away is carried by the turbid glacier torrent to considerable distances, and deposited as an alluvium wherever the agitated waters find a resting-place.

Thus the *débris* of the ordinary modern glacier is effectively separated into two or more very distinct deposits; the moraine at the glacier foot consisting of rock fragments of considerable size with very little sand or clay or other fine deposit between them, and a distant deposit of totally different character, consisting of gravel, sand, clay, or mud, according to the length and conditions of its journey. The "chips," as they have been well called, are thus separated from what I may designate the *filings* or *sawdust* of the glacier.

The filings from the existing glaciers of the Bernese Alps are gradually filling up the lake-basins of Geneva and Constance, repairing the breaches made by the erosive action of their gigantic predecessors; those of the southern slope of the Alps are doing a large share in filling up the Adriatic; while the chips of all merely rest upon the glacier beds forming the comparatively insignificant terminal moraine deposits.

The same in Scandinavia. The Storelv of the Jostedal is fed by the melting of the Krondal, Nygaard, Bjornestegs, and soldal glaciers. It has filled up a branch of the deep Sogne fjord, forming an extensive fertile plain at the mouth of its wild valley, and is depositing another subaqueous plain beyond, while the moraines of the glaciers are but inconsiderable and comparatively insignificant heaps of loose boulders, spread out on the present and former shores of the above-named glaciers, which are overflows from one side of the great *nevé*, the Jostedal Sneefond. All of these glaciers flow down small lateral valleys, spread out, and disappear in the main valley, which has now no glacier of its own, though it was formerly glaciated throughout.

What must have been the condition of this and the other great Scandinavian valleys when such was the case? To answer this question rationally we must consider the meteorological conditions of that period. Either the climate must have been much colder, or the amount of precipitation vastly greater than at present, in order to produce the general glaciation that rounded the mountains up to a height of some thousands of feet above the present sea-level. Probably both factors co-operated to effect this vast glaciation, the climate colder, and the snow-fall also greater. The whole of Scandinavia, or as much as then stood above the sea, must have been a *nevé* or sneefond on which the annual snow-fall exceeded the annual thaw.

This is the case at present on the largest *nevé* of Europe, the 500 square miles of the great plateau of the Jostedals and Nordfjords Sneefond, on all the overflowing *nevé* or snow-fields of the Alps above the snow-line; over the greater part of Greenland; and (as the structure of the southern icebergs prove) everywhere within the great Antarctic ice barrier.

What, then, must happen when the snow-line comes down, or nearly down, to the sea-level? It is evident that the out-thrust glaciers, the overflow down the valleys, cannot come to an end like the present Swiss and Scandinavian glaciers, by the direct melting action of the sun. They may be somewhat thinned from below by the heat of the earth, and that generated by their own friction on the rocks, but these must be quite inadequate to overcome the perpetual accumulation due to the snow-fall upon their own surface and the vast overflow from the great snow-fields above. They must go on and on, ever increasing, until they meet some new condition of climate or some other powerful agent of dissipation—something that can effectively melt them.

This agent is very near at hand in the case of the Scandinavian valleys and those of Scotland. It is the sea. I think I may safely say that the valley glaciers of these countries during the great ice age *must* have reached the sea, and there have terminated their existence, just as the Antarctic glaciers terminate at the present Antarctic ice-wall.

What must happen when a glacier is thus thrust out to sea? This question is usually answered by assuming that it slides along the bottom until it reaches such a depth that flotation commences and then it breaks off or "calves" as icebergs. This view is strongly expressed by Mr. Geikie (p. 47) when he says that—"The seaward portion of an Arctic glacier cannot by any possibility be floated up without sundering its connection with the frozen mass behind. So long as the bulk of the glacier much exceeds the depth of the sea, the ice will of course rest upon the bed of the fjord or bay without being subjected to any strain or tension. But when the glacier creeps outwards to greater depths, then the superior specific gravity of the sea-water will tend to press the ice upward. That ice, however, is a hard continuous mass, with sufficient cohesion to oppose for a time this pressure, and hence the glacier crawls on to a depth far beyond the point at which, had it been free, it would have risen to the surface and floated. If at this great depth the whole mass of the glacier could be buoyed up without breaking off, it would certainly go to prove that the ice of Arctic regions, unlike ice anywhere else, had the property of yielding to mechanical strain without rupturing. But the great tension to which it is subjected takes effect in the usual way, and the ice yields, not by bending and stretching, but by breaking." Mr. Geikie illustrates this by a diagram showing the "calving" of an iceberg.

In spite of my respect for Mr. Geikie as a geological authority, I have no hesitation in contradicting some of the physical assumptions included in the above.

Ice has no such rigidity as here stated. It *does* possess in a high degree "the property of yielding to mechanical strain without rupturing." We need not go far for evidence of this. Everybody who has skated or seen others skating on ice that is but just thick enough to "bear" must have felt or seen it yield to the mechanical strain of the skater's weight. Under these conditions it not only bends under him, but it afterwards yields to the reaction of the water below, rising and falling in visible undulations, demonstrating most unequivocally a considerable degree of flexibility. It may be said that in this case the flexibility is due to the thinness of the ice; but this argument is unsound, inasmuch as the manifestation of such flexibility does not depend upon absolute thickness or thinness, but upon the relation of thickness to superficial extension. If a thin sheet of ice can be bent to a given arc, a thick sheet may be bent in the same degree, but the thicker ice demands a greater radius and proportionate extension of circumference. But we have direct evidence that ice of great thickness—actual glaciers—may bend to a considerable curvature before breaking. This is seen very strikingly when the uncrevassed ice-sheet of a slightly inclined *nevé* suddenly reaches a precipice and is thrust over it. If Mr. Geikie were right, the projecting cornice thus formed should stand straight out, and then, when the transverse strain due to the weight of this rigid overhang exceeded the resistance of tenacity, it should break off short, exposing a face at right angles to the general surface of the supported body of ice. Had Mr. Geikie ever seen and carefully observed such an overhang or cornice of ice, I suspect that the above-quoted passage would not have been written.

Some very fine examples of such ice-cornices are well seen from the ridge separating the Handspikjen Fjelde from the head of the Jostedal, where a view of the great *nevé* or sneefond is obtained. This side of the *nevé* terminates in precipitous rock-walls; at the foot of one of these is a dreary lake, the Styggevand. The overflow of the *nevé* here forms great bending sheets that reach a short way down, and then break off and drop as small icebergs into the lake.[17]

The ordinary course of glaciers affords abundant illustrations of the plasticity of such masses of ice. They spread out where the valley widens, contract where the valley narrows, and follow all the convexities or concavities of the axial line of its bed. If the bending thus enforced exceeds a certain degree of abruptness crevasses are formed, but a considerable bending occurs before the rupture is effected, and crevasses of considerable magnitude are commonly formed without severing one part of a glacier from another. They are usually V-shaped, in vertical section, and in many the rupture does not reach the bottom of the glacier. Very rarely indeed does a crevasse cross the whole breadth of a glacier in such a manner as to completely separate, even temporarily, the lower from the upper part of the glacier.

If a glacier can thus bend *downwards* without "sundering its connection with the frozen mass behind," surely it may bend upwards in a corresponding degree, either with or without the formation of crevasses, according to the thickness of the ice and the degree of curvature.

A glacier reaching the sea by a very steep incline would probably break off, in accordance with Mr. Geikie's description, just as an Alpine glacier is ruptured fairly across when it makes a cascade over a suddenly precipitous bend of its path. One entering the sea at an inclination somewhat less precipitous than the minor limit of the effective rupture gradient would be crevassed in a contrary manner to the crevassing of Alpine glaciers. Its crevasses would gape downwards instead of upwards—have Λ-shaped instead of a V-shaped section.

---

[17] See "Through Norway with a Knapsack," chapters xi. and xii., for further descriptions of these.

With a still more moderate slope, the up-floating of the termination of the glacier, and a concurrent general up-lifting or upbending of the whole of its submerged portion might occur without even a partial rupture or crevasse formation occurring.

Let us now follow out some of the necessary results of these conditions of glacier existence and glacial prolongation. The first and most notable, by its contrast with ordinary glaciers, is the absence of lateral, medial, or terminal moraines. The larger masses of *débris*, the chippings that may have fallen from the exposed escarpments of the mountains upon the surface of the upper regions of the glacier, instead of remaining on the surface of the ice and standing above its general level by protecting the ice on which they rest from the general snow-thaw, would become buried by the upward accretion of the ice due to the unthawed stratum of each year's snow-fall.

The thinning agency at work upon such glaciers during their journey over the *terra firma* being the outflow of terrestrial heat and that due to their friction upon their beds, this thinning must all take place from below, and thus, as the glaciers proceed downwards, these rock fragments must be continually approaching the bottom instead of continually approaching the top, as in the case of modern Alpine glaciers flowing below the snow-line, and thawing from surface downwards.

It follows, therefore, that such glaciers could not deposit any moraines such as are in course of deposition by existing Alpine and Scandinavian glaciers.

What, then, must become of the chips and filings of these outfloating glaciers? They must be carried along with the ice *so long as that ice rests upon the land*; for this *débris* must consist partly of fragments imbedded in the ice, and partly of ground and re-ground excessively subdivided particles, that must either cake into what I may call ice-mud, and become a part of the glacier, or flow as liquid mud or turbid water beneath it, as with ordinary glaciers. The quantity of water being relatively small under the supposed conditions, the greater part would be carried forward to the sea by the ice rather than by the water.

An important consequence of this must be that the erosive power of these ancient glaciers was, *cæteris paribus*, greater than that of modern Alpine glaciers, especially if we accept those theories which ascribe an actual internal growth or regeneration of glaciers by the relegation below of some of the water resulting from the surface-thaw.

As the glacier with its lower accumulation advances into deeper and deeper water, its pressure upon its bed must progressively diminish until it reaches a line where it would just graze the bottom with a touch of feathery lightness. Somewhere before reaching this it would begin to deposit its burden on the sea-bottom, the commencement of this deposition being determined by the depth whereat the tenacity of the deposit, or its friction against the sea-bottom, or both combined, becomes sufficient to overpower the now-diminished pressure and forward thrusting, or erosive power of the glacier.

Further forward, in deeper water, where the ice becomes fairly floated above the original sea-bottom, a rapid under-thawing must occur by the action of the sea-water, and if any communication exists between this ice covered sea and the waters of warmer latitudes this thawing must be increased by the currents that would necessarily be formed by the interchange of water of varying specific gravities. Deposition would thus take place in this deeper water, continually shallowing it or bringing up the sea-bottom nearer to the ice-bottom.

This raising of the sea-bottom must occur not only here, but farther back, i.e., from the limit at which deposition commenced. This neutral ground, whereat the depth is just sufficient to allow the ice to rest lightly on its own deposit and slide over it without either sweeping it forward or depositing any more upon it, becomes an interesting critical region, subject to continuous forward extension during the lifetime of the glacier, as the deposition beyond it must continually raise the sea-bottom until it reaches the critical depth at which the deposition must cease. This would constitute what I may designate the normal depth of the glaciated sea, or the depth towards which it would be continually tending, during a great glacial epoch, by the formation of a submarine bank or plain of glacier deposit, over which the glacier would slide without either grinding it lower by erosion or raising it higher by deposition.

But what must be the nature of this deposit? It is evident that it cannot be a mere moraine consisting only of the larger fragments of rock such as are now deposited at the foot of glaciers that die out before reaching the sea. Neither can it correspond to the glacial silt which is washed away and separated from these larger fragments by glacial streams, and deposited at the outspreadings of glacier torrents and rivers. It will correspond to neither the assorted gravel, sand, nor mud of these alluvial deposits, but must be an agglomeration of all the infusible solid matter the glacier is capable of carrying.

It must contain, in heterogeneous admixture, the great boulders, the lesser rock fragments, the gravel chips, the sand, and the slimy mud; these settling down quietly in the cold, gloomy waters, overshadowed by the great ice-sheet, must form just such an agglomeration as we find in the boulder clay and tills, and lie just in those places where these deposits abound, provided the relative level of land and sea during the glacial epoch were suitable.

I should make one additional remark relative to the composition of this deposit, viz., that under the conditions supposed, the original material detached from the rocks around the upper portions of the glaciers would suffer a far greater degree of attrition at the glacier bottom than it obtains in modern Alpine glaciers, inasmuch as in these it is removed by the glacier torrent when it has attained a certain degree of fineness, while in the greater glaciers of the glacial epoch it would be carried much further in association with the solid ice, and be subjected to more grinding and regrinding against the bottom. Hence a larger proportion of slimy mud would be formed, capable of finally induring into stiff clay such as forms the matrix of the till and boulder clay.

The long journey of the bottom *débris* stratum of the glacier, and its final deposition when in a state of neutral equilibrium between its own tendency to repose and the forward thrust of the glacier, would obviously tend to arrange the larger fragments of rock in the manner in which they are found imbedded in the till, i.e., the oblong fragments lying with their longer axes and their best marked striæ in the direction of the motion of the glacier. The "*striated pavements*" of the till are thus easily explained; they are the surface upon which the ice advanced when its deposits had reached the critical or neutral height. Such a pavement would continually extend outwards.

The only sorting of the material likely to occur under these conditions would be that due to the earlier deposition and entanglement of the larger fragments, thus producing a more stony deposit nearer inland, just as Mr. Geikie describes the actual deposits of till where, "generally speaking, the stones are most numerous in the till of hilly districts; while at the lower levels of the country the clayey character of the mass is upon the whole more pronounced." These "hilly districts," upon the supposition of greater submergence, would be the near shore regions, and the lower levels the deeper sea where the glacier floated freely.

The following is Mr. Geikie's description of the distribution of the till (page 13):—"It is in the lower-lying districts of the country where till appears in greatest force. Wide areas of the central counties are covered up with it continuously, to a depth varying from two or three feet up to one hundred feet and more. But as we follow it towards the mountain regions it becomes thinner and more interrupted—the naked rock ever and anon peering through, until at last we find only a few shreds and patches lying here and there in sheltered hollows of the hills. Throughout the Northern Highlands it occurs but rarely, and only in little isolated patches. It is not until we get way from the steep rocky declivities and narrow glens and gorges, and enter upon the broader valleys that open out from the base of the highland mountains to the low-lying districts beyond, that we meet with any considerable deposits of stony clay. The higher districts of the Southern Uplands are almost equally free from any covering of till."

This description is precisely the same as I must have written, had I so far continued my imaginary sketch of the results of ancient glaciation as to picture what must remain after the glaciers had all melted away, and the sea had receded sufficiently to expose their submarine deposits.

Throughout the above I have assumed a considerable submergence of the land as compared with the present sea-level on the coasts of Scotland, Scandinavia, etc.

The universality of the terraces in all the Norwegian valleys opening westward proves a submergence of *at least* 600 or 700 feet. When I first visited Norway in 1856, I accepted the usual description of these as alluvial deposits; was looking for glacial vestiges in the form of moraines, and thus quite failed to observe the true nature of these vast accumulations, which was obvious enough when I re-examined them in the light of more recent information. Some few are alluvial, but they are exceptional and of minor magnitude. As an example of such alluvial terraces I may mention those near the mouth of the Romsdal, that are well seen from the Aak Hotel, and which a Russian prince, or other soldier merely endowed with military eyes, might easily mistake for artificial earthworks erected for the defence of the valley.

In this case, as in the others where the terraces are alluvial, the valley is a narrow one, occupied by a relatively wide river loaded with recent glacial *débris*. It evidently filled the valley during the period of glacial recession.

The ordinary wider valleys, with a river that has cut a narrow channel through the outspread terrace-flats, display a different formation. Near the mouth of such valleys I have seen cuttings of more than a hundred feet in depth, through an unbroken terrace of most characteristic till, with other traces rising above it. This is the ordinary constitution of the *lower portions* of most of the Scandinavian terraces.

These terraces are commonly topped with quite a different stratum, which at first I regarded as a subsequent alluvial or estuarine deposit, but further examination suggested another explanation of the origin of some portions of this superficial stratum, to which I shall refer hereafter.

Such terraces prove a rise of sea or depression of land, during the glacial epoch, to the extent of 600 feet as a *minimum*, while the well-known deposits of Arctic shells at Moel Tryfaen and the accompanying drift have led Prof. Ramsay to estimate "the probable amount of submergence during some part of the glacial period at about 2300 feet."[18]

---

[18] Lyell, "Elements of Geology," p. 159.

It would be out of place here to reproduce the data upon which geologists have based their rather divergent opinions respecting the actual extent of the submergence of the western coast of North Europe. All agree that a great submergence occurred, but differ only as to its extent, their estimates varying between 1,000 and 3,000 feet.

There is one important consideration that must not be overlooked, viz., that—if my view of the submarine origin of the till be correct—the mere submergence of the land at the glacial period does not measure the difference between the depth of the sea at that and the present time, seeing that the deposits from the glaciers must have shallowed it very materially.

It is only after contemplating thoroughly the present form of the granitic and metamorphic hills of Scandinavia,—hills that are always angular when subjected only to subaerial weathering,—that one can form an adequate conception of the magnitude of this shallowing deposit. The rounding, shaving, grinding, planing, and universal abrasion everywhere displayed appear to me to justify the conclusion that if the sea were now raised to the level of the terraces, i.e., 600 feet higher than at present, the mass of matter abraded from the original Scandinavian mountains, and lying under the sea, would exceed the whole mass of mountain left standing above it.

The first question suggested by reading Mr. Geikie's book was whether the terraces are wholly or partially formed of till, and more especially whether their lower portions are thus composed. This, as already stated, was easily answered by the almost unanimous reply of all the many Norwegian valleys I traversed. Any tourist may verify this. The next question was whether this same till extends below the sea. This was not so easily answered by the means at my disposal, as I travelled hastily round the coast from Stavanger via the North Cape to the frontier of Russian Lapland in ordinary passenger steam-packets, which made their stoppages to suit other requirements than mine. Still, I was able to land at many stations, and found, wherever there was a gently sloping strand at the mouth of an estuary, or of a valley whose river had already deposited its suspended matter (a common case hereabouts, where so many rivers terminate in long estuaries or open out into bag-shaped lakes near the coast), and where the bottom had not been modified by secondary glaciation, that the receding tide displayed a sea-bottom of till, covered with a thin stratum of loose stones and shells. In some cases the till was so bare that it appeared like a stiff mud deposited but yesterday.

At Bodö, an arctic coast station on the north side of the mouth of the Salten fjord (lat. 67° 20´), where the packets make a long halt, is a very characteristic example of this; a deposit of very tough till forming an extensive plain just on the sea-level. The tide rises over this, and the waves break upon it, forming a sort of beach by washing away some of the finer material, and leaving the stones behind. The ground being so nearly level, the reach of the tide is very great, and thus a large area is exposed at low tide. Continuous with this, and beyond the limit of high tide, is an extensive inland plain covered with coarse grass and weeds growing directly upon the surface of the original flat pavement of till.

There is no river at Bodö; the sea is clear, leaves no appreciable deposit, and the degree of denudation of the clayey matrix of the till is very much smaller than might be expected. The limit of high water is plainly shown by a beach of shells and stones, but at low tide the ground over which the sea has receded is a bare and scarcely modified surface of till. I have observed the same at low water at many other arctic stations. In the Tromsö Sund there are shallows at some distance from the shore which are just covered with water at low tide. I landed and waded on these, and found the bottom to consist of till covered with a thin layer of shells, odd fragments of earthenware, and other rubbish thrown overboard from vessels. It is evident that breakers of considerable magnitude are necessary for the loosening of this tough compact deposit—that it is very slightly, if at all, affected by the mere flow of running water.

I specify these instances as characteristic and easy of verification, as the packets all stop at these stations; but a yachtsman sailing at leisure amidst the glorious coast scenery of the Arctic Ocean might multiply such observations a hundredfold by stopping wherever such strands are indicated in passing. I saw a multitude of these in places where I was unable to go ashore and examine them.

A further question in this direction suggested itself on the spot, viz., what is the nature of the "*banks*" which constitute the fishing-grounds of Norway, Iceland, Newfoundland, etc. They are submarine plains unquestionably—they must have a high degree of fertility in order to supply food for the hundreds of millions of voracious cod-fish, coal-fish, haddocks, hallibut, etc., that people them. These large fishes all *feed on the bottom*, their chief food being mollusca and crustacea, which must find, either directly or indirectly, some pasture of vegetable origin. The banks are, in fact, great meadows or feeding grounds for the lower animals which support the higher.

From the Lofoten bank alone twenty millions of cod-fish are taken annually, besides those devoured by the vast multitude of sea-birds. Now this bank is situated precisely where, according to the above-stated view of the origin of the till, there should be a huge deposit. It occupies the Vest fjord, i.e., the opening between the mainland and the Lofoden Islands, extending from Moskenes, to Lodingen on Hindö, just where the culminating masses of the Kjolen Mountains must have poured their greatest glaciers into the sea by a westward course, and these glaciers must have been met by another stream pouring from the north, formed by the glaciers of Hindö and Senjenö, and both must have coalesced with a third flood pouring through the Ofoten fjord, the Tys fjord, etc., from the mainland. The Vest fjord is about sixty miles wide at its mouth, and narrows northward till it terminates in the Ofoten fjord, which forks into several branches eastward. A glance at a good map will show that here, according to my explanation of the origin of the till, there should be the greatest of all the submarine plains of till which the ancient Scandinavian glaciers have produced, and of which the plains of till I saw on the coast at Bodö (which lies just to the mouth of the Vest fjord, where the Salten fjord flows into it), are but the slightly inclined continuation.

Some idea of this bank may be formed from the fact that outside of the Lofodens the sea is 100 to 200 fathoms in depth, that it suddenly shoals up to 16 or 20 fathoms on the east side of these rocks, and this shallow plain extends across the whole 50 or 60 miles between these islands and the mainland.[19] It must not be supposed the fjords or inlets of Scandinavia are *usually* shallower than the open sea; the contrary is commonly the case, especially with the narrowest and those which run farthest inland. They are *very much* deeper than the open sea.

If space permitted I could show that the great Storregen bank, opposite Aalesund and Molde, where the Stor fjord, Mold fjord, etc., were the former outlets of the glaciers from the highest of all the Scandinavian mountains, and the several banks of Finmark, etc., from

---

[19] The celebrated "Maelström" is one of the currents that flow down the submarine incline between these islands when the tide is falling. Although I have ridiculed some of the accounts of this now innocent stream, I am not prepared to assert that it was always as mild as at present. If the ancient glaciers were stopped suddenly, as they may well have been, by the rocky barrier of Mosken, between Vaerö and Moskenesö, and they then suddenly concluded their deposition of till, a precipice must have been formed between this and the deep sea outside the islands, down which the sea would pitch when the tide was falling, and thus form some dangerous eddies. This cascade would gradually obliterate itself by wearing down the precipitous wall to an inclined plane such as at present exists, and down which the existing current flows.

which, in the aggregate, are taken another 20 or 30 millions of cod-fish annually, are all situated just where theoretically they ought to be found. The same is the case with the great bank of Newfoundland and the banks around Iceland, which are annually visited by large numbers of French fishermen from Dunkerque, Boulogne, and other ports.

Whenever the packet halted over these banks during our coasting trip we demonstrated their fertility by casting a line or two over the bulwark. No bait was required, merely a double hook with a flat shank attached to a heavy leaden plummet. The line was sunk till the lead touched the bottom, a few jerks were given, and then a tug was felt: the line was hauled in with a cod-fish or halibut hooked, not inside the mouth, but externally by the gill-plates, the back, the tail, or otherwise. The mere jerking of a hook near the bottom was sufficient to bring it in contact with some of the population. There is a very prolific bank lying between the North Cape and Nordkyn, where the Porsanger and Laxe fjords unite their openings. Here we were able, with only three lines, to cover the fore-deck of the packet with struggling victims in the course of short halts of fifteen to thirty minutes. Not having any sounding apparatus by which to fairly test the nature of the sea-bottom in these places, I cannot offer any direct proof that it was composed of till. By dropping the lead I could *feel* it sufficiently to be certain that it was not rock in any case, but a soft deposit, and the marks upon the bottom of the lead, so far as they went, afforded evidence in favor of its clayey character. A further investigation of this would be very interesting.

But the most striking—I may say astounding—evidence of the fertility of these banks, one which appeals most powerfully to the senses, is the marvelous colony of sea-birds at Sverholtklubben, the headland between the two last-named fjords. I dare not estimate the numbers that rose from the rocks and darkened the sky when we blew the steam-whistle in passing. I doubt whether there is any other spot in the world where an equal amount of animal life is permanently concentrated. All these feed on fish, and an examination of the map will show why—in accordance with the above speculations—they should have chosen Sverholtklubben as the best fishing-ground on the arctic face of Europe.

I am fully conscious of the main difficulty that stands in the way of my explanation of the formation of the till, viz., that of finding sufficient water to float the ice, and should have given it up had I accepted Mr. Geikie's estimate of the thickness of the great ice-sheet of the great ice age.

He says (page 186) that "The ice which covered the low grounds of Scotland during the early cold stages of the glacial epoch was certainly more than 2000 feet in thickness, and it must have been even deeper than this between the mainland and the Outer Hebrides. To cause such a mass to float, the sea around Scotland would require to become deeper than now by 1400 or 1500 feet at least."

I am unable to understand by what means Mr. Geikie measured this depth of the ice which covered these low grounds, except by assuming that its surface was level with that of the upper ice-marks of the hills beyond. The following passage on page 63 seems to indicate that he really has measured it thus:—

"Now the scratches may be traced from the islands and the coast-line up to an elevation of at least 3,500 feet; so that ice must have covered the country to that height at least. In the Highlands the tide of ice streamed out from the central elevations down all the main straths and glens; and by measuring the height attained by the smoothed and rounded rocks we are enabled to estimate roughly the probable thickness of the old ice-sheet. But it can only be a rough estimate, for so long a time has elapsed since the ice disappeared, the rain and frost together have so split up and worn down the rocks of these highland mountains that much of the smoothing and polishing has vanished. But although the finer marks of the ice-chisel have thus frequently been obliterated, yet the broader effects remain conspicuous enough. From an extensive examination of these we gather that the ice could not have been less, and was probably more than 3,000 feet thick in its deepest parts."

Page 80 he says: "Bearing in mind the vast thickness reached by the Scotch ice-sheet, it becomes very evident that the ice would flow along the bottom of the sea with as much ease as it poured across the land, and every island would be surmounted and crushed, and scored and polished just as readily as the hills of the mainland were."

Mr. Geikie describes the Scandinavian ice-sheet in similar terms, but ascribes to it a still greater thickness. He says (page 404)—"The whole country has been moulded and rubbed and polished by an immense sheet of ice, which could hardly have been less than 6,000 or even 7,000 feet thick," and he maintains that this spread over the sea and coalesced with the ice-sheet of Scotland.

My recollection of the Lofoden Islands, which from their position afford an excellent crucial test of this question, led me to believe that their configuration presented a direct refutation of Mr. Geikie's remarkable inference; but a mere recollection of scenery being too vague, a second visit was especially desirable in reference to this point. The result of the special observations I made during this second visit fully confirmed the impression derived from memory.

I found in the first place that all along the coast from Stavanger to the Varanger fjord every rock *near the shore* is glaciated; among the thousands of low-lying ridges that peer above the water to various heights none near the mainland are angular. The general character of these is shown in the sketch of "My Sea Serpent," in the last edition of "Through Norway with a Knapsack."

The rocks which constitute the extreme outlying limits of the Lofoden group, and which are between 60 and 70 miles from the shore, although mineralogically corresponding with those near the shore, are totally different in their conformation, as the sketch of three characteristic specimens plainly shows. Mr. Everest very aptly compares them to shark's teeth. Proceeding northward, these rocks gradually progress in magnitude, until they become mountains of 3,000 to 4,000 feet in height; their outspread bases form large islands, and the Vest fjord gradually narrows.

The remarkably angular and jagged character of these rocks when weathered in the air renders it very easy to trace the limits of glaciation on viewing them at a distance. The outermost and smallest rocks show from a distance no signs of glaciation. If submerged, the ice of the great ice age was then enough to float over without touching them; if they stood above the sea, as at present, they suffered no more glaciation than would be produced by such an ice-sheet as that of the "paleocrystic" ice recently found by Captain Nares on the north of Greenland. Progressing northward, the glaciation begins to become visible, running up to about 100 feet above the sea-level on the islands lying westward and southward of Ost Vaagen. Further northward along the coast of Ost Vaagen and Hindö, the level gradually rises to about 500 feet on the northern portion of Ost Vaagen, and up to more than 1,000 feet on Hindö, while on the mainland it reaches 3,000 to 4,000 feet.

A remarkable case of such variation, or descent of ice-level, as the ice-sheet proceeded seaward, is shown at Tromsö. This small oblong island (lat. 69° 40´), on which is the capital town of Finmark, lies between the mainland and the large mountainous island of Kvalö, with a long sea-channel on each side, the Tromösund and the Sandesund; the total width of these two channels and the island itself being about four or five miles. The general line of glaciation from the mainland crosses the broad side of these channels and the island, which has evidently been buried and ground down to its present moderate height of two or three hundred feet. Both of the channels are till-paved. On the east or inland side the mountains near the coast are glaciated to their summits—are simply *roches moutonnées*, over which the reindeer of the Tromsdal Lapps range and feed. On the west the mountains are dark, pyramidal, non-glaciated peaks, with long vertical snow-streaks marking their angular masses.

The contrast is very striking when seen from the highest part of the island, and is clearly due to a decline in the thickness of the ice-sheet in the course of its journey across this narrow channel. Speaking roughly from my estimation, I should say that this thinning or lowering of the limits of glaciation exceeds 500 feet between the opposite sides of the channel, which, allowing for the hill slopes, is a distance of about 6 miles. This very small inclination would bring a glacier of 3,000 feet in thickness on the shore down to the sea-level in an outward course of 30 miles, or about half the distance between the mainland and the outer rocks of the Lofodens shown in the engraving.

I am quite at a loss to understand the reasoning upon which Mr. Geikie bases his firm conviction respecting the depth of the ice-sheet on the low grounds of Scotland and Scandinavia. He seems to assume that the glaciers of the great ice age had little or no superficial down slope corresponding to the inclination of the base on which they rested. I have considerable hesitation in attributing this assumption to Mr. Geikie, and would rather suppose that I have misunderstood him, as it is a conclusion so completely refuted by all we know of glacier phenomena and the physical laws concerned in their production; but the passages I have quoted, and several others, are explicit and decided.

Those geologists who contend for the former existence of a great polar ice-cap radiating outwards and spreading into the temperate zones, might adopt this mode of measuring its thickness, but Mr. Geikie rejects this hypothesis, and shows by his map of "The Principal Lines of Glacial Erosion in Sweden, Norway, and Finland," that the glaciation of the extreme north of Europe proceeded from south to north; that the ice was formed on land, and proceeded seawards in all directions.

I may add to this testimony that presented by the North Cape, Sverholt, Nordkyn, and the rest of the magnificent precipitous headlands that constitute the characteristic feature of the arctic-face of Europe. They stand forth defiantly as a phalanx of giant heralds proclaiming aloud the fallacy of this idea of southward glacial radiation; and in concurrence with the structure and striation of the great glacier troughs that lie between them, and the planed table-land at their summits, they establish the fact that during the greatest glaciation of the glacial epoch the ice-streams were formed on land and flowed out to sea, just as they now do at Greenland, or other parts of the world where the snow line touches or nearly approaches the level of the sea.

All such streams must have followed the slope of the hill-sides upon which they rested and down which they flowed, and thus the upper limits of glaciation afford no measure whatever of the thickness of the ice upon "the low grounds of Scotland," or of any other glaciated country. As an example, I may refer to Mont Blanc. In climbing this mountain the journey from the lower ice-wall of the Glacier de Bessons up to the *bergschrund* above the *Grand Plateau* is over one continuous ice-field, the level of the upper part of which is more than 10,000 feet above its terminal ice-wall. Thus, if we take the height of the striations or smoothings of the upper *nevé* above the low grounds on which the ice-sheet rests, and adopt Mr. Geikie's reasoning, the lower ice-wall of the Glacier de Bessons should be 10,000 feet thick. Its actual thickness, as nearly as I can remember, is about 10 or 12 feet.

Every other known glacier presents the same testimony. The drawing of a Greenland glacier opposite page 47 of Mr. Geikie's book shows the same under arctic conditions, and where the ice-wall terminates in the sea.

I have not visited the Hebrides, but the curious analogy of their position to that of the Lofodens suggests the desirability of similar observations to those I have made in the latter. If the ice between the mainland and the Outer Hebrides was, as Mr. Geikie maintains, "certainly more than 2000 feet in thickness," and this stretched across to Ireland, besides uniting with the still thicker ice-sheet of Scandinavia, these islands should all be glaciated, especially the smaller rocks. If I am right, the smaller outlying islands, those south of Barra, should, like the corresponding rocks of the Lofodens, display no evidence of having been overswept by a deep "*mer de glace.*"

I admit the probability of an ice-sheet extending as Mr. Geikie describes, but maintain that it thinned out rapidly seaward, and there became a mere ice-floe, such as now impedes the navigation of Smith's Sound and other portions of the Arctic Ocean. The Orkneys and Shetlands, with which I am also unacquainted, must afford similar crucial instances, always taking into account the fact that the larger islands may have been independently glaciated by the accumulations due to their own glacial resources. It is the small rocks standing at considerable distance from the shores of larger masses of land that supply the required test-conditions.

From the above it will be seen that I agree with Mr. Geikie in regarding the till as a "*moraine profonde*," but differ as to the mode and place of its deposition. He argues that it was formed under glaciers of the thickness he describes, while their whole weight rested upon it.

This appears to me to be physically impossible. If such glaciers are capable of eroding solid rocks, the slimy mud of their own deposits could not possibly have resisted them. The only case where this might have happened is where a mountain-wall has blocked the further downward progress of a glacier, or in pockets, or steep hollows which a glacier might have bridged over and filled up; but such pockets are by no means the characteristic localities of till, though the till of Switzerland may possibly show examples of the first case. The great depth of the inland lakes of Norway, their bottoms being usually far below that of the present sea-bottom, is in direct contradiction of this.[20] They should, before all places, be filled with till, if the till were a ground moraine formed on land; but all we know of them confirms the belief that the glaciers deepened them by erosion instead of shallowing them by deposition.

Mr. Geikie's able defence of Ramsay's theory of lake-basin erosion is curiously inconsistent with his arguments in favor of the ground moraine.

---

[20] The largest of the Norwegian lakes, the Mjosen, is 1550 feet deep, and its surface 385 feet above the sea-level. Its bottom is about 1000 feet lower than the sea outside, or 500 to 800 feet below the bottom of the Christiana Fjord. The fjords, generally speaking, are very much shallower near their mouths than further inland, as though their depth had been determined by the thickness of the glaciers flowing down them, and the consequent limits of flotation and deposition.

I fully concur with Mr. Geikie's arguments against the iceberg theory of the formation of the till. This, I think, he has completely refuted.

Before concluding I must say a few words on those curious lenticular beds of sand and gravel in the till which appear so very puzzling. A simple explanation is suggested in connection with the above-sketched view of the formation of the till. All glaciers, whether in arctic or temperate climates, are washed by streamlets during summer, and these commonly terminate in the form of a stream or cascade pouring down a *"moulin"*—a well bored by themselves and reaching the bottom of the glacier. Now what must be the action of such a downflow of water upon my supposed submarine bed of till just grazing the bottom of the glacier? Obviously, to wash away the fine clayey particles, and leave behind the coarser sand or gravel. It must form just such a basin or lenticular cavity as Mr. Geikie describes. The oblong shape of these, their longer axis coinciding with the general course of the glacier, would be produced by the onward progress of the moulin. The accordance of their other features with this explanation will be seen on reading Mr. Geikie's description (pp. 18, 19, etc).

The general absence of marine animals and their occasional exceptional occurrence in the intercalated beds is just what might be expected under the conditions I have sketched. In the gloomy subglacial depths of the sea, drenched with continual supplies of fresh water and cooled below the freezing-point by the action of salt water on the ice, ordinary marine life would be impossible; while, on the other hand, any recession of the glacial limit would restore the conditions of arctic animal life, to be again obliterated with the renewed outward growth of the floating skirts of the inland ice-mantle.

But I must now refrain from the further discussion of these and other collateral details, but hope to return to them in another paper.

In "Through Norway with Ladies" I have touched lightly upon some of these, and have more particularly described some curious and very extensive evidences of secondary glaciation that quite escaped my attention on my first visit, and which, too, have been equally overlooked by other observers. In the above I have endeavored to keep as nearly as possible to the main subject of the origin of the till and the character of the ancient ice-sheet.

Chapter 14

THE BAROMETER AND THE WEATHER

The barometer was invented by Torricelli, an Italian philosopher of the seventeenth century. It consists essentially of a long tube open at one end and closed at the other, and partly filled with mercury; but instead of being filled like ordinary vessels, with the open end or mouth upwards and the closed end or bottom downwards, the barometer-tube is inverted, and has its open mouth downwards. This open mouth is either dipped into a little cup of mercury or bent a little upwards.

Why does not the mercury run out of this lower open end and overflow the little cup when it is inverted after being filled?

The answer to this question includes the whole mystery and principle of the barometer. The mercury does not fall down because something pushes it up and supports it with a certain degree of pressure, and that something is the atmosphere which extends all round the world, and presses downwards and sideways and upwards—in every direction, in fact—with a force equal to its weight, i.e., with a pressure equal to about 15 lbs. on every square inch. A column or perpendicular square stick of air one inch thick each way, and extending from the surface of the sea up to the top of the atmosphere, weighs about 15 lbs.; other columns or sticks next to it on all sides weigh the same, and so on with every portion; and all these are for ever squeezing down and against each other, and, being fluid, transmit their pressure in every direction, and against the earth and everything upon it, and therefore upon the mercury of the barometer-tube.

We have supposed the air to be made up of columns or sticks of air one inch each way, but might have taken any other size, and the weight and pressure would be proportionate. Now mercury, bulk for bulk, is so much heavier than air, that a stick or column of this liquid metal about 30 inches high weighs as much as a stick or column of air of same thickness reaching from the surface of the earth to the top of the atmosphere; therefore, the 30-inch stick of mercury balances the pressure of the many miles of atmosphere, and is supported by it. Thus the column of mercury may be used to counterbalance the atmosphere and show us its weight; and such a column of mercury is a barometer, or "weight measure." The word *barometer* is compounded of the two Greek words—*baros*, weight, and *metron*, a measure.

If you take a glass tube a yard long, stopped at one end and open at the other, fill it with mercury, stop the open end with your thumb, then invert the tube and just dip the open end in a little cup of mercury, some of the mercury in the tube will fall into the cup, but not all; only six inches will fall, the other 30 inches will remain, with an empty space between it and the stopped end of the tube. When you have done this you will have made a rude barometer. If you prop up the tube, and watch it carefully from day to day, you will find that the height of the column of mercury will continually vary. If you live at the sea-level, or thereabouts, it will sometimes rise more than 30 inches above the level of the mercury in the cup, and frequently fall below that height. If you live on the top of a high mountain, or on any high ground, it will never reach 30 inches, will still be variable, its average height less than if you lived on lower ground; and the higher you go the less will be this average height of the mercury.

The reason of this is easily understood. When we ascend a mountain we leave some portion of the atmosphere below us, and of course less remains above; this smaller quantity must have less weight and press the mercury less forcibly. If the barometer tells the truth, it must show this difference; and it does so with such accuracy that by means of a barometer, or rather of two barometers—one at the foot of the mountain and one at its summit—we may, by their difference, measure the height of the mountain provided we know the rules for making the requisite calculations.

The old-fashioned barometer, with a large dial-face and hands like a clock, is called the "wheel barometer," because the mercury, in rising and falling, moves a little glass float resting upon the mercury of the open bent end of the tube; to this float and its counterpoise a fine cord is attached; and this cord goes round a little grooved wheel to which the hands are attached. Thus the rising and falling of the mercury moves the float, the float-cord turns the wheel, and the wheel moves the hand that points to the words and figures on the dial. When this hand moves towards the right, or in the direction of an advancing clock-hand, the barometer is rising; when it goes backwards, or opposite to the clock-hand movement, the mercury is falling. By opening the little door at the back of such a barometer, the above-described mechanism is seen. In doing this, or otherwise moving your barometer, be careful always to keep it upright.

It sometimes happens to these wheel barometers that they, suddenly cease to act; and in most cases the owner of the barometer may save the trouble and expense of sending it to the optician by observing whether the cord has slipped from the little wheel, and if so, simply replacing it in the groove upon its edge. If, however, the mischief is caused by the tube being broken, which is seen at once by the mercury having run out, the case is serious, and demands professional aid.

The upright barometer, which shows the surface of the mercury itself, is the most accurate instrument, provided it is carefully read. This form of instrument is always used in meteorological observatories, where minute corrections are made for the expansion and contraction which variations of temperature produce upon the length of the mercury without altering its weight, and for the small fluctuations in the level of the mercury cistern. With such instruments, fitted with an apparatus called a "vernier" the height of the mercury may be read to hundredths of an inch.

The necessity for the 30 inches of mercury renders the mercurial barometer a rather cumbrous instrument: it must be more than 30 inches long, and is liable to derangement from the spilling of the mercury. On this account portable barometers of totally different construction have been invented. The "aneroid" barometer is one of these—the only one that is practically used to any extent. It contains a metal box partly filled with air; one face of the box is corrugated, and so thin that it can rise and fall like a stretched covering of india-rubber. As the pressure of the outside air varies it does rise and fall, and by a beautifully-delicate apparatus this rising and falling is magnified and represented upon the dial. Such barometers are made small enough to be carried in the pocket, and are very useful for measuring the heights of mountains; but they are not quite so accurate as the mercurial barometer, and are therefore not used for rigidly scientific measurements; but for all ordinary purposes they are accurate enough, provided they are occasionally compared with a standard mercurial barometer, and adjusted by means of a watch-key axis provided for that purpose, and seen on the back of the instrument. They are sufficiently delicate to tell the traveller in a railway whether he is ascending or descending an incline, and will indicate the difference of height between the upper and lower rooms of a three-story house. With due allowance for variations of level, the traveler may use them as weather indicators; especially as it is the direction in which the barometer is moving (whether rising or falling) rather than its absolute height that indicates changes of weather. Thus by placing the aneroid in his room on reaching his hotel at night, carefully marking its height then and there, and comparing this with another observation made on the following morning, he may use it as a weather-glass in spite of hill and dale.

Water barometers have been made on the same principle as the mercury barometer; but as water is 13½ times lighter, bulk for bulk, than mercury, the height of the column must be 13½ times 30 inches, or, allowing for variations, not less than 34 feet. This, of course, is very cumbrous; the evaporation of the water presents another considerable difficulty,[21] still such a barometer is a very interesting instrument, as it shows the atmospheric fluctuations on 13½ times the scale of the ordinary barometer. A range of about five feet is thus obtained; and not only the great waves, but even the comparatively small ripples of the atmospheric ocean are displayed by it. In stormy weather it may be seen to rise and fall and pulsate like a living creature, so sensitively does it respond to every atmospheric fluctuation.

But why should the height of the barometer vary while it remains in the same place?

If the quantity of air surrounding the earth remains the same, and if the barometer measures its weight correctly, why should the barometer vary?

Does the atmosphere grow bigger and smaller, lighter and heavier, from time to time?

These are fair questions, and they bring us at once to some of the chief uses of the barometer. The atmosphere is a great gaseous ocean surrounding the earth, and we are creeping about on the bottom of this ocean. It has its tides and billows and whirling eddies, but all these are vastly greater than those of the watery ocean. At one time we are under the crest or rounded portion of a mighty atmospheric wave, at another the hollow between two such waves is over our heads, and thus the depth of atmosphere, or quantity of air, above us is variable. This variation is the combined result of many co-operating causes. In the first place, there are great atmospheric tides, caused, like those of the sea, by the attraction of the sun and moon; but these do not *directly* affect the barometer, because the attracting body supports whatever it lifts. Variations of temperature also produce important fluctuations in the height and density of the atmosphere, some of which are indicated by the barometer—others are not. Thus a mere expansion or contraction of *dry* air, increasing the depth or the density of the atmospheric ocean, would not affect the barometer, as mere expansion and contraction only alter the *bulk* without affecting the *weight* of the air. But our atmosphere consists not only of the permanent gases, nitrogen and oxygen; it contains besides these and carbonic acid, a considerable quantity of gaseous matter, which is not permanent, but which may be a gas at

---

[21] This has been recently overcome to a great extent by using glycerine instead of water.

one moment—contributing its whole weight to that of the general atmosphere—and at another moment some of it may be condensed into liquid particles that fall through it more or less rapidly, and thus contribute nothing to its weight.

What, then, is this variable constituent that sometimes adds to the weight of the atmosphere and the consequent height of the barometer, and at others may suddenly cease to afford its full contribution to atmospheric pressure?

It is simply water, which, as we all know, exists as solid, liquid, or gas, according to the temperature and pressure to which it is exposed. We all know that steam when it first issues from the spout of a tea-kettle is a transparent gas, or true vapor, but that presently, by contact with the cool air, it becomes white, cloudy matter, or minute particles of water; and that, if these are still further cooled, they will become hoar-frost or snow, or solid ice. Artificial hoar-frost and snow may be formed by throwing a jet of steam into very cold, frosty air. If you take a tin canister or other metal vessel, fill it with a mixture of salt with pounded ice or snow, and then hold the outside of the canister against a jet of steam, such as issues from the spout of a tea-kettle, a snowy deposit of hoar-frost will coat the outside of the tin. Now let us consider what takes place when a warm south-westerly wind, that has swept over the tropical regions of the Atlantic ocean, reaches the comparatively cold shores of Britain. It is cooled thereby, and some of its gaseous water is condensed—forming mists, clouds, rain, hoar-frost or snow. The greater part of this forms and falls on the western coasts, on Cornwall, Ireland, the Western Highlands of Scotland. Ireland gets the lion's share of this humidity, and hence her "emerald" verdure. The western slope of a mountain, in like manner, receives more rain than the side facing the east.

How does this condensation affect the barometer?

It must evidently cause it to fall, inasmuch as the air must be lightened to the exact extent of all that is taken out of it and precipitated. But the precipitation is not completed immediately the condensation occurs. It takes some time for the minute cloudy particles to gather into rain drops and fall to the earth, while the effect of the condensation upon the barometer is instantaneous; the air begins to grow lighter immediately the gas is converted into cloud or mist, and the barometer falls just at the same time and same rate as this is produced; but the rain comes some time afterwards. Hence the use of the barometer as a "weather glass." When intelligently and properly used it is very valuable in this capacity; but, like most things, it may easily be misunderstood and misused.

The most common error in the use of the barometer is that to which people are naturally led by the words engraved upon it, "Stormy, Much Rain, Rain, Change, Fair, Set Fair," etc. A direct and absolute blunder or falsehood is usually short-lived, and deceives but few people; but a false statement, with a certain amount of superficial truth, may survive for ages, and deceive whole generations. Now this latter is just the character of the weather signs that are engraved on our popular barometers; they are unsound and deceptive, but not utterly baseless.

Stormy, *Much Rain*, and *Rain* are marked against the low readings of the barometer, and *Very Dry*, *Set Fair*, and *Fair* against the higher readings. A low barometer is not a reliable sign of wet or stormy weather, neither is a high barometer to be depended upon for expecting fine weather; and yet it is true that we are more likely to have fine weather with a high than with a low barometer, and also the liability to rain and storms is greater with a low than with a high barometer.

The best indications of the weather are those derived from the direction in which the barometer is moving—whether rising or falling—rather than its mere absolute height.

A sudden and considerable fall is an almost certain indication of strong winds and stormy weather. This is the most reliable of the prophetic warnings of the barometer, and the most useful, inasmuch as it affords the mariner just the warning he requires when lying off a dangerous coast, or otherwise in peril by a coming gale. Many a good ship has been saved by intelligent attention to the barometer, and by running into haven, or away from a rocky shore when the barometer has fallen with unusual rapidity.

The next in order of reliability is the indication afforded by a steady and continuous fall after a long period of fine weather. This is usually followed by a decided change of weather, and the greater the fall the more violent the change. If the fall is slow, and continues steadily for a long time, the change is likely to be less sudden but more permanent, i.e., the rain will probably arrive after some time, and then continue steadily for a long period.

In like manner, a steady, regular rise, going on for some days in the midst of wet weather, may be regarded as a hopeful indication of coming continuous fine weather—the more gradual and steady the rise, the longer is the fine weather likely to last.

The least reliable of all the barometric changes is a sudden rise. In winter it may be followed by hard and sudden frost, in summer by sultry weather and thunder-storms. All that may be safely said of such sudden rise is, that it indicates a change of some sort.

The barometer is usually high with N.E. winds, and low with S.W. winds. The preceding explanations show the reason of this. In a given place the extreme range of variation is from 2 to 2½ inches.

It has been proposed that the following rules should be engraved on barometer-plates instead of the usual words:—

1st. Generally, the rising of the mercury indicates the approach of fair weather; the falling of it shows the approach of foul weather.
2d. In sultry weather, the fall of the barometer indicates coming thunder. In winter, the rise of the mercury indicates frost. In frost, its fall indicates thaw, and its rise indicates snow.
3d. Whatever change in the weather suddenly follows a change in the barometer, may be expected to last but a short time.
4th. If fair weather continues for several days, during which the mercury continually falls, a long succession of foul weather will probably ensue; and again, if foul weather continues for several days, while the mercury continually rises, a long succession of fair weather will probably follow.
5th. A fluctuating and unsettled state of the mercurial column indicates changeable weather.

As the barometer is subject to slight diurnal variations, irrespective of those atmospheric changes which affect the weather, it is desirable in making comparative observations to do so at fixed hours of the day. Nine or ten in the morning and same hour in the evening are good times for observations that are to be recorded. These are about the hours of daily maxima or highest readings due to regular diurnal variation.

The true reading of the barometer is the height at which it would stand if placed at the level of the sea at high tide; but, as barometers are always placed more or less above this level, a correction for elevation is necessary. When the height of the place is known this correction may be made by adding one tenth of an inch to the actual reading for every 85 feet of elevation up to 510 feet; the same for every 90 feet between 510 and 1140 feet, for every 95 feet between 1140 and 1900 feet, and for every 100 feet above this and within our mountain limits. This simple and easy rule is sufficiently accurate for practical purposes. Thus, a barometer on Bray Head, or any place 800 feet above the sea, would require a correction of six-tenths for the first 510 feet, and a little more than three-tenths more for the remaining 290 feet. Therefore, if such a barometer registered the pressure at 29-1/10, the proper sea-level reading would be a little above 30 inches.

The most important prognostications of the barometer are those afforded by what is called the "barometric gradient or incline," showing the up-hill and down-hill direction of the atmospheric inequalities; but this can only be ascertained by comparing the state of the barometer at different stations at the same time. Thus, if the barometer is one-fourth of an inch higher at Dublin than at Galway, and the intermediate stations show intermediate heights, there must be an atmospheric down-hill gradient from Dublin to Galway; Dublin must be under the upper and Galway under the lower portion of a great atmospheric wave or current. It is evident that when there is thus more air over Dublin than over Galway, there must follow (if nothing else interferes) a flow of air from Dublin towards Galway. It is also evident that, in order to tell what else may interfere, we must know the atmospheric gradients beyond and around both Dublin and Galway, and for considerable distances.

We are now beginning to obtain such information by organizing meteorological stations and observatories, and transmitting the results of simultaneous observations by means of the electric telegraph to certain head-quarters.

The subject is occupying much attention, and the managers of those splendid monuments of British energy—our daily newspapers—are publishing daily weather charts, and therefore a few simple explanations of the origin, nature, and significance of such charts will doubtless be appreciated by our readers.

The grand modern improvement of the barometer, the thermometer, the anemometer, the pluviometer, etc., is that of making them "self-registering." We are told that Cadmus invented the art of writing, and we honor his memory accordingly. But he ventured no further than teaching human beings to write. Modern meteorologists have gone much further; they have taught the winds and the rains and the subtle heavings of the invisible air to keep their own diaries, to write their own histories on paper that is laid before them, with pencils that are placed in their fleshless, boneless, and shapeless fingers. This achievement is wrought by comparatively simple means. The paper is wound upon an upright drum or cylinder, and this cylinder is made to revolve by clock-work, in such a manner that a certain breadth travels on during the twenty-four hours. This breadth of paper is divided by vertical lines into twenty-four parts, each of which passes onward in one hour. Connected with the barometer is a pencil which, by means of a spring, presses lightly upon the revolving sheet, and this pencil, while thus pressing, rises and falls with the mercury. It is obvious that, in this manner, a line will be drawn as the paper moves. If the mercury is stationary, the line will be horizontal—only indicating the movement of the drum; if the mercury falls, the line will slope downwards; if it rises, it will incline upwards. By ruling horizontal lines upon the paper, representing inches, tenths, and smaller fractions, if desired, the whole history of the barometrical movements will be graphically recorded by the waving or zigzag lines thus drawn by the atmosphere itself.

The subjoined copy of the *Daily Telegraph* Barometer Chart represents, on a small scale, a four days' history of barometrical movements:

The large figures at the side (29 and 30) represent inches; the smaller figures tenths of inches.

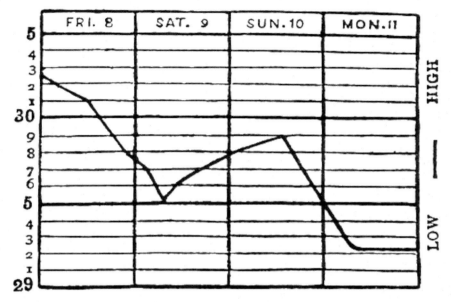

The pressure of the wind is similarly pictured by means of a large vane which turns with the wind, and to the windward face of which a flat board or plate of metal, one foot square, is attached perpendicularly. As the wind strikes this it presses against it with a force corresponding to a certain number of pounds, ounces, and fractions of an ounce. A spring like that of an ordinary spring letter-balance is compressed in proportion to this pressure. This movement of the spring is transmitted mechanically to another pencil like the above described, working against the same drum; thus another history is written on the same paper—the horizontal lines now representing fractions of pounds of pressure, instead of fractions of inches of mercury.

It has been found that if a semi-globular cup of thin metal is exposed to the wind, the pressure upon the round or convex side of the hemisphere is equal to two thirds of that upon the hollow or concave side. By placing four such cups upon cross-arms, and the arms on a pivot, the wind, from whatever quarter it may come, will always blow them round with their convex faces foremost; and they will move with one third of the actual velocity of the wind. By a simple clock-work arrangement, these arms move another pencil, in such a manner that it strikes the paper hammer-fashion every time the wind has completed a journey of one mile, or other given distance; and thus a series of dots upon the revolving paper records the velocity of the wind according to their distances apart. As the pressure of the wind is governed by two factors, viz., the density and velocity of the moving air, the relations between the barometer curve, the pressure curve, and the velocity dots, are very interesting.

The direction of the wind is written by a pencil fixed to a quick worm—a screw-thread upon the axis of the vane. As the vane turns round—N., E., S., or W.—it screws the pencil up or down, and thus the horizontal lines first described as registering tenths of inches of barometric pressure do duty as showing the points of the compass from which the wind is blowing; and, by reference to the zigzag line drawn by this pencil of the wind, its direction at any particular time of day may be ascertained as certified by its own sign-manual.

The wind-gauge is called an anemometer. Connected with this is the pluviometer, or rain-gauge—an upright vessel with an open mouth of measured area—say 100 square inches. This receives the rain that falls. By means of a pipe the water is conveyed to a vessel having a surface of—say one square inch. By this arrangement, when sufficient rain has fallen to cover the surface of the earth to the depth of one hundredth of an inch, the little vessel below will contain water one inch in depth. By balancing this vessel at the end of a long arm, it is made to preponderate gradually as the weight of water it receives increases, and finally, when filled, it tips over altogether, empties itself, and then rises to its starting place in equilibrium. To the other end of this arm a pencil is attached, which inscribes all these movements on the revolving paper, and thus tells the history of the rainfall. The line is zigzag while the rain is falling, and horizontal while the weather is fair. The amount of inclination of the zigzag line measures the depth of rain by means of the same ruled lines on the paper as measure the height of the barometer, etc. Every time the measuring vessel tips over a perpendicular line is drawn, and the pencil resumes its starting level. The papers containing these autographs of the elements may, of course, be kept as permanent records for reference whenever needed, or the results may be tabulated in other forms.

There are many modifications in the details of these self-registering instruments. In some of them photography is made to do a part of the work. The above description indicates the main principles of their construction, without attempting to enter upon minute details.

Meteorological observatories are provided with these instruments, and all nations worthy of the name of civilized co-operate with more or less efficiency in providing and endowing such establishments. They are placed in suitable localities, and communicate with each other, and with certain head-quarters, by means of the electric telegraph. One of these head-quarters is the Meteorological Office, at No. 116 Victoria Street, Westminster, S.W., which daily receives the results of the observations taken at about fifty stations on the British Islands and the Continent. The chief observations are made simultaneously—at 8 A.M.—and telegraphed in cypher to London, where they usually arrive before 10 A.M. As they come in they are marked down in their proper places upon a large chart, and when this chart is sufficiently completed, a condensed or abstract copy is made containing as much information as may be included in the small newspaper charts. This is copied mechanically on a reduced scale on a slab on which the outline chart has been already engraved. This engraving completed, casts are made in fusible metal with the black lines in relief, for printing with ordinary type, and the casts are set up with the ordinary newspaper types, and printed with the letterpress matter.

The engravings overleaf are taken from two of the newspaper weather charts for the dates of October 5th and 6th. They are enlarged and printed more clearly than the originals, with an explanation of signs at foot of the charts.

It will be observed that, in the chart for October 5th, an isobar of 29.2 runs up in a N.E. direction from between the Orkney and Shetland islands, crosses the North Sea, strikes the coast of Norway near Bergen, and then proceeds onwards towards Throndhjem. An isobar of 29.5 crosses Scotland, following very nearly the line of the Grampians, enters the North Sea about Aberdeen, and crosses to Christiansund; then runs up the Skager Rack and Christiania Fjord towards Christiania. Another isobar of 29.8 crosses Ireland through Connaught to Dublin, onward across England by Liverpool and the Humber, over the North Sea, and through Sleswig to the Baltic. These three are nearly parallel; but now we find another isobar—that of 30.2—taking quite a different course, by starting from the Bay of Biscay about Nantes; running on towards Paris and Strasbourg, and then bending sharp round, as though frightened by the Germans, and retreating to the Gulf of Lyons by an opposite course to that on which it started. On the following day all has changed; the northern isobars are running down south-eastwards instead of north-east, and are remarkably parallel. In the left-hand upper corner of this chart is a note that *"our west, north, and eastern coasts were warned yesterday."* Why was this? It was mainly because the barometric *gradient* or incline was so steep. On the 5th there was one inch of difference between the Orkneys and the Bay of Biscay, or between Bergen and Paris, while the barometer was still falling in Norway and at the same moment rising in Ireland and France. On the following day these movements culminated in a gradient of 1.4—nearly one and a half inches—between Cornwall and the ancient capital of Norway.

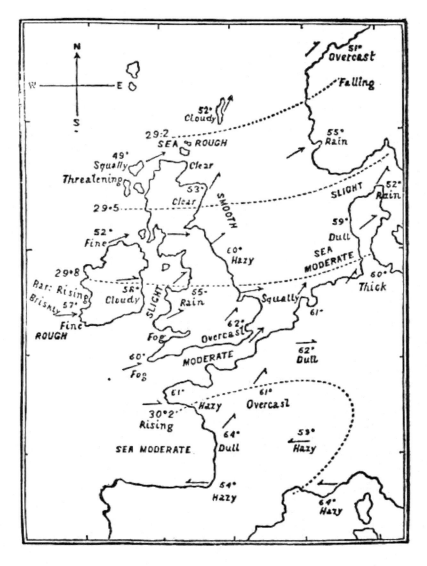

Weather Chart, October 5, 1875.

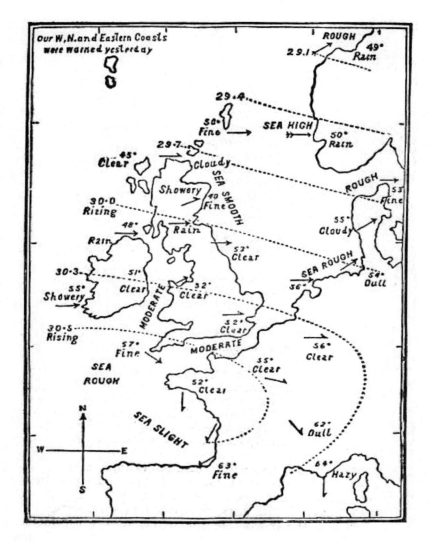

Weather Chart, October 6, 1875.

Explanation of Weather Chart.

In these charts the state of the sea—whether "rough," "smooth," "moderate," "slight," etc., is marked in capital letters; and the state of the weather—as "clear," "dull," "cloudy," "showery," etc., in small letters. The direction of the wind is indicated by the arrows. Unlike the arrows of a vane, these do not point towards the direction from which the wind is coming, but are *flying* arrows represented as moving *with* the wind, and consequently pointing to *where the wind is going*. The force of the wind is represented in five degrees of strength. 1st. A *calm*, by a horizontal line and zero—0 thus 0; 2nd. A *light wind*, by an arrow with one barb and no feathers ____\\; 3rd. A *fresh to strong breeze*, by an arrow with two barbs and no feathers ———>; 4th. A *gale*, by an arrow with two feathers >———>; and 5th. A *violent gale*, by an arrow with four feathers >>———>. The temperature—in the shade—is marked in figures with a small circle to the right, indicating degrees—as 60°. These figures stand in the places where the observations are made. The other figures—usually with decimals, and placed at the end of the dotted lines—give the height of the barometer—the dotted line showing where this particular height remained the same at the time of observation. These dotted lines are called "isobars," or *equal weights*—the weight or over-head pressure of the atmosphere being the same all along the line.

What must follow from this condition of the atmosphere? Clearly a great flow or rush of air from the south towards the comparatively vacuous regions of the north. The gases of our atmosphere, like the waters of the ocean, are always struggling to find their level, and thereby the winds are produced. The air flows from all sides towards the lowest isobar. But what, then, must be the course of the wind? Will it be in straight lines towards this point? If so, a strange conflict must result when all these currents meet from opposite directions. What will follow from this conflict? A skillful physicist can work out this problem mathematically, but we are not all mathematicians, some of us are not able to follow his formulæ, and, therefore, will do better by resorting to simple observation of other analogous and familiar phenomena. A funnel or any vessel with a hole in the bottom will answer our purpose. Let us fill such a vessel with water, then open the hole, and see what will be the course of the water when it is struggling to flow from all sides to the one point of vacuity. It will very soon establish a vortex or whirlpool, i.e., the water instead of flowing directly by straight lines from the sides to the centre of the funnel, will take a roundabout, spiral course, and thus screw its way down the outlet of the funnel.

This is just what occurs when the air is rushing to fill a comparatively vacuous atmospheric space. It moves in a spiral; and in the Northern Hemisphere this spiral always turns in the same way, viz., in the opposite direction to the hands of a clock when flowing inwards, and *vice versâ*, or *with* the clock hands, when the air is overflowing from a centre of high pressure.

In the chart for October 5th both these cases are illustrated. North of Dublin there is a curvature of isobars and an inrush of winds towards a northward low pressure, or vacuous region; while south of Dublin the isobar tends sharply round a high-pressure focus, and the overflowing wind is correspondingly reversed in direction, as shown by the arrows.

The next chart, for October 6th, shows that the overflow has spread northwards as far as Dublin, and the high-pressure focus has also moved northwards. It follows from this that if you know the barometric gradient, and stand with your left hand to the region of low barometer and your right hand to that of the high barometer, the wind will blow against your back, i.e., you will face the direction of the wind, or of those flying arrows on the chart. This interesting and important generalization is called "Buys Ballot's Law." In spite of the proverbial fickleness of the winds this simple law is rarely infringed, though it may require a slight modification of statement—inasmuch as the wind does not move in *circles* round the vacuous space, but in spirals, and thus it blows not quite square to the back, but rather obliquely, or a little on the right side. This is shown by the arrows in the charts, and is most strikingly displayed in the chart for October 6th, between the isobars of 30.3 and 30.5. To take, in Ireland, the position required by Buys Ballot's Law, one must have stood facing the east, and accordingly, the westerly wind would then blow upon one's back. In Paris, at the same moment, the position would be facing south-east, and the wind was curving round accordingly. Further south—at Bordeaux or the Pyrenees—the position becomes almost reversed, i.e., facing south-west, and the wind is reversed in equal degree.

Here, then, on these days we had the chief conditions of wind and rain, a steep and increasing barometric gradient, and a flow over our islands of humid air from the south and west regions of the great Atlantic. Strong winds and heavy rains did follow accordingly; and the prophetic warnings of the Meteorological Office, which are conveyed by means of signals displayed on prominent parts of the coast, were fulfilled.

Mr. Scott, the Director of the Meteorological Office, tells us that "The degree of success that has attended our warnings in these islands, on the average of the last two years, has been that over 45 per cent have been followed by severe gales; and over 33 per cent in addition have been followed by wind too strong for fishing-boats and yachts, though in themselves not severe gales; this gives a total percentage of success of nearly 80."

In winter the movements of the air are more decided, and the changes are often so rapid that the warning sometimes comes too late. With increased means—i.e., more money to cover additional work, and more stations—better results might be obtained. The United States expend 50,000*l*. a year in weather telegraphy, exclusive of salaries, while the United Kingdom only devotes 3,000*l*. a year to the same purpose. The difficulties on our side of the Atlantic are greater than on the American coasts, on account of the greater changeableness of our weather—mainly due to the more irregular distribution of land and water on this side. This, however, instead of discouraging national effort, should be regarded as a reason for increasing it. The greater the changes, the greater is the need for warnings, and the greater the difficulty the greater should be the effort. With our multitude of coastguard stations and naval men without employment, we ought to surpass all the world in such a work as this.

Those among our readers who are sufficiently interested in this subject to devote a little time to it, may make a very interesting weather scrap-book by cutting out the newspaper chart for each day, pasting it in a suitable album, and appending their own remarks on the weather at the date of publication, i.e., the day after the chart observations are made. Such an album would be far more interesting than the postage stamp and monogram albums that are so abundant.

Parents who desire their children to acquire habits of systematic observation, and to cultivate an intelligent interest in natural phenomena, will do well to supply such albums to their sons or daughters, and to hand over to them the daily paper for this purpose.

The Meteorological Office supplies by post copies of "Daily Weather Reports" to any subscriber who pays five shillings per quarter in advance; such subscriptions payable to Robt. H. Scott, Esq., Director Meteorological Office, 116 Victoria Street, Westminster, S.W.

These daily reports are printed on a large double sheet, on one half of which are four charts, representing separately the four records which are included in the one smaller newspaper chart—viz., those of the barometer, the thermometer, the rain-gauge, and the anemometer. On the other half of the sheet is a detailed separate tabular statement of the results of observations made at the following stations:

Haparanda
Hernösand
Stockholm
Wisby
Christiansund
Skudesnaes
Oxö (Christiansund)
Skagen (The Skaw)
Fanö
Cuxhaven

Sumburgh Head
Stornoway
Thurso
Wick
Nairn
Aberdeen
Leith
Shields
York
Scarborough
Nottingham
Ardrossan
Greencastle
Donaghadee
Kingstown
Holyhead
Liverpool
Valencia
Roche's Point
Pembroke
Portishead
Scilly
Plymouth
Hurst Castle
Dover
London
Oxford
Cambridge
Yarmouth
The Helder
Cape Griznez
Brest
L'Orient
Rochefort
Biarritz
Corunna
Brussels
Charleville
Paris
Lyons
Toulon

ON WINDS AND CURRENTS, FROM THE ADMIRALTY PHYSICAL ATLAS

In the Northern Hemisphere the effect of the veering of the wind on the barometer is according to the following law:

With East, South-east, and South winds, the barometer falls.
With South-west winds, the barometer ceases to fall and begins to rise.
With West, North-west, and North winds, the barometer rises.
With North-east winds, the barometer ceases to rise and begins to fall.

In the Northern Hemisphere the thermometer rises with East, South-east, and South winds; with a South-west wind it ceases to rise and begins to fall; it falls with West, North-west, and North winds; and with a North-east wind it ceases to fall and begins to rise.

Chapter 15

THE CHEMISTRY OF BOG RECLAMATION

The mode of proceeding for the reclamation of bog-land at Kylemore is first to remove the excess of water by "the big drain and the secondary drains," which must be cut deep enough to go right down to the gravel below. These are supplemented by the "sheep drains," or surface-drains, which are about twenty inches wide at top, and narrow downwards to six inches at bottom. They run parallel to each other, with a space of about ten yards between, and cost one penny per six yards.

This first step having been made, the bog is left for two years, during which it drains, consolidates, and sinks somewhat. If the bog is deep, the turf, which has now become valuable by consolidation, should be cut.

After this it is left about two years longer, with the drains still open. Then the drains are cleared and deepened, and a wedge-shaped sod, too wide to reach the bottom, is rammed in so as to leave below it a permanent tubular covered drain, which is thus made without the aid of any tiles or other outside material. The drainage is now completed, and the surface prepared for the important operation of dressing with lime, which, as the people expressively say, "boils the bog," and converts it into a soil suitable for direct agricultural operations.

Potatoes and turnips may now be set in "lazy bed" ridges. Mr. Mitchell Henry says, "Good herbage will grow on the bog thus treated; but as much as possible should at once be put into root-crops, with farm-yard manure for potatoes and turnips. The more lime you give the better will be your crop; and treated thus there is no doubt that even during the first year land so reclaimed will yield remunerative crops." And further, that "after being broken up a second time the land materially improves, and becomes doubly valuable." Also that he has no doubt that "all bog-lands may be thus reclaimed, but it is uphill work, and not remunerative to attempt the reclamation of bogs that are more than four feet in depth."

There is another and a simpler method of dealing with bogs—viz., setting them into narrow ridges; cutting broad trenches between the ridges; piling the turf cut out from these trenches into little heaps a few feet apart, burning them, and spreading the ashes over the ridges. This is rather largely practiced on the coast of Donegal, in conjunction with sea-weed manuring, and is prohibited in other parts of Ireland as prejudicial to the interests of the landlord.

We shall now proceed to the philosophy of these processes.

First, the drainage. Everybody in Ireland knows that the bog holds water like a sponge, and in such quantities that ordinary vegetation is rotted by the excess of moisture. There is good reason to believe that the ancient forests, which once occupied the sites of most of the Irish bogs, were in some cases destroyed by the rotting of their stems and roots in the excess of vegetable soil formed by generations upon generations of fallen leaves, which, in a humid climate like that of Ireland, could never become drained or air-dried.

But this is not all. There is rotting and rotting. When the rotting of vegetable matter goes on under certain conditions it is highly favorable to the growth of other vegetation, even of the vegetation of the same kind of plants as those supplying the rotting material. Thus, rotten and rotting straw is a good manure for wheat; and the modern scientific vine-grower carefully places the dressing of his vines about their roots, in order that they may rot, and supply the necessary salts for future growth. The same applies generally; rotting cabbage-leaves supply the best of manure for cabbages; rotting rhubarb-leaves for rhubarb; rose-leaves for rose-trees; and so on throughout the vegetable kingdom.

Why, then, should the bog-rotting be so exceptionally malignant? As I am not aware that any answer has been given to this question, I will venture upon one of my own. It appears to be mainly due to the excess of moisture preventing that slow combustion of vegetable carbon which occurs wherever vegetable matter is heaped together and *slightly* moistened. We see this going on in steaming dung-hills; in hayricks that have been stacked when imperfectly dried; in the spontaneous combustion of damp cotton in the holds of ships, and in factories where cotton-waste has been carelessly heaped; and in cucumber-frames and the other "hot-beds" of the gardener.

In ordinary soils this combustion goes on more slowly, but no less effectively, than in these cases. In doing so it maintains a certain degree of warmth about the roots of the plants that grow there, and *gradually* sets free the soluble salts which the rotting vegetables contain, and supplies them to the growing plants as manure, at the same time forming the humus so essential to vegetation.

A great excess of water, such as soddens the bog, prevents this, and also carries away any small quantity of soluble nutritious salts the soil may contain. Thus, instead of being warmed and nourished by slight humidity, and consequent oxidation, the bog soil is chilled and starved by excess of water.

The absolute necessity of the first operation—that of drainage—is thus rendered obvious; and I suspect that the need of four years' rest, upon which Mr. MacAlister insists, is somehow connected with a certain degree of slow combustion that accompanies and partially causes the consolidation of the bog. I have not yet had an opportunity of testing this by inserting thermometers in bogs under different conditions, but hope to do so.

The liming next demands explanation. Mr. Henry says that "it leaves the soil sweetened by the neutralization of its acids."

In order to test this theory I have digested (i.e., soaked) various samples of turf cut from Irish bogs in distilled water, filtered off the water, and examined it. I find that when this soaking has gone far enough to give the water a coloring similar to that which stands in ordinary bogs, the acidity is very decided—quite sufficiently so to justify this neutralization theory as a partial explanation. There is little reason to doubt that the lime is further effective in enriching the soil; or, in the case of pure bogs, that it forms the soil by disintegrating and decomposing the fibrous vegetable matter, and thus rendering it capable of assimilation by the crops.

Another effect which the lime must produce is the liberation of free ammonia from any fixed salts that may exist in the bog.

The bog-burning method of reclamation is easily explained. In the first place, the excessive vegetable encumbrance is reduced in quantity, and the remaining ashes supply the surface of the bog on which they rest with the non-volatile salts that originally existed in the burnt portions of the bog. In other words, they concentrate in a small space the salts that were formerly distributed too sparsely through the whole of the turf which was burnt.

As there are great differences in the composition of different bogs, especially in this matter of mineral ash, it is evident that the success of this method must be very variable, according to the locality.

On discussing this method with Mr. MacAlister (Mr. Henry's steward, under whose superintendence these reclamation works are carried out), he informed me that the bogs on the Kylemore estate yield a very small amount of ash—a mere impalpable powder that a light breath might blow away; that it was practically valueless, excepting from the turf taken at nearly the base of the bog. The ash I examined where the bog-burning is extensively practiced in Donegal, was quite different from this. The quantity was far greater, and its substance more granular and gritty. It, in fact, formed an important stratum, when spread over the surface of the ridges. These differences of composition may account for the differences of opinion and practice which prevail in different districts. It affords a far more rational explanation than the assumption that all such contradictions arise from local stupidities.

There is one evil, however, which is common to all bog-burning as compared with liming—it must waste the ammoniacal salts, as they are volatile, and are driven away into the air by the heat of combustion. Somebody may get them when the rain washes them down to the earth's surface again; but the burner himself obtains a very small share in this way.

We may therefore conclude that where lime is near at hand, bog-burning is a rude and wasteful, a viciously indolent mode of reclamation. It is only desirable where limestone is so distant that the expense of carriage renders lime practically unattainable, and where the bog itself is rich in mineral matter, and so deep and distant from a fuel demand, that it may be burned to waste without any practical sacrifice. Under such conditions it may be better to burn the bog than leave it in hopeless and worthless desolation.

I cannot conclude without again adverting to the importance of this subject, and affirming with the utmost emphasis, that the true Irish patriot is not the political orator, but he who by practical efforts, either as capitalist, laborer, or teacher, promotes the reclamation of the soil of Ireland, or otherwise develops the sadly neglected natural resources of the country.

With Mr. Mitchell Henry's permission I append to the above his own description of the results of his experiment, originally communicated in a letter to the *Times*; at the same time thanking him for his kind reception of a stranger at Kylemore Castle, and the facilities he afforded me for studying the subject on the spot.

"The interesting account you lately published of the extensive reclamations of His Grace the Duke of Sutherland, under the title of 'An Agricultural Experiment,' has been copied into very many newspapers, and must have afforded a welcome relief to thousands of readers glad to turn for a time from the terrible narratives that come to us from the east. If you will allow me, I should like to supplement your narrative by a rapid sketch of what has been done here during the last few years, on a much humbler scale, in the case of land similar, and some of it almost identical, with that in Sutherlandshire.

"The twelve *corps d'armée* under the Duke's command, in the shape of the twelve steam-engines and their ploughs, engaged in subduing the stubborn resistance of the unreclaimed wilds of Sutherlandshire, suggest to the mind the triumphs of great warriors, and fill us with admiration—not always excited by the details of great battle; but, as great battles can be fought seldom, and only by gigantic armies and at prodigious expense, so reclamation on such a scale is far beyond the opportunities or the means of most of us; while many may, perhaps, be encouraged to attempt work similar to that which has been successfully carried out here.

"And, first of all, a word as to the all-important matter of cost. Does it pay?

"Including farm-buildings and roads, the reclamations here have cost on an average 13*l*. an acre, which, at 5 per cent, means an annual rent-charge of 13*s*., to which is to be added a sum of from 1*s*. to 3*s*., the full annual value of the unreclaimed land. It is obvious that if we start with an outlay of 30*l*. *plus* the 1*s*. to 3*s*. of original rent, such an amount would usually be found prohibitory; but, on the other hand, excellent profits may be made if the expenditure is so kept down that the annual rent is not more than from 15*s*. to 18*s*. per acre. Before entering into further details, let me say that I claim no credit for originality in what has been done. The like has been effected on numerous properties in Ireland in bygone days, and is daily being carried out by the patient husbandman who year by year with his spade reclaims a little bit from the mountain side. And you must allow me emphatically to say that what has been done here economically and well would not have been done except for the prudence, patience, and thoughtful mind of my steward, Archibald MacAlister, a County Antrim man, descended from one of the race of Highland Catholic Scotch settlers, who have peopled the north of Ireland and added so much to its prosperity.

"The Pass of Kylemore, in which I live, is undoubtedly favorably situated for reclamation, for there is but little very deep bog, and there is abundance of limestone. In former ages it must have been an estuary of the sea, with a river flowing through it, now represented by a chain of lakes and the small rapid river Dowris. The subsoil is sand, gravel, and schist rock, with peat of various depths grown upon it. As by the elevation of the land the sea long ages ago was driven back, the mossy growth of peat commenced, followed by pine and yew trees, of which the trunks and roots are abundantly found; but, except over a space of about 400 acres, every tree that formerly clothed the hillsides has been cut down or has totally disappeared. The general result is that we have a pass several miles long, bounded on the north and south by a chain of rugged mountains of some 1500 or 1800 feet in height, while the east is blocked up by a picturesque chain running north and south, and separating the Joyce country from Connemara proper, the west being open to the Atlantic. The well-known Killery Bay, or Fiord, would, I doubt not, present an exact resemblance to Kylemore if the sea, which now flows up to its head, were driven out. There are miles of similar country in Ireland, waiting only for the industry of man, where, as here, there exist extensive stretches of undulating eskers, covered with heather growing on the light clay, with a basis of gravel or sand.

"A considerable difference exists between the reclamation of the flat parts, where the bog is pretty deep, and the hillsides, where there is little or no bog. Yet it is to be remembered that bog is nothing more than vegetable matter in a state of partial decomposition, and holding water like a sponge. The first thing is to remove the water by drains, some of which—that is, the big drain and the secondary drains—must go right down to the gravel below; but the other drains—called sheep-drains—need not, and, indeed, must not be cut so deep. The drains are cut wedge-shape by what are called Scotch tools, which employ three men—two to cut and one to hook out the sods; and all that is requisite to form a permanent drain is to replace the wedge-shaped sod, and ram it down between the walls of the drain, where it consolidates and forms a tube which will remain open for an indefinite number of years. We have them here as good as new, made twenty-five years ago; and at Chat Moss, in Lancashire, they are much older. After land has been thus drained—but not too much drained, or it will become dry turf—the surface begins to sink; what was tumid settles down, and in the course of a few months the land itself becomes depressed on the surface and much consolidated. Next it is to be dug by spade-labor or ploughed. We use oxen largely for this purpose, and, strange to say, the best workers we find to be a cross with the Alderney, the result being a light, wiry little animal, which goes gayly over the ground, is easy to feed, and is very tractable. The oxen are trained by the old wooden neck-yoke; but, when well broken, work in collars, which seem more easy to them. Horses on very soft land work well in wooden pattens. After the land has been broken up, a good dressing of lime is to be applied to it, and this, in the expressive language of the people here, 'boils the bog'—that is, the lime causes the vegetable matter, formerly half decomposed, to become converted into excellent manure. This leaves the soil sweetened by the neutralization of its acids, and in a condition pretty easily broken up by the chain-harrow; or, what is better still, by Randall's American revolving harrow.

"Good herbage will grow on bog thus treated, but as much as possible should at once be put into root-crops, with farmyard manure for potatoes and turnips. The more lime you give the better will be your crop, and, treated thus, there is no doubt that even during the first year, land so reclaimed will yield remunerative crops. People ask, 'But will not the whole thing go back to bog?' Of course it will if not kept under proper rotation, which we find to be one of five years—namely, roots followed by oats, laid down with clover and grass seed, which remains for two years. After being broken up a second time, the land materially improves and becomes doubly valuable. I have no doubt that all bog-lands may be thus reclaimed, but it is up-hill work and not remunerative to attempt the reclamation of bogs that are more than four feet in depth.

"And here I will make a remark as to the effects of drainage in a wet country. By no means does the whole effect result from raising the temperature of the soil; there is something else as important, and that is the supply of ammonia, brought down from the skies in the rain, which, with other fertilizing matter, is caught, detained, and absorbed in the soil. A well-drained field becomes, in fact, just like a water-meadow over which a river flows for a part of a year; and thus the very wetness of the climate may be made to reduce the supply of ammoniacal manures, so expensive to buy.

"The porous, well-drained soil carries quickly off the superfluous moisture, while the ammonia is absorbed by the roots and leaves of the plants. An excessive bill for ammoniacal manures has been the ruin of many a farmer; and our aim in Ireland should be to secure good crops by thorough drainage and constant stirring of the soil, without much outlay for concentrated manures. At the same time I ought to remark that we have grown excellent potatoes by using 5*l.* worth per acre of superphosphate and nitrate of soda in cases in which our farmyard manure has fallen short.

"The reclamation of mountain-land as distinguished from bog-land can best be illustrated by a record of what has been accomplished on two farms here. Three years ago the leases of two upland farms fell in, and I took them into my own hands. The first consists of 600 acres, one-half a nearly level flat of deepish bog running alongside the river, the other half moor heath, which with difficulty supported a few sheep and cattle.

"There had never been any buildings on this land, nor had a spade ever been put into it; and the tenant, being unable to pay his rent of 15*l.* a year for the 600 acres, was glad to give it up for a moderate consideration. The first thing accomplished was to fence and drain thoroughly as before described, and the best half of the land was then divided into forty-acre fields. Exactly now two years ago—on September 15th—a little cottage and a stable for a pair of horses and a pair of bullocks was completed and tenanted by two men and a boy. They ploughed all the week and came home on Saturdays to draw their supply of food and fodder for the ensuing seven days, thus approximating very nearly to the position of settlers in a new country. We limed all the land we could, manured part of it with seaweed and part with the farm manure made by the horses and oxen which were at work, and cropped with roots such as turnips and potatoes. A good portion we sowed with oats out of the lea, but the most satisfactory crop we found to be rape and grasses mixed, for on the best of the land they form at once an excellent permanent pasture. We have now had two crops from this land; and I venture to say that the thirteen stacks of oats and hay gathered in in good condition, and the turnips and roots now growing, which are not excelled in the county Galway—except those of Lord Clancarty at Ballinasloe, who has grown 110 tons of turnips to the Irish acre, equal to upwards of 68 tons to the acre here—present a picture most gratifying and cheering in every way.

"The second farm, of 240 acres, which adjoins this, had a good building on it; but, having been let on lease at about 10s. an acre to a large grazier whose stock-in-trade was a horse, a saddle, and a pair of shears, had not been cultivated or improved.

"Similar proceedings on this farm have produced similar results; and, if now let in the market, I have no doubt that after two years of good treatment these farms would be let at 20s. an acre, and I do not despair of doubling this figure in the course of time.

"The exact weight of the turnip crop this season is, on raw bog, drained, limed, and cropped this year for the first time, 24 tons per acre; manure, seaweed. On land ploughed but not cropped, last year 23½ tons; mixed mineral manure. On land from which a crop of oats had previously been taken, 29 tons; manure, farmyard, with 3 cwt. per acre mineral manure.

"Last year my excellent steward, Mr. MacAlister, visited the Duke of Sutherland's reclamations in Scotland, and was kindly and hospitably received. He found the land and the procedure adopted almost identical, with the conviction that oxen and horses will suit us better at the present time than steam culture, chiefly on the score of economy. He also visited the Bridgewater Estate at Chat Moss, near Manchester, where so much has been done to bring the deep peat into cultivation, and he found the system that has been followed there for so many years to be like that described above, marl, however, being used in the place of lime."

At the time of my visit to Kylemore the hay crops were down and partly carried on the reclaimed bog-land above described. The contrast of its luxuriance with the dark and dreary desolation of the many estates I had seen during three summers' wanderings through Ireland added further proof of the infamy of the majority of Irish landlords, by showing what Ireland would have been had they done their duty.

Chapter 16

AERIAL EXPLORATION OF THE ARCTIC REGIONS

On our own hemisphere, and separated from our own coasts by only a few days' journey on our own element, there remains a blank circle of unexplored country above 800 miles in diameter. We have tried to cross it, and have not succeeded. Nothing further need be said in reply to those who ask, "Why should we start another Arctic Expedition?"

The records of previous attempts to penetrate this area of geographical mystery prove the existence of a formidable barrier of mountainous land, fringed by fjords or inlets, like those of Norway, some of which may be open, though much contracted northward, like the Vestfjord that lies between the Lofoden Islands and the mainland of Scandinavia. The majority evidently run inland like the ordinary Norwegian fjords or the Scotch firths, and terminate in land valleys that continue upwards to fjeld regions, or elevated humpy land which acts as a condenser to the vapor-laden air continually flowing towards the Pole from the warmer regions of the earth, and returning in lower streams when cooled. The vast quantities of water thus condensed fall upon these hills and table lands as snow crystals. What becomes of this everlasting deposit?

Unlike the water that rains on temperate hill-sides, it cannot all flow down to the sea as torrents and liquid rivers, but it does come down nevertheless, or long ere this it would have reached the highest clouds. It descends mainly as glaciers, which creep down slowly, but steadily and irresistibly, filling up the valleys on their way; and stretching outwards into the fjords and channels, which they block up with their cleft and chasmed crystalline angular masses that still creep outward to the sea until they float, and break off or "calve" as mountainous icebergs and smaller masses of ice.

These accumulations of ice thus *formed on land* constitute the chief obstructions that bar the channels and inlets fringing the unknown Polar area. The glacier fragments above described are cemented together in the winter time by the freezing of the water between them. An open frozen sea, pure and simple, instead of forming a barrier to arctic exploration, would supply a most desirable highway. It must not be supposed that, because the liquid ocean is ruffled by ripples, waves, and billows, a frozen sea would have a similar surface. The freezing of such a surface could only start at the calmest intervals, and the ice would shield the water from the action of the wave-making wind, and such a sea would become a charming skating rink, like the Gulf of Bothnia, the Swedish and Norwegian lakes, and certain fjords, which, in the winter time, become natural ice-paved highways, offering incomparable facilities for rapid locomotion. In spite of the darkness and the cold, winter is the traveling season in

Sweden and Lapland. The distance that can be made in a given time in summer with a wheeled vehicle on well-made post roads can be covered in half the time in a *pulk* or reindeer sledge drawn over the frozen lakes. From Spitzbergen to the Pole would be an easy run of five or six days if nothing but a simply frozen sea stood between them.

This primary physical fact, that arctic navigators have not been stopped by a merely frozen sea, but by a combination of glacier fragments with the frozen water of bays, and creeks, and fjords, should be better understood than it is at present; for when it is understood, the popular and fallacious notion that the difficulties of arctic progress are merely dependent on latitude, and must therefore increase with latitude, explodes.

> It is the physical configuration of the fringing zone of the arctic regions, not its mere latitude, that bars the way to the Pole.

I put this in italics because so much depends upon it—I may say that all depends upon it—for if this barrier can be scaled at any part we may come upon a region as easily traversed as that part of the Arctic Ocean lying between the North Cape and Spitzbergen, which is regularly navigated every summer by hardy Norsemen in little sailing sloops of 30 to 40 tons burden, and only six or eight pair of hands on board; or by overland traveling as easily as the Arctic winter journey between Tornea and Alten. This trip over the snow-covered mountains is done in five or six days, at the latter end of every November, by streams of visitors to the fair at Alten, in latitude 70°, 3½ degrees N. of the Arctic circle; its distance, 430 miles, is just about equal to that which stands between the North Pole and the northernmost reach of our previous Arctic expeditions. One or the other of the above-named conditions, or an enclosed frozen Polar ocean, is what probably exists beyond the broken fjord barrier hitherto explored; a continuation of such a barrier is, in fact, almost a physical impossibility; and therefore the Pole will be ultimately reached, not by a repetition of such weary struggles as those which ended in the very hasty retreat of our last expedition, but by a bound across about 400 miles of open or frozen Polar ocean, or a rapid sledge-run over snow-paved fields like those so merrily traversed in Arctic Norway by festive bonders and their families on their way to Yule-time dancing parties.

Reference to a map of the circumpolar regions, or, better, to a globe, will show that the continents of Europe, Asia, and America surround the Pole, and hang, as it were, downwards or southwards from a latitude of 70° and upwards. There is but one wide outlet for the accumulations of Polar ice, and that is between Norway and Greenland, with Iceland standing nearly midway. Davis's and Behring's Straits are the narrower openings; the first may be only a fjord, rather than an outlet. The ice-block, or crowding together and heaping up of the glacier fragments and bay ice, is thus explained.

Attempts of two kinds have been made to scale this icy barrier. Ships have sailed northwards, threading a dangerous course between the floating icebergs in the summer, and becoming fast bound in winter, when the narrow spaces of brackish water lying between these masses of land ice become frozen, and the "ice-foot" clinging to the shore stretches out seaward to meet that on the opposite side of the fjord or channel. The second method, usually adopted as supplementary to the first, is that of dragging sledges over these glacial accumulations. The pitiful rate of progress thus attainable is shown by the record of the last attempt, when Commander Markham achieved about one mile per day, and the labor of doing this was nearly fatal to his men. Any tourist who has crossed or ascended an Alpine glacier with only a knapsack to carry, can understand the difficulty of dragging a cartload of provisions, etc., over such accumulations of iceberg fragments and of sea-ice squeezed and crumbled up between them. It is evident that we must either find a natural breach in this Arctic barrier or devise some other means of scaling it.

The first of these efforts has been largely discussed by the advocates of rival routes. I will not go into this question at present, but only consider the alternative to all land routes and all water routes, viz.: that by the other available element—an aerial route—as proposed to be attempted in the new Arctic expedition projected by Commander Cheyne, and which he is determined to practically carry out, provided his own countrymen, or, failing them, others more worthy, will assist him with the necessary means of doing so.

To reach the Pole from the northernmost point already attained by our ships demands a journey of about 400 miles, the distance between London and Edinburgh. With a favorable wind, a balloon will do this in a few hours, On November 27, 1870, Captain Roher descended near Lysthuus, in Hitterdal (Norway), in the balloon "Ville d'Orleans," having made the journey from Paris in fifteen hours. The distance covered was about 900 miles, more than double the distance between the Pole and the accessible shores of Greenland.

On November 7, 1836, Messrs. Holland, Mason, and Green ascended from Vauxhall Gardens, at 1.30 P.M., with a *moderate breeze*, and descended eighteen hours afterwards "in the Duchy of Nassau, about two leagues from the town of Weilburg," the distance in a direct line being about 500 miles. A similar journey to this would carry Commander Cheyne from his ship to the North Pole, or thereabouts, while a fresh breeze like that enjoyed by Captain Roher would, in the same time, carry him clear across the whole of the circumpolar area to the neighborhood of Spitzbergen, and two or three hours more of similar proceeding would land him in Siberia or Finland, or even on the shores of Arctic Norway, where he could take the Vadsö or Hammerfest packet to meet one of Wilson's liners at Trondhjem or Bergen, and thus get from the North Pole to London in ten days.

Lest any of my readers should think that I am writing this at random, I will supply the particulars. I have before me the "Norges Communicationer" for the present summer season of 1880. Twice every week a passenger excursion steam packet sails round the North Cape each way, calling at no less than twenty stations on this Arctic face of Europe to land and embark passengers and goods. By taking that which stops at Gjesvaer (an island near the foot of the North Cape) on Saturday, or that which starts from Hammerfest on Sunday morning, Trondhjem is reached on Thursday, and Wilson's liner, the "Tasso," starts on the same day for Hull, "average passage seventy hours." Thus Hammerfest, the northernmost town in the world, is now but eight days from London, including a day's stop at Tromsö, the capital of Lapland, which is about 3 degrees N. of the Arctic circle, and within a week of London. At Captain Roher's rate of traveling Tromsö would be but twenty-three hours from the Pole.

These figures are, of course, only stated as *possibilities* on the supposition that all the conditions should be favorable, but by no means as *probable*.

What, then, are the *probabilities* and the amount of risk that will attend an attempt to reach the Pole by an aerial route?

I have considered the subject carefully, and discussed it with many people; the result of such reflection and conversation is a conviction that the prevalent popular estimate of the dangers of Commander Cheyne's project extravagantly exaggerates them on almost all contingencies. I do not affirm that there is no risk, or that the attempt should be made with only our present practical knowledge of the subject, but I do venture to maintain that, after making proper preliminary practical investigations at home, a judiciously conducted aerostatic dash for the Pole will be far less dangerous than the African explorations of Livingstone, Stanley, and others that have been accomplished and are proposed. And further, that a long balloon journey starting in summer-time from Smith's Sound, or other suitable Arctic station, would be less dangerous than a corresponding one started from London; that it would involve less risk than was incurred by Messrs. Holland, Mason, and Green, when they traveled from Vauxhall Gardens to Nassau.

The three principal dangers attending such a balloon journey are: 1st. The variability of the wind. 2d. The risk of being blown out about the open ocean beyond the reach of land. 3d. The utter helplessness of the aeronaut during all the hours of darkness. I will consider these seriatim in reference to Arctic ballooning *versus* Vauxhall or Crystal Palace ballooning.

As regards the first danger, Vauxhall and Sydenham are in a position of special disadvantage, and all the ideas we Englishmen may derive from our home ballooning experience must tend to exaggerate our common estimate of this danger, inasmuch as we are in the midst of the region of variable winds, and have a notoriously uncertain climate, due to this local exaggeration of the variability of atmospheric movements. If instead of lying between the latitudes of 50° and 60°, where the N.E. Polar winds just come in collision with the S.W. tropical currents, and thereby effect our national atmospheric stir-about, we were located between 10° and 30° (where the Canary Islands are, for example), our notions on the subject of balloon traveling would be curiously different. The steadily blowing trade-wind would long ere this have led us to establish balloon mails to Central and South America, and balloon passenger expresses for the benefit of fast-going people or luxurious victims of sea-sickness. To cross the Atlantic—three thousand miles—in forty-eight hours, would be attended with no other difficulty than the cost of the gas, and that of the return carriage of the empty balloon.

It is our exceptional meteorological position that has generated the popular expression "as uncertain as the wind." We are in the very centre of the region of meteorological uncertainties, and cannot go far, either northward or southward, without entering a zone of greater atmospheric regularity, where the direction of the wind at a given season may be predicted with more reliability than at home. The atmospheric movements in the Arctic regions appear to be remarkably regular and gentle during the summer and winter months, and irregular and boisterous in spring and autumn. A warm upper current flows from the tropics towards the Pole, and a cold lower one from the Arctic circle towards the equator. Commander Cheyne, who has practical experience of these Arctic expeditions, and has kept an elaborate log of the wind, etc., which he has shown me, believes that, by the aid of pilot balloons to indicate the currents at various heights, and by availing himself of these currents, he may reach the Pole and return to his ship, or so near as to be able to reach it by traveling over the ice in light sledges that will be carried for that purpose. In making any estimate of the risk of Arctic aerostation, we must banish from our minds the preconceptions induced by our British experience of the uncertainties of the wind, and only consider the atmospheric actualities of the Polar regions, so far as we know them.

Let us now consider the second danger, viz., that of being blown out to sea and there remaining until the leakage of gas has destroyed the ascending power of the balloon, or till the stock of food is consumed. A glance at a map of the world will show how much smaller is the danger to the aeronaut who starts from the head of Baffin's Bay than that which was incurred by those who started from Vauxhall in the Nassau balloon, or by Captain Roher, who started from Paris. Both of these had the whole breadth of the Atlantic on the W. and S.W., and the North Sea and Arctic Ocean N. and N.E. The Arctic balloon, starting from Smith's Sound or thereabouts, with a wind from the South (and without such a wind the start would not, of course, be made), would, if the wind continued in the same direction, reach the Pole in a few hours; in seven or eight hours at Roher's speed; in fourteen or fifteen hours at the average rate made by the Nassau balloon in a "moderate breeze." Now look again at the map and see what surrounds them. Simply the continents of Europe, Asia, and America, by which the circumpolar area is nearly land-locked, with only two outlets, that between Norway and Greenland on one side, and the narrow channel of Behring's Straits on the other. The wider of these is broken by Spitzbergen and Iceland, both inhabited islands, where a balloon may descend and the aeronauts be hospitably received. Taking the 360 degrees of the zone between the 70th parallel of latitude and the Arctic circle, 320 are land-locked and only 40 open to the sea; therefore the chances of coming upon land at *any one* part of this zone is as 320 to 40; but with a choice of points for descent such as the aeronauts would have unless the wind blew precisely down the axis of the opening, the chances would be far greater. If the wind continued as at starting, they would be blown to Finland; a westerly deflection would land them in Siberia, easterly in Norway; a strong E. wind at the later stage of the trip would blow them back to Greenland.

In all the above I have supposed the aeronauts to be quite helpless, merely drifting at random with that portion of the atmosphere in which they happened to be immersed. This, however, need not be the case. Within certain limits they have a choice of winds, owing to the prevalence of upper and lower currents blowing in different and even in opposite directions. Suppose, for example, they find themselves N. of Spitzbergen, where "Parry's furthest" is marked on some of our maps, and that the wind is from the N.E., blowing them towards the Atlantic opening. They would then ascend or descend in search of a due N. or N. by W. wind that would blow them to Norway, or W.N.W. to Finland, or N.W. to Siberia, or due E. back to Greenland, from whence they might rejoin their ships. One or other of these would almost certainly be found. A little may be done in steering a balloon, but so very little that small reliance should be placed upon it. Only in a very light wind would it have a sensible effect, though in case of a "near shave" between landing, say at the Lofodens or Iceland, and being blown out to sea, it might just save them.

As already stated, Commander Cheyne believes in the possibility of returning to the ship, and bases his belief on the experiments he made from winter quarters in Northumberland Sound, where he inflated four balloons, attached to them proportionately different weights, and sent them up simultaneously. They were borne by diverse currents of air in *four different directions according to the different altitudes*, viz., N.W., N.E., S.E., and S.W., "thus proving that in this case balloons could be sent in any required direction by ascending to the requisite altitude. The war balloon experiments at Woolwich afford a practical confirmation of this important feature in aerostation." Cheyne proposes that one at least of the three balloons shall be a rover to cross the unknown area, and has been called a madman for suggesting this merely as an alternative or secondary route. I am still more lunatic, for I strongly hold the opinion that the easiest way for him to return to his ship will be to drift rapidly across to the first available inhabited land, thence come to England, and sail in another ship to rejoin his messmates; carrying with him his bird's-eye chart, that will demonstrate once for all the possibility or impossibility of circumnavigating Greenland, or of sailing, or sledging, or walking to the Pole.

The worst dilemma would be that presented by a dead calm, and it is not improbable that around the Pole there may be a region of calms similar to that about the Equator. Then the feather-paddle or other locomotive device worked by man-power would be indispensable. Better data than we at present possess are needed in order to tell accurately what may thus be done. Putting various estimates one against the other, it appears likely that five miles an hour may be made. Taking turn and turn about, two or three aeronauts could thus travel fully 100 miles per day, and return from the Pole to the ship in less than five days.

Or take the improbable case of a circular wind blowing round the Pole, as some have imagined. This would simply demand the working of the paddle always northwards in going to the Pole, and always southwards in returning. The resultant would be a spiral course winding inwards in the first case, and outwards in the second. The northward or southward progress would be just the same as in a calm if the wind were truly concentric to the Pole. Some rough approximation to such currents may exist, and might be dealt with on this principle.

Let us now consider the third danger, that of the darkness. The seriousness of this may be inferred from the following description of the journey of the Nassau balloon, published at the time: "It seemed to the aeronauts as if they were cleaving their way through an interminable mass of black marble in which they were imbedded, and which, solid a few inches before them, seemed to soften as they approached in order to admit them still further within its cold and dusky enclosure. In this way they proceeded blindly, as it may well be called, until about 3.30 A.M., when in the midst of the impenetrable darkness and profound stillness an unusual explosion issued from the machine above, followed by a violent rustling of the silk, and all the signs which might be supposed to accompany the bursting of the balloon. The car was violently shaken. A second and a third explosion followed in quick succession. The danger seemed immediate, when suddenly the balloon recovered her usual form and stillness. These alarming symptoms seemed to have been produced by collapsing of the balloon under the diminished temperature of the upper regions after sunset, and the silk forming into folds under the netting. Now, when the guide rope informed the voyagers that the balloon was too near the earth, ballast was thrown out, and the balloon rising rapidly into a thinner air experienced a diminution of pressure, and consequent expansion of the gas.

"The cold during the night ranged from a few degrees below to the freezing point. As morning advanced the rushing of waters was heard, and so little were the aeronauts aware of the course which they had been pursuing during the night, that they supposed themselves to have been thrown back upon the shores of the German Ocean, or about to enter the Baltic, whereas they were actually over the Rhine, not far from Coblentz."

All this blind drifting for hours, during which the balloon may be carried out to sea, and opportunities of safe descent may be lost, is averted in an Arctic balloon voyage, which would be made in the summer, when the sun never sets. There need be no break in the survey of the ground passed over, no difficulty in pricking upon a chart the course taken and the present position at any moment. With an horizon of 50 to 100 miles' radius the approach of such a danger as drifting to the open ocean would be perceived in ample time for descent, and as a glance at the map will show, this danger cannot occur until reaching the latitudes of inhabited regions.

The Arctic aeronauts will have another great advantage over those who ascend from any part of England. They can freely avail themselves of Mr. Green's simple but most important practical invention—the drag-rope. This is a long and rather heavy rope trailing on the ground. It performs two important functions. First, it checks the progress of the balloon, causing it to move less rapidly than the air in which it is immersed. The aeronaut thus gets a slight breeze equivalent to the difference between the velocity of the wind and that of the balloon's progress. He may use this as a fulcrum to effect a modicum of steerage.

The second and still more important use of the drag-rope is the very great economy of ballast it achieves. Suppose the rope to be 1000 feet long, its weight equal to 1 lb. for every ten feet, and the balloon to have an ascending power of 50 lbs. It is evident that under these conditions the balloon will retain a constant elevation of 500 feet above the ground below it, and that 500 feet of rope will trail upon the ground. Thus, if a mountain is reached no ballast need be thrown away in order to clear the summit, as the balloon will always lift its 500 feet of rope, and thus always rise with the up-slope and descend with the down-slope of hill and dale. The full use of this simple and valuable adjunct to aerial traveling is prevented in such a country as ours by the damage it might do below, and the temptation it affords to mischievous idiots near whom it may pass.

In the course of many conversations with various people on this subject I have been surprised at the number of educated men and women who have anticipated with something like a shudder the terrible cold to which the poor aeronauts will be exposed.

This popular delusion which pictures the Arctic regions as the abode of perpetual freezing, is so prevalent and general, that some explanation is demanded.

The special characteristic of Arctic climate is a cold and long winter and a short and *hot summer*. The winter is intensely cold simply because the sun never shines, and the summer is very hot because the sun is always above the horizon, and, unless hidden by clouds or mist, is continually shining. The summer heat of Siberia is intense, and the vegetable proportionately luxuriant. I have walked over a few thousand miles in the sunny South, but never was more oppressed with the heat than in walking up the Tromsdal to visit an encampment of Laplanders in the summer of 1856.

On the 17th July I noted the temperature on board the steam packet when we were about three degrees north of the Arctic circle. It stood at 77° well shaded in a saloon under the deck; it was 92° in the "rōk lugar," a little smoking saloon built on deck; and 108° in the sun on deck. This was out at sea, where the heat was less oppressive than on shore. The summers of Arctic Norway are very variable on account of the occasional prevalence of misty weather. The balloon would be above much of the mist, and would probably enjoy a more equable temperature during the twenty-four hours than in any part of the world where the sun sets at night.

I am aware that the above is not in accordance with the experience of the Arctic explorers who have summered in such places as Smith's Sound. I am now about to perpetrate something like a heresy by maintaining that the summer climate there experienced by these explorers is quite exceptional, is not due to the latitude, but to causes that have hitherto escaped the notice of the explorers themselves and of physical geographers generally. The following explanation will probably render my view of this subject intelligible:

As already stated, the barrier fringe that has stopped the progress of Arctic explorers is a broken mountainous shore down which is pouring a multitude of glaciers into the sea. The ice of these glaciers is, of course, fresh-water ice. Now, we know that when ice is mixed with salt water we obtain what is called "a freezing mixture"—a reduction of temperature far below the freezing point, due to the absorption of heat by the liquefaction of the ice. Thus the heat of the continuously shining summer sun *at this particular part of the Arctic region* is continuously absorbed by this powerful action, and a severity quite exceptional is thereby produced. Every observant tourist who has crossed an Alpine glacier on a hot summer day has felt the sudden change of climate that he encounters on stepping from *terra firma* on to the ice, and in which he remains immersed as long as he is on the glacier. How much greater must be this depression of temperature where the glacier ice is broken up and is floating in sea-water, to produce a vast area of freezing mixture, which would speedily bring the hottest blasts from the Sahara down to many degrees below the freezing point.

A similar cause retards the *beginning* of summer in Arctic Norway and in Finland and Siberia. So long as the winter snow remains unmelted, i.e., till about the middle or end of June, the air is kept cold, all the solar heat being expended in the work of thawing. This work finished, then the warming power of a non-setting sun becomes evident, and the continuously accumulating heat of his rays displays its remarkable effect on vegetable life, and everything capable of being warmed. These peculiarities of Arctic climate must become exaggerated as the Pole is approached, the winter cold still more intense, and the accumulation of summer heat still greater. In the neighborhood of the North Cape, where these contrasts astonish English visitors, where inland summer traveling becomes intolerable on account of the clouds of mosquitoes, the continuous sunshine only lasts from May 11 to August 1. At the North Pole the sun would visibly remain above the horizon during about seven months—from the first week in March to the first week in October (this includes the effect of refraction and the prolonged summer of the northern hemisphere due to the eccentricity of the earth's orbit).

This continuance of sunshine, in spite of the moderate altitude of the solar orb, may produce a very genial summer climate at the Pole. I say "may," because mere latitude is only one of the elements of climate, especially in high latitudes. Very much depends upon surface configuration and the distribution of land and water. The region in which our Arctic expedition ships have been ice-bound combines all the most unfavorable conditions of Arctic summer climate. It is extremely improbable that those conditions are maintained all the way to the Pole. We know the configuration of Arctic Europe and Arctic Asia, that they are masses of land spreading out northward round the Arctic circle and narrowing southward to angular terminations. The southward configuration and northward outspreading of North America are the same, but we cannot follow the northern portion to its boundary as we may that of Europe and Asia, both of which terminate in an Arctic Ocean. Greenland is remarkably like Scandinavia; Davis's Strait, Baffin's Bay, and Smith's Sound corresponding with the Baltic and the Gulf of Bothnia. The deep fjords of Greenland, like those of Scandinavia, are on its western side, and the present condition of Greenland corresponds to that of Norway during the milder period of the last glacial epoch. If the analogy is maintained a little further north than our explorers have yet reached we must come upon a Polar sea, just as we come upon the White Sea and the open Arctic Ocean if we simply travel between 400 and 500 miles due north from the head of the frozen Gulf of Bothnia.

Such a sea, if unencumbered with land ice, will supply the most favorable conditions for a genial arctic summer, especially if it be dotted with islands of moderate elevation, which the analogies of the known surroundings render so very probable. Such islands may be inhabited by people who cannot reach us on account of the barrier wall that has hitherto prevented us from discovering them. Some have even supposed that a Norwegian colony is there imprisoned. Certainly the early colonists of Greenland have disappeared, and their disappearance remains unexplained. They may have wandered northwards, mingled with the

Esquimaux, and have left descendants in this unknown world. If any of Franklin's crew crawled far enough they may still be with them, unable to return.

In reference to these possibilities it should be noted that a barrier fringe of mountainous land like that of Greenland and arctic America would act as a condensing ground upon the warm air flowing from the south, and would there accumulate the heavy snows and consequent glaciers, just as our western hills take so much of the rain from the vapor-laden winds of the Atlantic. The snowfall immediately round the Pole would thus be moderated, and the summer begin so much earlier.

I have already referred to the physical resemblances of Baffin's Bay, Smith's Sound, etc., to the Baltic, the Gulf of Bothnia, and Gulf of Finland. These are frozen every winter, but the Arctic Ocean due north of them is open all the winter, and every winter. The hardy Norse fishermen are gathering their chief harvest of cod fish in the open sea around and beyond the North Cape, Nordkyn, etc., at the very time when the Russian fleet is hopelessly frozen up in the Gulf of Finland. But how far due north of this frozen Baltic are these open-sea fishing banks? More than 14 degrees—more than double the distance that lies between the winter quarters of some of our ships in Smith's Sound and the Pole itself. This proves how greatly physical configuration and oceanic communication may oppose the climatic influences of mere latitude. If the analogy between Baffin's Bay and the Baltic is complete, a Polar sea will be found that is open in the summer at least.

On the other hand, it may be that ranges of mountains covered with perpetual snow, and valleys piled up with huge glacial accumulations, extend all the way to the Pole, and thus give to our globe an arctic ice-cap like that displayed on the planet Mars. This, however, is very improbable, for, if it were the case, we ought to find a circumpolar ice-wall like that of the antarctic regions; the Arctic Ocean beyond the North Cape should be crowded with icebergs instead of being open and iceless all the year round. With such a configuration the ice-wall should reach Spitzbergen and stretch across to Nova Zembla; but, instead of this, we have there such an open stretch of arctic water, that in the summer of 1876 Captain Kjelsen, of Tromsö, sailed in a whaler to lat. 81° 30´ without sighting ice. He was then but 510 geographical miles from the Pole, with open sea right away to his north horizon, and nobody can say how much farther.

These problems may all be solved by the proposed expedition. The men are ready and willing; one volunteer has even promised 1000*l.* on condition that he shall be allowed to have a seat in one of the balloons. All that is wanted are the necessary funds, and the amount required is but a small fraction of what is annually expended at our racecourses upon villainous concoctions of carbonic acid and methylated cider bearing the name of "champagne."

Arrangements are being made to start next May, but in the meantime many preliminary experiments are required. One of these, concerning which I have been boring Commander Cheyne and the committee, is a thorough and practical trial of the staying properties of hydrogen gas when confined in given silken or other fabrics saturated with given varnishes. We are still ignorant on this fundamental point. We know something about coal-gas, but little or nothing of the hydrogen, such as may be used in the foregoing expedition. Its exosmosis, as proved by Graham, depends upon its adhesion to the surface of the substance confining it. Every gas has its own speciality in this respect, and a membrane that confines a hydrocarbon like coal-gas may be very unsuitable for pure hydrogen, or *vice versâ*. Hydrogen passes through hard steel, carbonic oxide through red-hot iron plates, and so on with other gases.

They are guilty of most improbable proceedings in the matter of penetrating apparently impenetrable substances.

The safety of the aeronauts and the success of the aerial exploration primarily depends upon the length of time that the balloons can be kept afloat in the air.

A sort of humanitarian cry has been raised against this expedition, on the ground that unnaturally good people (of whom we now meet so many) should not be guilty of aiding and abetting a scheme that may cause the sacrifice of human life. These kind friends may be assured that, in spite of their scruples, the attempt will be made by men who share none of their fears, unless the preliminary experiments prove that a balloon cannot be kept up long enough. Therefore the best way to save their lives is to subscribe *at once* for the preliminary expense of making these trials, which will either discover means of traveling safely, or demonstrate the impossibility of such ballooning altogether. Such experiments will have considerable scientific value in themselves, and may solve other problems besides those of arctic exploration.

Why not apply balloons to African exploration or the crossing of Australia? The only reply to this is that we know too little of the practical possibilities of such a method of traveling when thus applied. Hitherto the balloon has only been a sensational toy. We know well enough that it cannot be steered in a predetermined *line*, i.e., from one *point* to another given *point*, but this is quite a different problem from sailing over a given *surface of considerable area*. This can be done to a certain extent, but we want to know definitely to what extent, and what are the limits of reliability and safety. With this knowledge, and its application by the brave and skillful men who are so eager to start, the solution of the Polar mystery assumes a new and far more hopeful phase than it has ever before presented.

The Anglo-American Arctic Expedition

Commander Cheyne has gone to America to seek the modest equipment that his own countrymen are unable to supply. He proposes now that his expedition shall be "Anglo-American." I have been asked to join an arctic council, to coöperate on this side, and have refused on anti-patriotic grounds. As a member of the former arctic committee, I was so much disgusted with the parsimony of our millionaires and the anti-geographical conduct of the Savile Row Mutual Admiration Society, that I heartily wish that in this matter our American grandchildren may "lick the Britishers quite complete." It will do us much good.

My views, expressed in the "Gentleman's Magazine" of July 1880, and repeated above, remain unchanged, except in the direction of confirmation and development. I still believe that an enthusiastic, practically trained, sturdy arctic veteran, who has endured hardship both at home and abroad, whose craving eagerness to reach the Pole amounts to a positive monomania, who lives for this object alone, and is ready to die for it, who will work at it purely for the work's sake—will be the right man in the right place when at the head of a modestly but efficiently equipped Polar expedition, especially if Lieutenant Schwatka is his second in command.

They will not require luxurious saloons, nor many cases of champagne; they will care but little for amateur theatricals; they will follow the naval traditions of the old British "sea-dogs" rather than those of our modern naval lap-dogs, and will not turn back after a first struggle with the cruel arctic ice, even though they should suppose it to be "paleocrystic."

MR. WALTER POWELL

Scientific aerostation has lost its most promising expert by the untimely death of Walter Powell. He was not a mere sensational ballooner, nor one of those dreamers who imagine they can invent flying machines, or steer balloons against the wind by mysterious electrical devices or by mechanical paddles, fan-wheels, or rudders.

He perfectly understood that a balloon is at the mercy of atmospheric currents and must drift with them, but nevertheless he regarded it as a most promising instrument for geographical research. I had a long conference with him on the subject in August last, when he told me that the main objects of the ascents he had already made, and should be making for some little time forward, were the acquisition of practical skill, and of further knowledge of atmospheric currents; after which he should make a dash at the Atlantic with the intent of crossing to America.

On my part, I repeated with further argument what I have already urged on page 113 of the "Gentleman's Magazine" for July, 1880, viz., the primary necessity of systematic experimental investigation of the rate of exosmosis (oozing out) of the gas from balloons made of different materials and variously varnished.

Professor Graham demonstrated that this molecular permeation of gases and liquids through membranes mechanically air-tight, depends upon the adhesive affinities of particular solids for other particular fluids, and these affinities vary immensely, their variations depending on chemical differences rather than upon mechanical impermeability. My project to attach captive balloons of small size to the roof of the Polytechnic Institution, holding them by a steelyard that should indicate the pull due to their ascending power, and the rate of its decline according to the composition of the membrane, was heartily approved by Mr. Powell, and, had the Polytechnic survived, would have been carried out, as it would have served the double purpose of scientific investigation and of sensational advertisement for the outside public.

If the aeronaut were quite clear on this point—could calculate accurately how long his balloon would float—he might venture with deliberate calculation on journeys that without such knowledge are mere exploits of blind daring.

The varnishes at present used are all permeable by hydrogen gas and hydrocarbon coal-gas, as might be expected, *à priori*, from the fact that they are themselves solid hydrocarbons, soluble in other liquid or gaseous hydrocarbons. Nothing, as far as I can learn, has yet been done with *silicic or boracic varnishes*,[22] which are theoretically impermeable by hydrogen and its carbon compounds; but whether they are practically so under ballooning conditions, and can be made sufficiently pliable and continuous, are questions only to be solved by

---

[22] Since the above was written I have made some experiments with a solution of shellac in borax (obtained by long boiling), and hereby claim the invention of its application to this purpose, in order to prevent anybody from patenting it. I shall not do so myself.

practical experiments of the kind above named. Now that the best man for making these experiments is gone, somebody else should undertake them. Unfortunately, they must of necessity be rather expensive.

Chapter 17

THE LIMITS OF OUR COAL SUPPLY[23]

Estimating the actual consumption of coal for home use in Great Britain at 110 millions of tons per annum, a rise of eight shillings per ton to consumers is equivalent to a tax of 44 millions per annum. These are the figures taken by Sir William Armstrong in his address at Newcastle last February. As the recent abnormal rise in the value of coal has amounted to more than this, consumers have been paying at some periods above a million per week as premium on fuel, even after making fair deduction for the rise of price necessarily due to the diminishing value of gold.

Are we, the consumers of coal, to write off all this as a dead loss, or have we gained any immediate or prospective advantage that may be deducted from the bad side of the account? I suspect that we shall gain sufficient to ultimately balance the loss, and, even after that, to leave something on the profit side.

The abundance of our fuel has engendered a shameful wastefulness that is curiously blind and inconsistent. As a typical example of this inconsistency, I may mention a characteristic incident. A party of young people were sitting at supper in the house of a colliery manager. Among them was the vicar of the parish, a very jovial and genial man, but most earnest withal in his vocation. Jokes and banterings were freely flung across the table, and no one enjoyed the fun more heartily than the vicar; but presently one unwary youth threw a fragment of bread-crust at his opposite neighbor, and thus provoked retaliation. The countenance of the vicar suddenly changed, and in stern clerical tones he rebuked the wickedness of thus wasting the bounties of the Almighty. A general silence followed, and a general sense of guilt prevailed among the revellers. At the same time, and in the same room, a blazing fire, in an ill-constructed open fire-place, was glaring reproachfully at all the guests, but no one heeded the immeasurably greater and utterly irreparable waste that was there proceeding. To every unit of heat that was fully utilized in warming the room, there were eight or nine passing up the chimney to waste their energies upon the senseless clouds and boundless outer atmosphere. A large proportion of the vicar's parishioners are colliers, in whose cottages huge fires blaze most wastefully all day, and are left to burn all night to save the trouble of re-lighting. The vicar diligently visits these cottages, and freely admonishes where he deems it necessary; yet he sees in this general waste of coal no corresponding sinfulness to that of wasting bread. Why is he so blind in one direction, while his moral vision is so microscopic

---

[23] Written during the coal famine of 1872–73.

in the other? Why are nearly all Englishmen and Englishwomen as inconsistent as the vicar in this respect?

There are doubtless several combining reasons for this, but I suspect that the principal one is the profound impression which we have inherited from the experience and traditions of the horrors of bread-famine. A score of proverbs express the important practical truth that we rarely appreciate any of our customary blessings until we have tasted the misery of losing them. Englishmen have tasted the consequences of approximate exhaustion of the national grain store, but have never been near to the exhaustion of the national supply of coal.

I therefore maintain most seriously that we need a severe coal famine, and if all the colliers of the United Kingdom were to combine for a simultaneous winter strike of about three or six months' duration, they might justly be regarded as unconscious patriotic martyrs, like soldiers slain upon a battle-field. The evils of such a thorough famine would be very sharp, and proportionally beneficent, but only temporary; there would not be time enough for manufacturing rivals to sink pits, and at once erect competing iron-works; but the whole world would partake of our calamity, and the attention of all mankind would be aroused to the sinfulness of wasting coal. Six months of compulsory wood and peat fuel, with total stoppage of iron supplies, would convince the people of these islands that waste of coal is even more sinful than waste of bread,—would lead us to reflect on the fact that our stock of coal is a definite and limited quantity that was placed in the present storehouse long before human beings came upon the earth; that every ton of coal that is wasted is lost for ever, and cannot be replaced by any human effort, while bread is a product of human industry, and *its* waste may be replaced by additional human labor; that the sin of bread-wasting does admit of agricultural atonement, while there is no form of practical repentance that can positively and directly replace a hundredweight of wasted coal.

Nothing short of the practical and impressive lesson of bitter want is likely to drive from our households that wretched fetish of British adoration, the open "Englishman's fireside." Reason seems powerless against the superstition of this form of fire-worship. Tell one of the idolaters that his household god is wasteful and extravagant, that five-sixths of the heat from his coal goes up the chimney, and he replies, "I don't care if it does; I can afford to pay for it. I like to *see* the fire, and have the right to waste what is my own." Tell him that healthful ventilation is impossible while the lower part of a room opens widely into a heated shaft, that forces currents of cold air through doors and window leakages, which unite to form a perpetual chilbrain stratum on the floor, and leaves all above the mantel-piece comparatively stagnant. Tell him that no such things as "draughts" should exist in a properly warmed and ventilated house, and that even with a thermometer at zero outside, every part of a well-ordered apartment should be equally habitable, instead of merely a semicircle about the hearth of the fire-worshiper; he shuts his ears, locks up his understanding, because his grandfather and grandmother believed that the open-mouthed chimney was the one and only true English means of ventilation.

But suppose we were to say, "You love a cheerful blaze, can afford to pay for it, and therefore care not how much coal you waste in obtaining it. We also love a cheerful blaze, but have a great aversion to coal-smoke and tarry vapors; and we find that we can make a beautiful fire, quite inoffensive even in the middle of the room, provided we feed it with stale quartern loaves. We know that such fuel is expensive, but can afford to pay for it, and choose to do so." Would he not be shocked at the sight of the blazing loaves, if this extravagance were carried out?

This popular inconsistency of disregarding the waste of a valuable and necessary commodity, of which the supply is limited and unrenewable, while we have such proper horror of wilfully wasting another similar commodity which can be annually replaced as long as man remains in living contact with the earth, will gradually pass away when rational attention is directed to the subject. If the recent very mild suggestion of a coal-famine does something towards placing coal on a similar pedestal of popular veneration to that which is held by the "staff of life," the million a week that it has cost the coal consumer will have been profitably invested.

Many who were formerly deaf to the exhortations of fuel economists are now beginning to listen. "*Forty shillings per ton*" has acted like an incantation upon the spirit of Count Rumford. After an oblivion of more than eighty years, his practical lessons have again sprung up among us. Some are already inquiring how he managed to roast 112 lbs. of beef at the Foundling Hospital with 22 lbs. of coal, and to use the residual heat for cooking the potatoes, and why it is that with all our boasted progress we do not now in the latter third of the nineteenth century, repeat that which he did in the eighteenth.

The fact that the consumption of coal in London during the first four months of 1873 has, in spite of increasing population, amounted to 49,707 tons less than the corresponding period of 1872, shows that some feeble attempts have been made to economize the domestic consumption of fuel. One very useful result of the recent scarcity of coal has been the awakening of a considerable amount of general interest in the work of stock-taking, a tedious process which improvident people are too apt to shirk, but which is quite indispensable to sound business proceedings, either of individuals or nations.

There are many discrepancies in the estimates that have been made of the total available quantity of British coal. The speculative nature of some of the data renders this inevitable, but all authorities appear to agree on one point, viz., that the amount of our supplies will not be determined by the actual total quantity of coal under our feet, but by the possibilities of reaching it. This is doubtless correct, but how will these possibilities be limited, and what is the extent or range of the limit? On both these points I venture to disagree with the eminent men who have so ably discussed this question. First, as regards the nature of the limit or barrier that will stop our further progress in coal-getting. This is generally stated to be the depth of the seams. The Royal Commissioners of 1870 based their tables of the quantity of available coal in the visible and concealed coal-fields upon the assumption that 4000 feet is the limit of possible working. This limit is the same that was taken by Mr. Hull ten years earlier. Mr. Hull, in the last edition of "The Coal Fields of Great Britain," p. 326, referring to Professor Ramsay's estimate, says, "These estimates are drawn up for depths down to 4000 feet below the surface, and even beyond this limit; but with this latter quantity it is scarcely necessary that we should concern ourselves." I shall presently show reasons for believing that the time may ultimately arrive when we *shall* concern ourselves with this deep coal, and actually get it; while, on the other hand, that remote epoch will be preceded by another period of practical approximate exhaustion of British coal supply, which is likely to arrive long before we reach a working depth of 4000 feet.

The Royal Commissioners estimate that within the limits of 4000 feet we have hundreds of square miles of attainable coal capable of yielding, after deducting 40 per cent for loss in getting, etc., 146,480 millions of tons; or, if we take this with Mr. Hull's deduction of one-twentieth for seams under two feet in thickness, there remains 139,000 millions of tons, which, at present rate of consumption, would last about 1200 years. But the rate of consumption is annually increasing, not merely on account of increasing population, but also from the fact that mechanical inventions are perpetually superseding hand labor, and the source of power in such cases, is usually derived from coal. This consideration induced Professor Jevons, in 1865, to estimate that between 1861 and 1871 the consumption would increase from 83,500,000 tons to 118,000,000 tons. Mr. Hunt's official return for 1871 shows that this estimate was a close approximation to the truth, the actual total for 1871 having been 117,352,028 tons. At this rate of an arithmetical increase of three and a half tons per annum, 139,000 millions of tons would last but 250 years. Mr. Hull, taking the actual increase at three millions of tons per annum, extends it to 276 years. Hitherto the annual increase has followed a geometrical, rather than arithmetical progress, and those who anticipate a continuance of this allow us a much shorter lease of our coal treasures. Mr. Price Williams maintains that the increase will proceed in a diminishing ratio like that of the increase of population; and upon this basis he has calculated that the annual consumption will amount to 274 millions of tons a hundred years hence, and the whole available stock of coal will last about 360 years.

The latest returns show, for 1872, an output of 123,546,758 tons, which, compared with 1871, gives a rate of increase of more than double the estimate of Mr. Hull, and indicate that prices have not yet risen sufficiently to check the geometrical rate of increase.[24] Mr Hull very justly points out the omission in those estimates which do not "take into account the diminishing ratio at which coal must be consumed when it becomes scarcer and more expensive;" but, on the other hand, he omits the opposite influence of increasing prices on production, which has been strikingly illustrated by the extraordinary number of new coal-mining enterprises that have been launched during the last six months. If we continue as we are now proceeding, a practical and permanent coal famine will be upon us within the lifetime of many of the present generation. By such a famine, I do not mean an actual exhaustion of our coal seams (which will never be effected), but such a scarcity and rise of prices as shall annihilate the most voracious of our coal-consuming industries, those which depend upon abundance of cheap coal, such as the manufacture of pig-iron, etc.[25]

The action of increasing prices has been but lightly considered hitherto, though its importance is paramount in determining the limits of our coal supply; I even venture so far as to affirm that it is not the depth of the coal seams, not the increasing temperature nor pressure as we proceed downwards, nor even thinness of seam, that will practically determine the limits of British coal-getting, but simply the price per ton at the pit's mouth.

In proof of this, I may appeal to actual practice. Mr. Hull and others have estimated the working limit of thinness at two feet, and agree in regarding thinner seams than this as unworkable. This is unquestionably correct so long as the getting is effected in the usual manner. A collier cannot lie down and hew a much thinner seam than this, if he works as colliers work at present. But the lead and copper miners succeed in working far thinner lodes, even down to the thickness of a few inches, and the gold-digger crushes the hardest component of the earth's crust to obtain barely visible grains of the precious metal. This extension of effort is entirely determined by market value. At a sufficiently high price the

---

[24] From 1870 to 1880 the amount has risen from 110,431,192 to 146,818,622 tons per annum, an average increase of 3,638,743 tons per annum.

[25] At the present time (1882) we are receiving the excessive supplies consequent upon the opening of new pits that, under the stimulus of high prices, were in the course of sinking when the above was written. Hence the present low prices. Presently the annual increase of consumption will overtake this increased supply, and another "coal famine" like that then existing will follow. This is not far distant.

two-feet limit of coal-getting would vanish, and the collier would work after the manner of the lead-miner.

We may safely apply the same reasoning to the limits of depth. The 4000 feet limit of the Royal Commissioners is *at present* unattainable, simply because the immediately prospective price of coal would not cover the cost of such deep sinking and working; but as prices go up, pits will go down, deeper and deeper still.

The obstacles which are assumed to determine the 4000 feet limit are increasing density due to greater pressure, and the elevation of temperature which proceeds as we go downwards. The first of these difficulties has, I suspect, been very much overstated, if not altogether misunderstood; though it is but fair to add that Mr. Hull, who most prominently dwells upon it, does so with all just and philosophic caution. He says that "it is impossible to speak with certainty of the effect of the accumulative weight of 3000 or 4000 feet of strata on mining operations. In all probability one effect would be to increase the density of the coal itself, and of its accompanying strata, so as to increase the difficulty of excavating," and he concludes by stating that "in the face of these two obstacles—temperature and pressure, ever increasing with the depth—I have considered it utopian to include in calculations having reference to coal supply any quantity, however considerable, which lies at a greater depth than 4000 feet. Beyond that depth I do not believe that it will be found practicable to penetrate. Nature rises up, and presents insurmountable barriers."[26]

On one point I differ entirely from Mr. Hull, viz., the conclusion that the increased "density of the coal itself and of its accompanying strata" will offer any serious obstacle. On the contrary, there is good reason to believe that such density is one of the essential conditions for working deep coal. Even at present depths of working, density and hardness of the accompanying strata is one of the most important aids to easy and cheap coal-getting. With a dense roof and floor the collier works vigorously and fearlessly, and he escapes the serious cost of timbering.

Those who have never been underground, and only read of colliery disasters, commonly regard the fire-damp and choke-damp as the collier's most deadly enemies, but the collier himself has quite as much dread of a rotten roof as of either of these: he knows by sad experience how much bruising, and maiming, and crushing of human limbs are due to the friability of the rock above his head. Mr. Hull quotes the case of the Dunkinfield colliery, where, at a depth of about 2500 feet, the pressure is "so resistless as to crush in circular arches of brick four feet thick," and to snap a cast-iron pillar in twain; but he does not give any account of the density of the accompanying strata at the place of these occurrences. I suspect that it was simply *a want of density* that allowed the superincumbent pressure to do such mischief. The circular arches of brick four feet thick were but poor substitutes for a roof of solid rock of 40 or 400 feet in thickness; an arch cut in such a rock would be all key-stone: and I may safely venture to affirm that if, in the deep sinkings of the future, we do encounter the increased density which Mr. Hull anticipates, this will be altogether advantageous. I fear, however, that it will not be so, that the chief difficulty of deep coal-mining will arise from occasional "running in" due to deficient density, and that this difficulty will occur in about the same proportion of cases as at present, but will operate more seriously at the greater depths.

A very interesting subject for investigation is hereby suggested. Do rocks of given composition and formation increase in density as they dip downwards; and if so, does this

---

[26] "The Coal Fields of Great Britain," pp. 447, 448.

increase of density follow any law by which we may determine whether their power of resisting superincumbent pressure increases in any approach to the ratio of the increasing pressure to which they are naturally subjected? If the increasing density and power of resistance reaches or exceeds this ratio, deep mining has nothing to fear from pressure. If they fall short of it, the difficulties arising from pressure may be serious. Friability, viscosity, and power of resisting a crushing strain must be considered in reference to this question.

Mr. Hull has collected a considerable amount of data bearing upon the rate of increase of temperature with depth. His conclusions give a greater rate of increase than is generally stated by geologists; but for the present argument I will accept, without prejudice, as the lawyers say, his basis of a range of 1° F. for 60 feet. According to this, the *rocks* will reach 99.6°, a little above blood-heat, at 3000 feet, and 116.3° at the supposed limit of 4000 feet. It is assumed by Mr. Hull, by the Commissioners, and most other authorities, that this rock temperature of 116° will limit the possibilities of coal-mining. At the average prices of the last three years, or the prospective prices of the next three years, this temperature may be, like difficulties of the thin seams, an insurmountable barrier; but I contend that at higher prices we may work coal at this, and even far higher, rock temperatures; that it matters not how high the thermometer rises as we descend, we shall still go lower and still get coal so long as prices rise with the mercury. Given this condition, and I have no doubt that coal may be worked where the rock temperature shall reach or even exceed 212°. I do not say that we shall actually work coal at such depths; but if we do not, the reason will be, not that the thermometer is too high, but that prices are too low; in other words, value, not temperature, will determine the working limits.

Mr. Leifchild, in the last number of the "Edinburgh Review," in discussing this question, tells us that "the normal heat of our blood is 98°, and fever heat commences at 100°, and the extreme limit of fever heat may be taken at 112°. Dr. Thudichum, a physician who has specially investigated this subject, has concluded from experiments on his own body at high temperatures, that at a heat of 140° no work whatever could be carried on, and that at a temperature of from 130° to 140° only a very small amount of labor, and that at short periods, was practicable; and further, that human labor daily, and at ordinary periods, is limited by 100° of temperature, as a fixed point, and then the air must be dry, for in moist air he did not think men could endure ordinary labor at a temperature exceeding 90°."

It may be presumptuous on my part to dispute the conclusions of a physician on such a subject, but I do so nevertheless, as the data required are simple practical facts such as are better obtained by furnace-working than by sick-room experience.

During the hottest days of the summer of 1868 I was engaged in making some experiments in the re-heating furnaces at Sir John Brown & Co.'s works, Sheffield, and carried a thermometer about with me which I suspended in various places where the men were working. At the place where I was chiefly engaged (a corner between two sets of furnaces), the thermometer, suspended in a position where it was not affected by direct radiations from the open furnaces, stood at 120° while the furnace doors were shut. The *radiant* heat to which the men themselves were exposed while making their greatest efforts in placing and removing the piles was far higher than this, but I cannot state it, not having placed the thermometer in the position of the men. In one of the Bessemer pits the thermometer reached 140°, and men worked there at a kind of labor demanding great muscular effort. It is true that during this same week the puddlers were compelled to leave their work; but the tremendous amount of concentrated exertion demanded of the puddler in front of a furnace, which, during the time of removing the balls, radiates a degree of heat quite sufficient to roast a sirloin of beef if placed in the position of the puddles hands, is beyond comparison with that which would be demanded of a collier working even at a depth giving a theoretical rock temperature of 212°, and aided by the coal-cutting and other machinery that sufficiently high prices would readily command. In some of the operations of glass-making, the ordinary summer working temperature is considerably above 100°, and the radiant neat to which the workmen are subjected far exceeds 212°. This is the case during a "pot setting," and in the ordinary work of flashing crown glass.

As regards the mere endurance of a high temperature, the well-known experiments of Blagden, Sir Joseph Banks, and others have shown that the human body can endure for short periods a temperature of 260° F., and upwards. My own experience of furnace-work, and of Turkish baths, quite satisfies me that I could do a fair day's work of six or eight hours in a temperature of 130° F., provided I were free from the encumbrances of clothing, and had access to abundance of tepid water. This in a still atmosphere; but with a moving current of dry air capable of promoting vigorous evaporation from the skin, I suspect that the temperature might be ten or fifteen degrees higher. I *enjoy* ordinary walking exercise in a well-ventilated Turkish bath at 150°, and can endure it at 180°.

In order to obtain further information on this point, I have written to Mr. Tyndall, the proprietor of the Turkish baths at Newington Butts. He is an architect, who has had considerable experience in the employment of workmen and in the construction of Turkish baths and other hot-air chambers. He says: "Shampooers work in my establishment from four to five hours at a time *in a moist atmosphere* at a temperature ranging from 105° to 110°. I have myself worked twenty hours out of twenty-four in one day in a temperature over 110°. Once for one half-hour I shampooed in 185°. At the enamel works in Pimlico, belonging to Mr. Mackenzie, men work daily in a heat of over 300°. The moment a man working in a 110° heat begins to drink alcohol, his tongue gets parched, and he is obliged to continue drinking while at work, and the brain gets so excited that he cannot do half the amount. I painted my skylights, taking me about four hours, at a temperature of about 145°; also the hottest room skylights, which took me one hour, coming out at intervals for "a cooler," at a temperature of 180°. I may add in conclusion, that a man can work well in a moist temperature of 110° if he perspires freely."

The following, by a writer whose testimony may be safely accepted, is extracted from an account of ordinary passenger ships of the Red Sea, in the "Illustrated News," of November 9, 1872: "The temperature in the stoke-hole was 145°. The floor of this warm region is close to the ship's keel, so it is very far below. There are twelve boilers, six on each side, each with a blazing furnace, which has to be opened at regular intervals to put in new coals, or to be poked up with long iron rods. This is the duty of the poor wretches who are doomed to this work. It is hard to believe that human beings could be got to labor under such conditions, yet such persons are to be found. The work of stoking or feeding the fires is usually done by Arabs, while the work of bringing the coal from the bunkers is done by sidi-wallahs or Negroes. At times some of the more intelligent of these *are promoted to the stoking*. The Negroes who do this kind of work come from Zanzibar. They are generally short men, with strong limbs, round bullet heads, and the very best of good nature in their dispositions. Some of them will work half an hour in such a place as the stoke-hole without a drop of perspiration on their dark skins. Others, particularly the Arabs, when it is so hot as it often is in the Red

Sea, have to be carried up in a fainting condition, and are restored to animation by dashing buckets of water over them as they lie on deck."

It must be remembered that the theoretical temperature of 116° at 4000 feet, the 133° at 5000 feet, or the 150° at 6000 feet, are the temperatures of *the undisturbed rock*; that this rock is a bad conductor of heat, whose surface may be considerably cooled by radiation and convection; and therefore we are by no means to regard the rock temperature as that of the air of the roads and workings of the deep coal-pits of the future.[27] It is true that the Royal Commissioners have collected many facts showing that the actual difference between the face of the rocks of certain pits and the air passing through them is but small; but these data are not directly applicable to the question under consideration for the three following reasons:

First. The comparisons are made between the temperature of the air and the actual temperature of the opened and already cooled strata, while the question to be solved is the difference between the theoretical temperature of the unopened earth depths and that of the air in roads and working's to be opened through them.

Second. The cooling effect of ventilation must (as the Commissioners themselves state) increase in a ratio which "somewhat exceeds the ratio of the difference between the temperature of the air and that of the surrounding surface with which it is in contact." Thus, the lower we proceed the more and more effectively cooling must a given amount of ventilation become.

The third, and by far the most important, reason is, that in the deep mining of the future, special means will be devised and applied to the purpose of lowering the temperature of the workings; that as the descending efforts of the collier increase with the ascending value of the coal, a new problem will be offered for solution, and the method of working coal will be altered accordingly. In the cases quoted by the Commissioners, the few degrees of cooling were effected by a system of ventilation that was devised to meet the requirements of respiration, and not for the purpose of cooling the mine.

It would be very presumptuous for anyone in 1873 to say how this special cooling will actually be effected, but I will nevertheless venture to indicate one or two principles which may be applied to the solution of the problem. First of all, it must be noted that very deep mines are usually dry; and there is good reason to believe that, before reaching the Commissioners' limit of 4000 feet, dry mining would be the common, and at and below 4000 feet the universal, case. At present we usually obtain coal from water-bearing strata, and all our arrangements are governed by this very serious contingency. With water removed, the whole system of coal-mining may be revolutionized, and thus the aspect of this problem of cooling the workings would become totally changed.

Those who are acquainted with the present practice of mining are aware that when an estate is taken, and about to be worked for coal, the first question to be decided is the dip of the measures, in order that the sinking may be made "on the deep" of the whole range. The pits are not sunk at that part of the same range where, at first sight, the coal appears the most accessible, but, on the contrary, at the deepest part. It is then carried on to some depth below the coal seam which is to be worked, in order to form a "sumpf" or receptacle from which the

---

[27] In a paper on the Comstock mines, read at the Pittsburg meeting of the American Institute of Mining Engineers in 1879, by Mr. John A. Church, the hot mine waters are described as reaching 158° Fahr. (so hot that men have been scalded to death by falling into them). The highest recorded *air temperature* there is 128°. These are silver mines, and vigorously worked in spite of this temperature and great humidity. A much higher temperature is endurable in *dry air*.

water may be wound or pumped. The necessity for this in water-bearing strata is obvious enough. If the collier began at the shallowest portion of his range, and attempted to proceed downwards, he would be "drowned out" unless he worked as a coal-diver rather than a coal-miner. By sinking in the deep he works upwards, away from the water, which all drains down to the sumpf, from which it is pumped.

The modern practice is to sink "a pair of pits," *both on the deep*, and within a short distance of each other. The object of the second is ventilation. By contrivances, which I need not here detail, the air is made to descend one of the pits, "the downcast shaft," then to traverse the roads and workings wherein ventilation is required, and return by a reverse route to the "upcast shaft," by which it ascends to the surface.

Thus it will be seen that, whenever the temperature of the roads and workings exceeds that of the outer atmosphere; the air currents have to be forced to travel through the mine in a direction contrary to their natural course. The cooler air of the downcast shaft has to climb the inclined roads, and then after attaining its maximum temperature in the fresh workings must *descend* the roads till it reaches the upcast shaft. The cool air must rise and the warmer air descend.

What, then, would be the course of the mining engineer when all the existing difficulties presented by water-bearing strata should be removed, and their place taken by a new and totally different obstacle, viz., high temperature? Obviously to reverse the present mode of working—to sink on the upper part of the range and drive downwards. In such a system of working the ventilation of the pit will be most powerfully aided or altogether effected by natural atmospheric currents. An upcast once determined by artificial means, it will thereafter proceed spontaneously, as the cold air of the downcast shaft will travel by a descending road to the workings, and then after becoming heated will simply obey the superior pressure of the heavy column behind, and proceed by an upward road to the upcast shaft. As the impelling force of the air current will be the difference between the weight of the cool column of air in the downcast shaft and roads and the warm column in the upcast, the available force of natural ventilation and cooling will increase just as demanded, i.e., it will increase with the depth of the workings and the heat of the rocks. A mining engineer who knows what is actually done with present arrangements, will see at once that with the above-stated advantages a gale of wind or even a hurricane might be directed through any particular roads or long-wall workings that were once opened. Let us suppose the depth to be 5000 feet, the rock temperature at starting 133°, and that of the outer air 60°, we should have a torrent of air, 73° cooler than the rocks, rushing furiously downwards, then past the face of the heated strata, and absorbing its heat to such an extent that the upcast shaft would pour forth a perpetual blast of hot air like a gigantic furnace chimney.

But this is not all; the heat and dryness of these deep workings of the future place at our disposal another and still more efficient cooling agency than even that of a hurricane of dry-air ventilation. In the first part of the sinking of the deep shafts the usual water-bearing strata would be encountered, and the ordinary means of "tubbing" or "coffering" would probably be adopted for temporary convenience during sinking. Doorways, however, would be left in the tubbing at suitable places for tapping at pleasure the wettest and most porous of the strata. Streams of cold water could thus be poured down the sides of the shaft, which, on reaching the bottom, would flow by a downhill road into the workings. The stream of air rushing by the same route and becoming heated in its course would powerfully assist the evaporation of the water. The deeper and hotter the pit, the more powerful would be these cooling agencies.

As the specific heat of water is about five times that of the coal-measure rocks, or the coal itself, every degree of heat communicated to each pound of water would abstract one degree from five pounds of rocks. But in the conversion of water at 60° into vapor at say 100°, the amount of heat absorbed is equivalent to that required to raise the same weight of water about 1000°, and thus the effective cooling power on the rock would be equivalent to 5000°.

The workings once opened (I assume as a matter of course that by this time pillar-and-stall working will be entirely abandoned for long-wall or something better), there would be no difficulty in thus pouring streams of water and torrents of air through the workings during the night, or at any suitable time preparatory to the operations of the miner, who long before the era of such deep workings will be merely the director of coal-cutting and loading machinery.

Given a sufficiently high price for coal at the pit's mouth to pay wages and supply the necessary fixed capital, I see no insuperable difficulty, *so far as mere temperature is concerned*, in working coal at double the depth of the Royal Commissioners' limit of possibility. At such a depth of 8000 feet the theoretical rock-temperature is 183°.

By the means above indicated, I have no doubt that this could be reduced to an *air* temperature below 110°—that at which Mr. Tyndall's shampooers ordinarily work. Of course the newly-exposed face of the coal would have its initial temperature of 183°; but this is a trivial heat compared to the red-hot radiant surfaces to which puddlers, shinglers, glassmakers, etc., are commonly exposed. Divested of the incumbrance of clothing, with the whole surface of the skin continuously fanned by a powerful stream of air—which, during working hours need be but partly saturated with vapor—a sturdy midland or north-countryman would work merrily enough at short hours and high wages, even though the newly-exposed face of coal reached 212°; for we must remember that this new coal-face would only correspond to the incomparably hotter furnace-doors and fires of the steamship stoke-holes.

The high temperature at 8000 or even 10,000 feet would present a really serious difficulty during the first opening of communications between the two pits. A spurt of brave effort would here be necessary, and if anybody doubts whether Englishmen could be found to make the effort, let him witness a "pot-setting" at a glass-house. Negro labor might be obtained if required, but my experience among English workmen leads me to believe that they will never allow Negroes or any others to beat them at home in any kind of work where the wages paid are proportionate to the effort demanded.

If I am right in the above estimates of working possibilities, our coal resources may be increased by about forty thousand millions of tons beyond the estimate of the Commissioners. To obtain such an additional quantity will certainly be worth an effort, and unless we suffer a far worse calamity than the loss of all our minerals, viz., a deterioration of British energy, the effort will assuredly be made.

I have said repeatedly that it is not physical difficulties but market value that will determine the limits of our coal-mining. This, like all other values, is of course determined by the relation between demand and supply. Fuel being one of the absolute necessaries of life, the demand for it must continue so long as the conditions of human existence remain as at present, and the outer limits of the possible value of coal will be determined by that of the next cheapest kind of fuel which is capable of superseding it.

We begin by working the best and most accessible seams, and while those remain in abundance the average value of coal will be determined by the cost of producing it under these easy conditions. Directly these most accessible seams cease to supply the whole demand, the market value rises until it becomes sufficient to cover the cost of working the less accessible; and the average value will be regulated not by the cost of working what remains of the first or easy mines, but by that of working the most difficult that must be worked in order to meet the demand. This is a simple case falling under the well-established economic law, that the natural or cost value of any commodity is determined by the cost value of the most costly portion of it. Thus, the only condition under which we can proceed to sink deeper and deeper, is a demand of sufficient energy to keep pace with the continually increasing cost of production. This condition can only be fulfilled when there is no competing source of cheaper production which is adequate to supply the demand.

The question then resolves itself into this: Is any source of supply likely to intervene that will prevent the value of coal from rising sufficiently to cover the cost of working the coal seams of 4000 feet and greater depth? Without entering upon the question of peat and wood fuel, both of which will for some uses undoubtedly come into competition with British coal as it rises in value, I believe that there are sound reasons for concluding that our London fireplaces, and those of other towns situated on the sea-coast and the banks of navigable rivers, will be supplied with transatlantic coal long before we reach the Commissioner's limit of 4000 feet. The highest prices of last winter, if steadily maintained, would be sufficient to bring about this important change. Temporary upward jerks of the price of coal have very little immediate effect upon supply, as the surveying, conveyance, boring, sinking, and fully opening of a new coal estate is a work of some years.

The Royal Commissioners estimate that the North-American coal-fields contain an untouched coal area equal to seventy times the whole of ours. Further investigation is likely to increase rather than diminish this estimate. An important portion of this vast source of supply is well situated for shipment, and may be easily worked at little cost. Hitherto, the American coal-fields have been greatly neglected, partly on account of the temptations to agricultural occupation which are afforded by the vast area of the American continent, and partly by the barbarous barriers of American politics. Large amounts of capital which, under the social operation of the laws of natural selection, would have been devoted to the unfolding of the vast mineral resources of the United States, are still wastefully invested in the maintenance of protectively nursed and sickly imitation of English manufactures. When the political civilization of the United States become sufficiently advanced to establish a national free-trade policy, this perverted capital will flow into its natural channels, and the citizens of the States will be supplied with the more highly elaborated industrial products at a cheaper rate than at present, by obtaining them in exchange for their superabundant raw material from those European countries where population is overflowing the raw material supplies.

When this time arrives, and it may come with the characteristic suddenness of American changes, the question of American *versus* English coal in the English markets will reduce itself to one of horizontal *versus* vertical difficulties. If at some future period the average depth of the Newcastle coal-pits becomes 3000 feet greater than those of the pits near the coast of the Atlantic or American lakes, and if the horizontal difficulties of 3000 miles of distance are less than the vertical difficulties of 3000 feet of depth, then coals will be carried from America to Newcastle. They will reach London and the towns on the South Coast before this, that is, when the vertical difficulties at Newcastle plus those of horizontal traction from Newcastle to the south, exceed those of eastward traction across the Atlantic.

As the cost of carriage increases in a far smaller ratio than the open ocean distance, there is good reason for concluding that the day when London houses will be warmed by American coal is not very far distant. We, in England, who have outgrown the pernicious folly of "protecting native industry" will heartily welcome so desirable a consummation. It will render unnecessary any further inquiry into the existence of London "coal rings" or combinations for restricted output among colliers or their employers. If any morbid impediments to the free action of the coal trade do exist, the stimulating and purgative influence of foreign competition will rapidly restore the trade to a healthy condition.

The effect of such introduction of American coal will not be to perpetually lock up our deep coal nor even to stop our gradual progress towards it. We shall merely proceed downwards at a much slower rate, for in America, as with ourselves, the easily accessible coal will be first worked, and as that becomes exhausted, the deeper, more remote, thinner, and inferior will only remain to be worked at continually increasing cost. When both our own and foreign coal cost more than peat, or wood, or other fuel, then and therefore will coal become quite inaccessible to us, and this will probably be the case long before we are stopped by the physical obstacles of depth, density, or high temperature.

As this rise of value must of necessity be gradual, and as the superseding of British by foreign coal, as well as the final disuse of coal, will gradually converge from the circumference towards the centres of supply, from places distant from coal-pits to those close around them, we shall have ample warning and opportunity for preparing for the social changes that the loss of the raw material will enforce.

The above-quoted writer, in the "Edinburgh Review," expresses in strong and unqualified terms an idea that is very prevalent in England and abroad: he says that, "The course of manufacturing supremacy of wealth and of power is directed by coal. That wonderful mineral, of the possession of which Englishmen have thought so little but wasted so much, is the modern realization of the philosopher's stone. This chemical result of primeval vegetation has been the means by its abundance of raising this country to an unprecedented height of prosperity, and its deficiency might have the effect of lowering it to slow decline."

*** "It raises up one people and casts down another; it makes railways on land and paths on the sea. It founds cities, it rules nations, it changes the course of empires."

The fallacy of these customary attributions of social potency to mere mineral matter is amply shown by facts that are previously stated by the reviewer himself. He tells us that "the coal-fields of China extend over an area of 400,000 square miles; and a good geologist, Baron Von Richthofen, has reported that he himself has found a coal-field in the province of Hunau covering an area of 21,700 square miles, which is nearly double our British coal area of 12,000 square miles. In the province of Shansi, the Baron discovered nearly 30,000 square miles of coal with unrivaled facilities for mining. But all these vast coal-fields, capable of supplying the whole world for some thousands of years to come, are lying unworked."

If "the course of manufacturing supremacy of wealth and of power" were directed by coal, then China, which possesses 33·3 times more of this directive force than Great Britain, and had so early a start in life, should be the supreme summit of the industrial world. If this solid hydrocarbon "raises up one people and casts down another," the Chinaman should be raised thirty-three times and three tenths higher than the Englishman; if it "makes railways on land and paths on the sea," the Chinese railways should be 33·3 times longer than ours, and the tonnage of their mercantile marine 33·3 times greater.

Every addition to our knowledge of the mineral resources of other parts of the world carries us nearer and nearer to the conclusion that the old idea of the superlative abundance of the natural mineral resources of England is a delusion. We are gradually discovering that, with the one exception of tin-stone, we have but little if any more than an average supply of useful ores and mineral fuel. It is a curious fact, and one upon which we may profitably ponder, that the poorest and the worst iron ores that have ever been commercially reduced, are those of South Staffordshire and the Cleveland district, and these are the two greatest iron-making centres of the world. There are no ores of copper, zinc, tin, nickel, or silver in the neighborhood of Birmingham, nor any golden sands upon the banks of the Rea, yet this town is the hardware metropolis of the world, the fatherland of gilding and plating, and is rapidly becoming supreme in the highest art of gold and silver work.

These, and a multitude of other analogous facts, abundantly refute the idea that the native minerals, the natural fertility, the navigable rivers, or the convenient seaports, determine the industrial and commercial supremacy of nations. The moral forces exerted by the individual human molecules are the true components which determine the resulting force and direction of national progress. It is the industry and skill of our workmen, the self-denial, the enterprise, and organizing ability of our capitalists, that has brought our coal so precociously to the surface and redirected for human advantage the buried energies of ancient sunbeams, while the fossil fuel of other lands has remained inert.

The foreigner who would see a sample of the source of British prosperity must not seek for it in a geological museum or among our subterranean rocks; let him rather stand on the Surrey side of London Bridge from 8 to 10 A.M. and contemplate the march of one of the battalions of our metropolitan industrial army, as it pours forth in an unceasing stream from the railway stations towards the City. An analysis of the moral forces which produce the earnest faces and rapid steps of these rank and file and officers of commerce will reveal the true elements of British greatness, rather than any laboratory dissection of our coal or ironstone.

Fuel and steam-power have been urgently required by all mankind. Englishmen supplied these wants. Their urgency was primary and they were first supplied, even though the bowels of the earth had to be penetrated in order to obtain them. In the present exceptional and precocious degree of exhaustion of our coal treasures, we have the *effect* not the *cause* of British industrial success.

If in a ruder age our greater industrial energy enabled us to take the lead in supplying the ruder demands of our fellow-creatures, why should not a higher culture of those same abundant energies qualify us to maintain our position and enable us to minister to the more refined and elaborate wants of a higher civilization? There are other necessary occupations quite as desirable as coal-digging, furnace-feeding, and cotton-spinning.

The approaching exhaustion of our coal supplies should therefore serve us as a warning for preparation. Britain will be forced to retire from the coal trade, and should accordingly prepare her sons for higher branches of business,—for those in which scientific knowledge and artistic training will replace mere muscular strength and mechanical skill. We have attained our present material prosperity mainly by our excellence in the use of steam-power; let us now struggle for supremacy in the practical application of brain-power.

We have time and opportunity for this. The exhaustion of our coal supplies will go on at a continually retarding pace—we shall always be approaching the end, but shall never absolutely reach it, as every step of approximation will diminish the rate of approach; like the everlasting process of reaching a given point by continually halving our distance from it.

First of all we shall cease to export coal; then we shall throw up the most voracious of our coal-consuming industries, such as the reduction of iron-ore in the blast-furnace; then copper-smelting and the manufacture of malleable iron and steel from the pig, and so on progressively. If we keep in view the natural course and order of such progress, and intelligently prepare for it, the loss of our coal need not in the smallest degree retard the progress of our national prosperity.

If, however, we act upon the belief that the advancement of a nation depends upon the mere accident of physical advantages, if we fold our arms and wait for Providence to supply us with a physical substitute for coal, we shall become Chinamen, minus the unworked coal of China.

If our educational efforts are conducted after the Chinese model; if we stultify the vigor and freshness of young brains by the weary, dull, and useless cramming of words and phrases; if we poison and pervert the growing intellect of British youth by feeding it upon the decayed carcases of dead languages, and on effete and musty literature, our progress will be proportionately Chinaward; but if we shake off that monkish inheritance which leads so many of us blindly to believe that the business of education is to produce scholars rather than men, and direct our educational efforts towards the requirements of the future rather than by the traditions of the past, we need have no fear that Great Britain will decline with the exhaustion of her coal-fields.

The teaching and training in schools and colleges must be directly and designedly preparatory to those of the workshop, the warehouse, and the office; for if our progress is to be worthy of our beginning, the moral and intellectual dignity of industry must be formally acknowledged and systematically sustained and advanced. Hitherto, we have been the first and the foremost in utilizing the fossil forces which the miner has unearthed; hereafter we must in like manner avail ourselves of the living forces the philosopher has revealed. Science must become as familiar among all classes of Englishmen as their household fuel. The youth of England must be trained to observe, generalize, and *investigate* the phenomena and forces of the world outside themselves; and also those moral forces within themselves, upon the right or wrong government of which the success or failure, the happiness or misery of their lives will depend.

With such teaching and training the future generations of England will make the best and most economical use of their coal while it lasts, and will still advance in material and moral prosperity in spite of its progressive exhaustion.

Chapter 18

"THE ENGLISHMAN'S FIRESIDE"

During the investment of Paris, the *Comptes Rendus* of the Acadamy of Sciences were mainly filled with papers on the construction and guidance of balloons; with the results of ingenious researches on methods of making milk and butter without the aid of cows; on the extraction of nutritious food from old boots, saddles, and other organic refuse; and other devices for rendering the general famine more endurable. In like manner, our present coal famine is directing an important amount of scientific, as well as commercial, attention to the subject of economizing coal and finding substitutes for it.

A few thoughtful men have shocked their fellow-sufferers very outrageously by wishing that coal may reach 3*l*. per ton, and remain at that price for a year or two. I confess that, in spite of my own empty coal-cellar and small income, I am one of those hard-hearted cool calculators, being confident that, even from the narrow point of view of my own outlay in fuel, the additional amount I should thus pay in the meantime would be a good investment, affording by an ample return in the saving due to consequent future cheapness.

Regarded from a national point of view, I am convinced that 3*l*. a ton in London, and corresponding prices in other districts, if thus maintained, would be an immense national blessing. I say this, being convinced that nothing short of pecuniary pains and penalties of ruinous severity will stir the blind prejudices of Englishmen, and force them to desist from their present stupid and sinful waste of the greatest mineral treasure of the island.

One of the grossest of our national manifestations of Conservative stupidity is our senseless idolatrous worship of that domestic fetish, "the Englishman's fireside." We sacrifice health, we sacrifice comfort, we begrime our towns and all they contain with sooty foulness, we expend an amount far exceeding the interest of the national debt, and discount our future prospects of national prosperity, in order that we may do what? Enjoy the favorite recreation of idiots. It is a well-known physiological fact that an absolute idiot, with a cranium measuring sixteen inches in circumference, will sit and stare at a blazing fire for hours and hours continuously, all the day long, except when feeding, and that this propensity varies with the degree of mental vacuity.

Few sights are more melancholy than the contemplation of a party of English fire-worshipers seated in a semicircle round the family fetish on a keen frosty day. They huddle together, roast their knees, and grill their faces, in order to escape the chilling blast that is brought in from all the chinks of leaky doors and windows by the very agent they employ, at so much cost, for the purpose of keeping the cold away. The bigger the fire the greater the draught, the hotter their faces the colder their backs, the greater the consumption of coal the more abundant the crop of chilblains, rheumatism, catarrh, and other well-deserved miseries.

The most ridiculous element of such an exhibition is the complacent self-delusion of the victims. They believe that their idol bestows upon them an amount of comfort unknown to other people, that it affords the most perfect and salubrious ventilation, and, above all, that it is a "cheerful" institution. The "cheerfulness" is, perhaps, the broadest part of the whole caricature, especially when we consider that, according to this theory of the cheerfulness of fire-gazing, the 16-inch idiot must be the most cheerful of all human beings.

The notion that our common fireplaces and chimneys afford an efficient means of ventilation, is almost too absurd for serious discussion. Everybody who has thought at all on the subject is aware that in cold weather the exhalations of the skin and lungs, the products of gas-burning, etc., are so much heated when given off that they rise to the upper part of the room (especially if any cold outer air is admitted), and should be removed from there before they cool again and descend. Now, our fireplace openings are just where they ought *not* to be for ventilation; they are at the lower part of the room, and thus their action consists in creating a current of cold air or "draught" from doors and windows, which cold current at once descends, and then runs along the floor, chilling our toes and provoking chilblains.

This cold fresh air having done its worst in the way of making us uncomfortable, passes directly up the chimney without doing us any service for purposes of respiration. Our mouths are usually above the level of the chimney opening, and thus we only breathe the vitiated atmosphere which it fails to remove.

Not only does the fire-opening fail to purify the air we breathe, it actually prevents the leakage of the lower part of the windows and doors from assisting in the removal of the upper stratum of vitiated air, for the strong up-draught of the chimney causes these openings to be fully occupied by an inflowing current of cold air, which at once descends, and then proceeds, as before stated, to the chimney. If the leakage is insufficient to supply the necessary amount of chilblain-making and bronchitis-producing draught, it has to enter by way of the chimney-pot in the form of occasional spasms of down-draught, accompanied by gusts of choking and blackening smoke. It is a fact not generally known, that smoky chimneys are especial English institutions, one of the peculiar manifestations of our very superior domestic comfortableness.

It is true that, in some of our rooms, an Arnott's ventilator opens into the upper part of the chimney, but this was intended by Dr. Arnott as an adjunct to his modification of the German stove, and such ventilator can only act efficiently where a stove is used. The pressure required to fairly open it can only be regularly obtained when the chimney is closed below, or its lower opening is limited to that of a stovepipe.

The mention of a German stove has upon an English fire-worshiper a similar effect to the sight of water upon a mad dog. Again and again, when I have spoken of the necessity of reforming our fireplaces, the first reply elicited has been, "What, would you have us use German stoves?" In every case where I have inquired of the exclaimer, "What sort of a thing is a German stove?" the answer has proved that the exclamation was but a manifestation of blind prejudice based upon total ignorance. These people who are so much shocked at the notion of introducing "German stoves" have no idea of the construction of the stoves which deservedly bear this title. Their notion of a German stove is one of those wretched iron boxes of purely English invention known to ironmongers as "shop stoves." These things get red hot, their red-hot surface frizzles the dust particles that float in the atmosphere and perfume the apartment accordingly. This, however disagreeable, is not very mischievous, perhaps the reverse, as many of these dust particles, which are revealed by a sunbeam, are composed of organic matter which, as Dr. Tyndall argues, may be carriers of infection. If we must inhale such things, it is better that we should breathe them cooked than take them raw.

The true cause of the headaches and other mischief which such stoves unquestionably induce is very little understood in this country. It has been falsely attributed to over-drying of the atmosphere, and accordingly evaporating pans and other contrivances have been attached to such stoves, but with little or no advantage. Other explanations are given, but the true one is that iron *when red hot is permeable by carbonic oxide*. This was proved by the researches of Professor Graham, who showed that this gas not only *can* pass through red-hot iron with singular facility, but actually *does* so whenever there is atmospheric air on one side and carbonic oxide on the other.

For the benefit of my non-chemical readers, I may explain that when any of our ordinary fuel is burned, there are two products of carbon combustion, one the result of complete combustion, the other of semi-combustion—carbonic acid and carbonic oxide—the former, though suffocating when breathed alone or in large proportion, is not otherwise poisonous, and has no disagreeable odor; it is in fact rather agreeable in small quantities, being the material of champagne bubbles and of those of other effervescing drinks. Carbonic oxide, the product of semi-combustion, is quite different. Breathed only in small quantities, it acts as a direct poison, producing peculiarly oppressive headaches. Besides this, it has a disagreeable odor. It thus resembles many other products of imperfect combustion, such as those which are familiar to everybody who has ever blown out a tallow candle, and left the red wick to its own devices.

On this account alone any kind of iron stove capable of becoming red-hot should be utterly condemned. If Englishmen did their traveling in North Europe in the winter, their self-conceit respecting the comfort of English houses would be cruelly lacerated, and none such would perpetrate the absurdity of applying the name of "German stove" to the iron fire-pots that are sold as stoves by English ironmongers.

As the Germans use so great a variety of stoves, it is scarcely correct to apply the title of German to any kind of stove, unless we limit ourselves to North Germany. There, and in Sweden, Denmark, Norway, and Russia, the construction of stoves becomes a specialty. The Russian stove is perhaps the most instructive to us, as it affords the greatest contrast to our barbarous device of a hole in the wall into which fuel is shoveled, and allowed to expend nine-tenths of its energies in heating the clouds, while only the residual ten per cent does anything towards warming the room. With the thermometer outside below zero, a house in Moscow or St. Petersburg is kept incomparably more warm and comfortable, and is *better* ventilated (though, perhaps, not so *much* ventilated) than a corresponding class of house in England, where the outside temperature is 20 or 30 degrees higher, and this with a consumption of about one-fourth of the fuel which is required for the production of British bronchitis.

This is done by, first of all, sacrificing the idiotic recreation of fire-gazing, then by admitting no air into the chimney but that which is used for the combustion of the fuel; thirdly, by sending as little as possible of the heat up the chimney; fourthly, by storing the heat obtained from the fuel in a suitable reservoir, and then allowing it gradually and steadily to radiate into the apartment from a large but not overheated surface.

The Russian stove by which these conditions are fulfilled is usually an ornamental, often a highly artistic, handsome article of furniture, made of fire-resisting porcelain, glazed and otherwise decorated outside. Internally it is divided by thick fire-clay walls into several upright chambers or flues, usually six. Some dry firewood is lighted in a suitable fireplace, and is supplied with only sufficient air to effect combustion, all of which enters below and passes fairly through the fuel. The products of combustion being thus undiluted with unnecessary cold air, are very highly heated, and in this state pass up compartment or flue No. 1; they are then deflected, and pass down No. 2; then up No. 3, then down No. 4, then up No. 5, then down No. 6. At the end of this long journey they have given up most of their heat to the 24 heat-absorbing surfaces of the fire-clay walls of the six flues.

When the interior of the stove is thus sufficiently heated, the fire-door and the communication with the chimney are closed, and the fire is at once extinguished, having now done its day's work; the interior of the stove has bottled up its calorific force, and holds it ready for emission into the apartment. This is effected by the natural properties of the walls of the earthenware reservoir. They are bad conductors and good radiators. The heat slowly passes through to the outside of the stove, is radiated into the apartment from a large and moderately-heated surface, which affords a genial and well-diffused temperature throughout.

There is no scorching in one little red-hot hole, or corner, or box, and freezing in the other parts of the room. There are no draughts, as the chimney is quite closed as soon as the heat reservoir is supplied. If one of these heat reservoirs is placed in the hall, where it may form a noble ornament and can easily communicate with an underground flue, it warms every part of the house, and enables the Russian to enjoy a luxurious temperate climate indoors in spite of arctic winter outside.

In a house thus warmed and free from draughts or blasts of cold air, ventilation becomes the simplest of problems. Nothing more is required than to provide an inlet and outlet in suitable places, and of suitable dimensions, when the difference between the specific gravity of the cold air without and warm air within does all the rest. Nothing is easier to arrange than to cause all the entering air to be warmed on its way by the hall stove, and to regulate the supply which each apartment shall receive from this general or main stream by adjusting its own upper outlet. In our English houses, with open chimneys, all such systematic, scientific ventilation is impossible, on account of the dominating, interfering, useless, and comfort-destroying currents produced by these wasteful air-shafts.

I should add that the Russian porcelain reservoirs may be constructed for a heat supply of a few hours or for a whole day, and I need say nothing further in refutation of the common British prejudice which confounds so admirable and truly scientific a contrivance with the iron fire-pot above referred to.

There is another kind of stove, which, for the sake of distinction, I may call Scandinavian, as it is commonly used in Norway, Sweden, and Denmark, besides some parts of North Germany. This is a tall, hollow iron pillar, of rectangular section, varying from three to six feet in width, and rising half-way to the ceiling of the room, and sometimes higher. A fire is lighted at the lower part, and the products of combustion, in their way upwards, meet with horizontal iron plates, which deflect them first to the right, then to the left, and thus compel them to make a long serpentine journey before they reach the chimney. By this means they give off their heat to the large surface of iron plate, and enter the chimney at a comparatively low temperature. The heat is radiated into the apartment from the large metal surface, no part of which approaches a red-heat. A further economy is commonly effected by placing this iron pillar in the wall separating two rooms, so that one of its faces is in each room. Thus two rooms are heated by one fire. One of these may be the kitchen, and the same fire that prepares the food may be used to warm the dining-room. The fire-worshiper is of course deprived of his "cheerful" occupation of staring at the coals, and he also loses his playthings, as neither poker, tongs, nor coal-scuttle are included in the furniture of an apartment thus heated. People differently constituted consider that an escape from the dust, dirt, and clatter of these is a decided advantage.

Of course these stoves of our northern neighbors are costly—may be very costly when highly ornamental. The stove of a Norwegian "bonder," or peasant proprietor, costs nearly half as much as the two-roomed wooden house in which it is erected, but the saving it effects renders it a good investment. It would cost 100*l.* or 200*l.* to fit up an English mansion with suitable porcelain stoves of the Russian pattern, but a saving of 20*l.* a year in fuel would yield a good return as regards mere cost, while the gain in comfort and healthfulness would be so great that, once enjoyed and understood, such outlay would be willingly made by all who could afford it, even if no money saving were effected.

Only last week I was discussing this question in a railway carriage, where one of my fellow-passengers was an intelligent Holsteiner. He confirmed the heresy by which I had shocked the others, in exulting in the high price of coal, and wishing it to continue. He told us that when wood was abundant in his country, fuel was used as barbarously, as wastefully, and as inefficiently as it now is here, but that the deforesting of the land, and the great cost of fuel, forced upon them a radical reform, the result of which is that they now have their houses better warmed, and at a less cost than when fuel was obtainable at one fourth of its present cost.

Such will be the case with us also if we can but maintain the present coal famine during one or two more winters, especially if we should have the further advantage of some very severe weather in the meantime. Hence the cruel wishes above expressed. The coal famine would scarcely be necessary if we had Russian winters, for in such case our houses, instead of being as they are, merely the most uncomfortable in North Europe, would be quite uninhabitable. With our mild winters we require the utmost severity of fuel prices to civilize our warming and ventilating devices.

Chapter 19

"BAILY'S BEADS"

TO THE EDITOR OF THE *TIMES*

Sir,—The curious breaking up of the thin annular rim of the sun which is uncovered just before and just after totality, or which surrounds the moon during an annular eclipse, has been but occasionally observed, and some scepticism as to the accuracy of Baily's observations has lately arisen. Having attempted an explanation of the "beads," I have looked with much interest for the reports of the eclipse of 1870, for, if I am right, they ought to have been well seen on this occasion. This has been the case. We are informed that both Lord Lindsay and the Rev. S. J. Perry have observed them, and that Lord Lindsay has set aside all doubts respecting their reality by securing a photographic record of their appearance.

My explanation is that they are simply sun-spots seen in profile—spots just caught in the fact of turning the sun's edge. All observers are now agreed as to the soundness of Galileo's original description of the spots—that they are huge cavities, great rifts of the luminous surface of the sun, many thousands of miles in diameter, and probably some thousand miles deep. Let us suppose the case of a spot—say, 2,000 miles deep and 10,000 miles across (Sir W. Herschel has measured spots of 50,000 miles diameter). When such a spot in the course of the sun's rotation reaches that part which forms the visible edge of the sun, it must, if rendered visible, be seen as a notch; but what will be the depth of such a notch? Only about 1-430th of the sun's diameter. But the apparent depth would be much less as the edge or rim of the spot next to the observer would cut off more or less of its actually visible depth, this amount depending upon the lateral or east and west diameter of the spot and its position at the time of observation.

Thus, the visible depth of such a notch would rarely exceed one thousandth of the sun's apparent diameter, or might be much less. The sun being globular, the edge which is visible to us is but our horizon of his fiery ocean, which we see athwart the intervening surface as it gradually bends away from our view. So small an indent upon this edge would, under ordinary circumstances of observation, be rendered quite invisible by the irradiation of the vast globular surface of the glaring photosphere, upon which it would visually encroach.

If, however, this body of glare could be screened off, and only a line of the sun's edge, less than one thousandth of his diameter, remain visible, the notch would appear as a distinct break in this curved line of light. If a group of spots, or a great irregular spot with several umbræ, were at such a time situated upon the sun's edge, the appearance of a series of such notches or breaks leaving intermediate detachments of the visible ring of the photosphere would be the necessary result, and thus would be presented exactly the appearance described as "Baily's beads."

I have been led to anticipate a display of these beads during the late eclipse by the fact that some days preceding it a fine group of spots—visible to the naked eye through a London fog—were traveling towards the eastern edge of the sun, and should have reached the limb at about the time of the eclipse. The beads were observed by the Rev. S. J. Perry just where I expected them to appear. I have not yet learnt on which side of the sun they were observed and photographed by Lord Lindsay.

Baily's first observation of the beads was made during the annular eclipse of May 15, 1836. That year, like 1870, was remarkable for a great display of sun-spots. As in 1870, they were then visible to the naked eye. I well remember my own boyish excitement when, a few weeks before the eclipse of 1836, I discovered a spot upon the reddened face of the setting sun—a thing I had read about, and supposed that only great astronomers were privileged to see. The richness of this sun-spot period is strongly impressed on my memory by the fact that I continued painfully watching the dazzling sun, literally "watching and weeping," up to the Sunday of the eclipse, on which day also I saw a large spot through my bit of smoked glass.

The previous records of these appearances of fracture of the thin line of light are those of Halley, in his memoir on the total eclipse of 1715, and Maclauren's on that of 1737. Both of these correspond to great spot periods; the intervals between 1715, 1737, 1836, and 1870 are all divisible by eleven. The observed period of sun-spot occurrence is eleven years and a small fraction.

I am anxiously awaiting the arrival of Lord Lindsay's long-exposure photographs of the corona, for if they represent the varying degrees of splendor of this solar appendage, the explanations offered in Chapter xii. of my essay on "The Fuel of the Sun" will be very severely tested by them.

Yours respectfully,
W. Mattieu Williams.

Woodside Green, Croydon, January 4, 1871.

Chapter 20

THE COLORING OF GREEN TEA

The following is a copy of my report to the *Grocer* on a sample of the ingredients actually used by the Chinese for coloring of tea, which sample was sent to the *Grocer* office by a reliable correspondent at Shanghai (November, 1873). I reprint it because the subject has a general interest and is commonly misunderstood:

I have examined the blue and the yellowish-white powders received from the office, and find that the blue is not indigo, as your Shanghai correspondent very naturally supposes, but is an ordinary commercial sample of Prussian blue. It is not so bright as some of our English samples, and by mere casual observation may easily be mistaken for indigo. Prussian blue is a well-known compound of iron, cyanogen, and potassium. Commercial samples usually contain a little clayey or other earthy impurities, which is the case with this Chinese sample. There are two kinds of Prussian blue—the insoluble, and the basic or soluble. The Chinese sample is insoluble.

This is important, seeing that we do not eat our tea-leaves, but merely drink an infusion of them; and thus even the very small quantity which faces the tea-leaf remains with the spent leaves, and is not swallowed by the tea-drinker, who therefore need have no fear of being poisoned by this ornamental adulterant.

Its insolubility is obvious, from the fact that green tea does not give a blue infusion, which would be the case if the Prussian blue were dissolved.

There are some curious facts bearing on this subject and connected with the history of the manufacture of Prussian blue. Messrs. Bramwell, of Newcastle-on-Tyne, who may be called the fathers of this branch of industry, established their works about a century ago. It was first sold at two guineas per lb.; in 1815 it had fallen to 10*s.* 6*d.*, in 1820 to 2*s.* 6*d.*, then down to 1*s.* 9*d.* in 1850. I see by the Price Current of the *Oil Trade Review* that the price has recently been somewhat higher.

In the early days of the trade a large portion of Messrs. Bramwell's produce was exported to China. The Chinese then appear to have been the best customers of the British manufacturers of this article. Presently, however, the Chinese demand entirely ceased, and it was discovered that a common Chinese sailor, who had learned something of the importation of this pigment to his native country, came to England in an East Indiaman, visited, or more probably obtained employment at a Prussian blue manufactory, learned the process, and, on his return to China, started there a manufactury of his own, which was so successful that in a short time the whole of the Chinese demand was supplied by native manufacture; and thus ended our export trade. Those who think the Chinese are an unteachable and unimprovable people may reflect on this little history.

The yellowish powder is precisely what your Shanghai correspondent supposes. It is steatite, or "soapstone." This name is very deceptive, and coupled with the greasy or unctuous feel of the substance, naturally leads to the supposition that it is really as it appears, an oleaginous substance. This, however, is not the case. It is a compound of silicia, magnesia, and water, with which are sometimes associated a little clay and oxide of iron. Like most magnesian minerals, it has a curiously smooth or slippery surface, and hence its name. It nearly resembles meerschaum, the smoothness of which all smokers understand.

When soapstone is powdered and rubbed over a moderately rough surface, it adheres, and forms a shining film; just as another unctuous mineral, graphite (the "black-lead" of the housemaid), covers and polishes ironwork. On this account, soapstone is used in some lubricating compounds, for giving the finishing polish to enameled cards, and for other similar purposes.

With a statement of these properties before us, and the interesting description of the process by your Shanghai correspondent, the whole riddle of green-tea coloring and facing is solved. The Prussian blue and soapstone being mixed together when dry in the manner described, the soapstone adheres to the surface of the particles of blue, and imparts to them not only a pale greenish color, but also its own unctuous, adhesive, and polishing properties. The mixture being well stirred in with the tea-leaves, covers them with this facing, and thus gives both the color and peculiar pearly lustre characteristic of some kinds of green tea. I should add that the soapstone, like the other ingredient, is insoluble, and therefore perfectly harmless.

Considering the object to be attained, it is evident from the above that John Chinaman understands his business, and needs no lessons from European chemists. It would puzzle all the Fellows of the Chemical Society, though they combined their efforts for the purpose, to devise a more effective, cheap, simple, and harmless method of satisfying the foolish demand for unnaturally colored tea-leaves.

When the tea-drinking public are sufficiently intelligent to prefer naturally colored leaves to the ornamental stuff they now select, Mr. Chinaman will assuredly be glad enough to discontinue the addition of the Prussian blue, which costs him so much more per pound than his tea-leaves, and will save him the trouble of the painting and varnishing now in demand.

In the meantime, it is satisfactory to know that, although a few silly people may be deceived, nobody is poisoned by this practice of coloring green tea. I say "a few silly people," for there can be only a few, and those very silly indeed, who judge of their tea by its appearance rather than by the quality of the infusion it produces.

With these facts before us it is not difficult to trace the origin of the oft-repeated and contradicted statement that copper is used in coloring green tea. One of the essential ingredients in the manufacture of Prussian blue is sulphate of iron, the common commercial name which is "green copperas." It is often supposed to contain copper, but this is not the case.

Your Shanghai correspondent overrates the market value of soapstone when he supposes that Chinese wax may be used as a cheap substitute. In many places—as, for instance, the "Lizard" district of Cornwall—great veins of this mineral occur, which, if needed, might be quarried in vast abundance, and at very little cost on account of its softness. The romantic scenery of Kynance Cove, its caverns, its natural arches, the "Devil's Bellows," the "Devil's

Post-office," the "Devil's Cauldrons," and other fantastic formations of this part of the coast, attributed to his Satanic Majesty or the Druids, are the natural results of the waves beating away the veins of soft soapstone, and leaving the deformed skeleton rocks of harder serpentine behind.

Chapter 21

"IRON FILINGS" IN TEA

I have watched the progress of the tea controversy and the other public performances of the public analysts with considerable interest; it might have been with amusement, but for the melancholy degradation of chemical science which they involve.

Among the absurdities and exaggerations which for some years past have been so industriously trumpeted forth by the pseudo-chemists who trade upon the adulteration panic and consequent demand for chemical certificates of purity, the continually repeated statements concerning the use of iron filings as a fraudulent adulterant of tea take a prominent place. I need scarcely remark that, in order to form such an adulterant, the quantity added must be sufficiently great to render its addition commercially profitable to an extent commensurate with the trouble involved.

The gentlemen who, since the passing of the Adulteration Act, have by some kind of inspiration suddenly become full-blown chemists, have certified to wilful adulteration of tea with iron filings, and have obtained *convictions* on such certificates, when, according to their own statement, the quantity contained has not exceeded 5 per cent in the cheapest qualities of tea. Now, the price of such tea to the Chinaman tea-grower, who is supposed to add these iron filings, is about fourpence to sixpence per pound; and we are asked to believe that he will fraudulently deteriorate the market value of his commodity for the sake of this additional 1-20th of weight. Supposing that he could obtain his iron filings at twopence per pound, his total gain would thus be about 1-10th of a penny per pound. But can he obtain such iron filings in the quantity required at such a price? A little reflection on a few figures will render it evident that he cannot, and that such adulteration is utterly impossible.

I find by reference to *The Grocer* of November 8th, that the total deliveries of tea into the port of London during the first ten months of 1872 were 142,429,337 lbs., and during the corresponding period of 1873, 139,092,409 lbs. Of this about 8½ millions of pounds in 1873, and 10 millions of pounds in 1872, were green, the rest black. This gives in round numbers about 160 millions of pounds of black tea per annum, of which above 140 millions come from China. As the Russians are greater tea-drinkers than ourselves—the Americans and British colonists are at least equally addicted to the beverage, and other nations consume some quantity—the total exports from China may be safely estimated to reach 400 or 500 millions of pounds.

Let us take the smaller figure, and suppose that only one fourth of this is adulterated, to the extent of 5 per cent, with iron filings. How much would be required? Just five millions of pounds per annum.

It must be remembered that *coarse* filings could not possibly be used; they would show themselves at once to the naked eye as rusty lamps, and would shake down to the bottom of the chest; neither could borings, nor turnings, nor plane-shavings be used. Nothing but *fine* filings would answer the supposed purpose. I venture to assert that if the China tea-growers were to put the whole world under contribution for their supposed supply of fine iron filings, this quantity could not be obtained.

Let anyone who doubts this borrow a blacksmith's vice, a fine file, and a piece of soft iron, then take off his coat and try how much labor will be required to produce a single ounce of filings, and also bear in mind that fine files are but very little used in the manufacture of iron. As the price of a commodity rises when the demand exceeds the supply the Chinaman would have to pay far more for his adulterant than for the leaves to be adulterated. As Chinese tea-growers are not public analysts, we have no right to suppose that they would perpetrate any such foolishness.

The investigations recently made by Mr. Alfred Bird, of Birmingham, show that the iron found in tea-leaves is not in the metallic state, but in the condition of oxide; and he confirms the conclusions of Zöller, quoted by Mr. J. A. Wanklyn in the *Chemical News* of October 10th—viz., that compounds of iron naturally exist in genuine tea. It appears, however, that the ash of many samples of *black* tea contains more iron than naturally belongs to the plant; and, accepting Mr. Bird's statement, that this exists in the leaf as oxide mixed with small siliceous and micaceous particles I think we may find a reasonable explanation of its presence without adopting the puerile theory of the adulteration maniac, who, in his endeavor to prove that everybody who buys or sells anything is a swindler, has at once assumed the impossible addition of iron filings as a makeweight.

In the first place we must remember that the commodity in demand is *black* tea, and that ordinary leaves dried in an ordinary manner are not black, but brown. Tea-leaves, however, contain a large quantity of tannin, a portion of which is, when heated in the leaves, rapidly convertible into gallo-tannic or tannic acid. Thus a sample of tea rich in iron would, when heated in the drying process, become, by the combination of this tannic acid with the iron it contains, much darker than ordinary leaves or than other teas grown upon less ferruginous soils and containing less iron.

This being the case, and a commercial demand for *black* tea having become established, the tea-grower would naturally seek to improve the color of his tea, especially of those samples naturally poor in iron, and a ready mode of doing this is offered by stirring in among the leaves while drying a small additional dose of oxide of iron, if he can find an oxide in such a form that it will spread over the surface of the leaf as a thin film. Now, it happens that the Chinaman has lying under his feet an abundance of material admirably adapted for this purpose—viz., red hæmatite, some varieties of which are as soft and unctuous as graphite, and will spread over his tea-leaves exactly in the manner required. The micaceous and siliceous particles found by Mr. Bird are just what should be found in addition to oxide of iron, if such hæmatite were used.

The film of oxide thus easily applied, and subjected to the action of the exuding and decomposing extractive matter of the heated leaves, would form the desired black dye or "facing."

The knotty question of whether this is or is not an adulteration is one that I leave to lawyers to decide, or for those debating societies that discuss such interesting questions as whether an umbrella is an article of dress. If it is an adulteration, and, as already admitted, is not at all injurious to health, then all other operations of dyeing are also adulterations; for the other dyers, like the Chinaman, add certain impurities to their goods—the silk, wool, or cotton—in order to alter their natural appearance, and to give them the false facing which their customers demand, but with this difference, if I am right in the above explanation: that in darkening tea nothing more is done but to increase the proportion of one of its natural ingredients, and to intensify its natural color; while in the dyeing of silk, cotton, or wool, ingredients are added which are quite foreign and unnatural, and the natural color of the substance is altogether falsified.

The above appeared in the *Chemical News* November 21, 1873, when the adulteration in question was generally believed to be commonly perpetrated, and many unfortunate shopkeepers had been and were still being summoned to appear at Petty Sessions, etc., and publicly branded as fraudulent adulterators on the evidence of the newly-fledged public analysts, who confidently asserted that they found such filings mixed with the tea. Some discussion followed in subsequent numbers of the *Chemical News*; but it only brought out the fact that "finely divided iron" exists in considerable quantities in Sheffield,—may be "begged," as Mr. Alfred H. Allen (an able analytical chemist, resident in Sheffield,) said. The fact that such finely divided iron is thus without commercial value still further confirms my conclusion that it is not used for the adulteration of tea. If it were, its collection would be a regular business, and truck-loads would be transmitted from Sheffield to London, the great centre of tea-importation. No evidence of any commercial transactions in iron filings or iron dust for such purposes came forward in reply to my challenge.

The practical result of the controversy is that iron filings are no longer to be found in the analytical reports of the adulteration of tea.

Chapter 22

CONCERT-ROOM ACOUSTICS

The acoustics of public buildings are now occupying considerable attention in London. The vast audiences which any kind of sensational performance in the huge metropolis is capable of attracting, is forcing the subject upon all who cater for public amusement or instruction. There was probably no building in London, or anywhere else, more utterly unfit for musical performances than the Crystal Palace in its original condition; but, nevertheless, the Handel Festival of last week was a great success. I attended the first of these immense gatherings, and this last; but nothing of the kind intermediate, and, therefore, am the better able to make comparisons.

My recollections of the first were so very unsatisfactory that I gladly evaded the grand rehearsal of Friday week, and went to the "Messiah" on Monday with an astronomical treatise in my pocket, in order that my time should not be altogether wasted. Being seated at the further end of the transept, in a gallery above the level of the general ridge-and-furrow roof of the nave, the plump little Birmingham tenor, who rose to sing the first solo, appeared, under the combined optical conditions of distance and vertical foreshortening, like a chubby cheese-mite viewed through a binocular microscope. Taking it for granted that his message of comfort could not possibly reach my ear, I determined to anticipate the exhortation by settling down for a comfortable reading of a chapter or two, but was surprised to find I could hear every note, both of recitative and air.

It thus became obvious that the alterations that have gradually grown since the time when Clara Novello's voice was the only one that could be heard across the transept are worthy of study; that the advertised success of the "velarium" is something more than mere puffery. I accordingly used my eyes as well as my ears, and made a few notes which may be interesting to musical and architectural, as well as to scientific readers.

Sound, like light, heat, and all other radiations, loses its intensity as it is outwardly dispersed, is enfeebled in the ratio of the squares of distance; thus at twenty feet from the singer the loudness of the sound is one fourth of that at ten feet, at thirty feet one ninth, at forty feet one sixteenth, at fifty feet one twenty-fifth, and so on; that is, supposing the singer or other source of sound is surrounded on all sides by free, open, and still air.

But this condition is never fulfilled in practice, excepting, perhaps, by Simeon Stylites when he preached to the multitude from the top of his column. If Mr. Vernon Rigby had stood on the top of one of his native South Staffordshire chimney-shafts, of the same height above the ground as the upper press gallery of the Crystal Palace is above the front of the orchestra, and I had stood on the open ground at the same distance away and below him, his solo of "Comfort ye, my People" would have been utterly inaudible.

What, then, is the reason of this great difference of effect at equal distances? If we can answer this question, we shall know something about the acoustics of concert-rooms.

The uninitiated reader will at once begin by saying that "sound rises." This is almost universally believed, and yet it is a great mistake, as commonly understood. Sound radiates equally in every direction—downwards, upwards, north, south, east, or west, unless some special directive agency is used. The directive agency commonly used is a reflecting or reverberating surface.

Thus the voice of the singer travels forward more abundantly than backward, because he uses the roof, and, to some extent, the walls and floor of his mouth, as a sound reflector. The roof of his mouth being made of concave plates of bone with a thin velarium of integument stretched tightly over them, supplies a model sound reflector; and I strongly recommend every architect who has to build a concert or lecture-room, or theatre, to study the roof of his own mouth, and imitate it as nearly as he can in the roof of his building.

The great Italian singing masters of the old school, who, like the father of Persiani, could manufacture a great voice out of average raw material, studied the physiology of the vocal organs, and one of their first instructions to their pupils was that they should sing against the roof of the mouth, then throw the head back and open the mouth, so that the sound should reverberate forwards, clear of the teeth and lips. For the first year or two the pupil had to sing only "la, la," for several hours per day, until the faculty of doing this effectually and habitually was acquired.

The popular notion that sound rises has probably originated from the fact that in our common experience the sounds are produced near to some kind of floor, which reflects the sounds upwards, and thus adds the reflected sound to that which is directly transmitted, and thereby the general result is materially augmented.

But if we would economize sound most effectively, we must have not only a reflecting floor, but also a reflecting roof and reflecting walls on all sides of the concert room. These are the conditions that were wanting in the original structure of the Crystal Palace transept, for then the sound of the singer's voice could travel upwards to that lofty arch and sidewise in all directions, almost as freely as in the open air.

This defect has been remedied to a very great extent by the velarium stretched across from the springing of the great arch of glass and iron, and forming a ceiling to the concert-room part of the building. Besides this, a wall of drapery is stretched across each side of the transept, and the orchestra has its special walls, roof, and back. There are other minor arrangements for effecting lateral reverberation; that is, for returning the sound into the auditorium proper instead of allowing it to wander feebly throughout the building.

The general result of these arrangements is to render that portion of the building in which the reserved seats are placed a really luxurious and efficient concert-room, of magnificent proportions; but, very unfortunately and inevitably, these conditions, which are so favorable for the happy eight or nine thousand who can afford reserved seats, render the position of the other half-dozen thousand outsiders more disappointing and vexatious than ever. For my own part I would rather spend a holiday afternoon in the mild atmosphere and the quiet, soothing gloom of a coal-pit than be teased and irritated by a strained listening to the indefinite roar of a grand choir, and the occasional dying vibrations of Sims Reeves' "top A."

I have in the above advocated reverberation as a remedy for diffusion of sound. This may, perhaps, appear rather startling to some musicians who have a well-founded dread of echoes, and who read the words *echo* and *reverberation* as synonymous. This requires a little explanation. As light is transmitted, reflected, and absorbed in the same manner as sound, and as light is visible—or, rather, renders objects visible—I will illustrate my meaning by means of light.

Let us suppose three apartments of equal size and same shape, one having its walls covered with mirrors, the second with white paper, and the third with black woollen cloth, and all lighted with central chandeliers of equal brilliancy. The first and second will be much lighter than the third, but they will be illuminated very differently.

In the first, there will be a repetition of chandeliers in the mirrored walls, each wall definitely reflecting the image of each particular light. In the second room there will be reflection also, and economy of light, but no reflection of definite images; the apartment will appear to be filled with a general and well-diffused luminosity, rendering every object distinctly visible, and there will be no deep shadows anywhere.

In scientific language, we shall have, in the first room, *regular reflection*; in the second, *scattering reflection*; in the third room we should have comparative gloom, owing to the *absorption* of the light by the black cloth.

We may easily suppose the parallels of these in the case of sound. If the velarium and side walls of the transept and orchestra were made of sheet iron, or smooth, bare, unbroken vibrating wooden boards, we should have a certain amount of *regular* reflection of sound or echo. Just as we should see the particular lights of the chandelier reflected in the first room, so should we hear the particular notes of the singer or player echoed by such regularly vibrating walls and ceiling.

If, again, the velarium and side drapery of the transept and orchestra had been thick, soft woollen cloths, the sound, like the light, would have been absorbed or "muffled," and, though very clear, it would be weak and insufficient.

The reader will now ask—What, then, is the right material for such velarium and walls? I cannot pretend to say what is the best possible, believing that it has yet to be discovered. The best yet known, and attainable at moderate expense, is common canvas or calico, washed or painted over with a mixture of size and lime, or other attainable material that will fill up the pores of the fabric, and give it a moderately smooth face or surface. Thus prepared, it is found to reflect sound, as paper, ground glass, etc., reflect light, by scattering reverberation, but without definite echo.

It will now be understood how the velarium acted in rendering the solos so clearly audible at the great height and distance of the Upper Press Gallery. Instead of being wasted by diffusion in the great vault above, they were stopped and reflected by the velarium, but not so reflected as to produce disagreeable repetition notes, just audible at particular points, as the lights of the mirror reflections of the chandeliers would be.

Flat surfaces reflect radially, while concave surfaces with certain curves reflect sound, light, heat, etc., in parallel lines. The walls and roof of a music-hall should scatter their reflections on all sides, and, therefore, should be flat, or nearly so, excepting at the angles, which should be curved or hollowed. From the orchestra the sound is chiefly required to be projected forward as from the singer's mouth; and, therefore, an orchestra should have curved walls and roof.

Space will not permit a dissertation here on the particular curve required. This has, I believe, been carefully calculated in constructing the Crystal Palace orchestra. Viewed from a distance, the whole orchestra is curiously like a huge wide-opened mouth that only requires to close a little and open a little more, according to the articulations of the choir, to represent the vocal effort of one gigantic throat.

There is, I think, one fault in the shape of this mouth. It extends too far laterally in proportion to its perpendicular dimensions. The angles of the mouth are too acute; the choir extends too far on each side. The singers should be packed more like those of the Birmingham Festival Choir.

There is an acoustic limit to the magnitude of choirs. Sound travels at about 1100 feet per second, and thus, if one of the singers of a choir is 110 feet nearer than another singer to any particular auditor, the near singer will be heard one-tenth of a second before the more distant, though they actually sing exactly together. In rapid staccato passages this would produce serious confusion, though in such music as most of Handel's it would be scarcely observable.

Some observations which I have made convince me that the actual choir of the Handel Festivals has reached, if not exceeded, the acoustic limits even for Handel's music, and decidedly exceeds the limits permissible for Mendelsshon and most other composers.

I found that when standing on the floor of the building in front of the orchestra, and on one side, I could plainly distinguish the wave of difference of time due to the traveling of the sound, and in all the passages which required to be taken up smartly and simultaneously by the opposite sides of the choir, the effect was very disagreeable.

The defect, however, was not observable from the press gallery, which is placed as nearly as may be to the focus of the orchestral curve, so that radial lines drawn from the auditor to different parts of the orchestra do not differ so much in length as to effect perceptible differences in the moment at which the different sounds reach the ear.

My conclusion, therefore, is that if any amendment is to be made in the numbers of the Handel Festival choir, it should rather be done by a reduction than an increase; that the four thousand voices should rather be reduced to three thousand than increased to five thousand. With greater severity of selection as regards quality, power, and training of each individual voice, and with better packing, the three thousand would be more effective than the four thousand.

Chapter 23

SCIENCE AND SPIRITUALISM

A rather startling paper in the current number of the "Quarterly Journal of Science," from the pen of William Crookes, F.R.S. (who is well known in the scientific world by his discovery of the metal thallium, his investigations of its properties and those of its compounds, besides many other important researches, and also as the able and spirited editor of the *Chemical News*), is now the subject of much scientific gossip and discussion.

Mr. Crookes has for some time past been engaged in investigating some of the phenomena which are attributed on one hand to the agency of spiritual visitors, and on the other side to vulgar conjuring. Nobody acquainted with Mr. Crookes can doubt his ability to conduct such an investigation, or will hesitate for a moment in concluding that he has done so with philosophical impartiality, though many think it quite possible that he may have been deceived. None, however, can yet say how.

For my own part, I abstain from any conclusion in the meantime, until I have time and opportunity to witness a repetition of some of these experiments, and submitting them to certain tests which appear to me desirable. Though struggling against a predisposition to prejudge, and to conclude that the phenomena are the results of some very skillful conjuring, I very profoundly respect the moral courage that Mr. Crookes has displayed in thus publicly grappling with a subject which has been soiled by contact with so many dirty fingers. Nothing but a pure love of truth, overpowering every selfish consideration, could have induced Mr. Crookes to imperil his hard-earned scientific reputation by stepping thus boldly on such very perilous ground.

It is only fair, at the outset, to state that Mr. Crookes is not what is called "a spiritualist." This I infer, both from what he has published and from conversation I have had with him on the subject. He has witnessed some of the "physical manifestations," and, while admitting that many of these may be produced by the jugglery of impostors, he has concluded that others cannot be thus explained; but, nevertheless, does not accept the spiritual theory which attributes them to the efforts of departed human souls.

He suspects that the living human being may have the power of exerting some degree of force or influence upon bodies external to himself—may, for instance, be able to counteract or increase the gravitation of substances by an effort of the will. He calls this power the "psychic force," and supposes that some persons are able to manifest it much more powerfully than others, and thus explains the performances of those "mediums" who are not mere impostors.

There is nothing in this hypothesis which the sternest, the most sceptical, and least imaginative of physical philosophers may not unhesitatingly investigate, provided some first-sight evidence of its possibility is presented to him. We know that the Torpedo, the Gymnotus, the Silurus Electricus, and other fishes, can, by an effort of the will, act upon bodies external to themselves. Faraday showed that the electric eel exhibited some years ago at the Adelaide Gallery was able, by an effort of its will, to make a magnetic needle suddenly turn thirty degrees aside from its usual polar position; that this same animal could—still by an effort of will—overpower the gravitation of pieces of gold leaf, cause them to be uplifted and outstretched from their pendent position, could decompose iodide of potassium, and perform many other "physical manifestations," simply by a voluntary nervous effort, and without calling in the aid of any souls of other departed eels.

Before this gymnotus was publicly exhibited it was deposited at a French hotel in the neighborhood of Leicester Square. A burly fishmonger's man, named Wren, brought in the daily supply of fish to the establishment, when some of the servants told him they had an eel so large that he would be afraid to pick it up. He laughed at the idea of being afraid of an eel, and when taken to the tub boldly plunged in both hands to seize the fish. A hideous roar followed this attempt. Wren had experienced a demonstration of the "psychic force" of the electrical eel, and his terror so largely exaggerated the actual violence of the shock, that he believed for the remainder of his life that he was permanently injured by it. He had periodical spasms across the chest, which could only be removed by taking a half-quartern of gin. As he was continually narrating his adventure to public-house audiences, and always had a spasm on concluding, which his hearers usually contributed to relieve, the poor fellow's life was actually shortened by the shock from the gymnotus.

The experiments which Mr. Crookes relates in support of his psychic force hypothesis are as follows:—In the first place he contrived an apparatus for testing Mr. Home's alleged power of modifying the gravitation of bodies. As Mr. Home requires to lay his hands, or at least his finger-ends, upon the body to be influenced, Mr. Crookes attached one end of a long board to a suspended spring steelyard of delicate construction; the other end of the board rested on a fulcrum in such a manner that one half of the weight of the board was supported by the fulcrum and the other half by the steelyard. The weight of the board thus suspended was carefully noted, and then Mr. Home put his fingers upon that end of the board immediately resting on the fulcrum in such a manner that he could not by simple pressure affect the dependent end of the board.

Dr. Huggins, the eminent astronomer, was present, and also Serjeant Cox, besides Mr. Crookes. They all watched Mr. Home, the board, and the steelyard; they observed first a vibration and fluctuation of the index, and finally that the steelyard indicated an increase of weight amounting to about three pounds. Mr. Crookes tried to produce the same effect by mechanical pressure exerted in a similar manner, but failed to do so. The details of the experiment are fully described and illustrated by an engraving.

Another and still more striking experiment is described. Mr. Crookes purchased a new accordion from Messrs. Wheatstone, and himself constructed a wire cage open at top and bottom, and large enough for the accordion to be suspended within it by holding it over the open top, while the bottom of the cage rested on the floor. The accordion was then handed to Mr. Home, who held it with one hand by the wooden framework of the bottom of the instrument, as shown in an illustrative drawing. The keys were thus hanging downwards and the bellows distended by the weight of the instrument thus pendent. It was then held so that it should be entirely surrounded by the wire-work of the cage, and the results were, as before, watched keenly by Mr. Crookes, Dr. Huggins, and Serjeant Cox. After a while the instrument began to wave about, then the bellows contracted, and the lower part (i.e., the key-board end) rose a little, presently sounds were produced, and finally the instrument played a tune upon itself in obedience, as Mr. Crookes supposes, to the psychic force which Mr. Home exerted upon it.

Before the publication of the paper describing these experiments a proof was sent to both Dr. Huggins and Serjeant Cox, and each has written a letter testifying to its accuracy, which letters are printed with the paper in the "Quarterly Journal of Science."

Here, then, we have the testimony of an eminent lawyer, accustomed to sifting evidence, that of the most distinguished of experimental astronomers, the man whose discoveries in celestial physics have justly excited the admiration of the whole civilized world; and besides these, of another Fellow of the Royal Society, who has been severely trained in "putting nature to the torture" by means of the most subtle devices of the modern physical and chemical laboratory.

Such testimony must not be treated lightly. It would be simple impertinence for any man dogmatically to assert that these have been deceived merely because he is unconvinced.

Though one of the unconvinced myself, I would not dare to regard the investigations of these gentlemen with any other than the profoundest respect. Still a suggestion occurs to me which may appear very brutal, but I make it nevertheless. It is this:—That the testimony of another witness—of an expert of quite a different school—should have been added. I mean such a man as Döbler, Houdin, or the Wizard of the North. He might possibly have detected something which escaped the scrutiny of the legitimate scientific experimentalist.

There is one serious defect in the accordion experiment. The cage is represented in the engraving as placed under a table; Mr. Home holds the instrument in his hand, which is concealed by the table, and it does not appear that either Mr. Crookes, Dr. Huggins, or Serjeant Cox placed themselves under the table during the concertina performance, and thus neither of them saw Mr. Home's hand. Such, at least, appears from the description and the engraving. A story being commonly circulated respecting some of Mr. Home's experiments in Russia, according to which he failed entirely when a glass table was provided instead of a wooden one, it would be well, if only in justice to Mr. Home, to get rid of the table altogether.

It is very desirable that these experiments should be continued, for two distinct reasons; first, as a matter of ordinary investigation for philosophical purposes, and, secondly, as a means of demolishing the most degrading superstition of this generation.

If Mr. Crookes succeeds in demonstrating the existence of the psychic force and reducing it to law—as it must be reducible if it is a force—then the ground will be cut from under the feet of spiritualism, just as the old superstitions, which attributed thunder and lightning to Divine anger, were finally demolished by Franklin's kite. If, on the other hand, the arch-medium, Mr. Home, is proved to be a common conjuror, then surely the dupes of the smaller "mediumistic" fry will have their eyes opened, provided the cerebral disturbance which spiritualism so often induces has not gone so far as to render them incurable lunatics.

It is very likely that I shall be accused of gross uncharitableness in thus applying the term lunatic to "those who differ from me," and therefore state that I have sad and sufficient reasons for doing so.

The first spiritualist I ever knew, and with whom I had many conferences on the subject many years ago, was a lady of most estimable qualities, great intellectual attainments, and distinguished literary reputation. I watched the beginning and the gradual progress of her spiritual "investigations," as she called them, and witnessed the melancholy end—shocking delusions, intellectual shipwreck, and confirmed, incurable insanity, directly and unmistakably produced by the action of these hideous superstitions upon an active, excitable imagination.

I well remember the growing symptoms of this case, have seen their characteristic features repeated in others, and have now before me some melancholy cases where the same changes, the same decline of intellect and growth of ravenous credulity, is progressing with most painfully visible distinctness.

The necessity for some strong remedy is the more urgent, inasmuch as the diabolical machinery of the spiritual impostors has been so much improved of late. The lady whose case I first referred to had reached the highest stage of spiritualistic development—viz., the lunatic asylum—before "dark séances" had been invented, or, at any rate, before they were introduced into this country. When the conditions of these séances are considered, it is not at all surprising that persons of excitable temperament, especially women, should be morbidly affected by them.

We are endowed with certain faculties, and placed in a world wherein we may exercise them healthfully upon their legitimate objects. Such exercise, properly limited, promotes the growth and vigor of our faculties; but if we pervert them by directing them to illegitimate objects, we gradually become mad. God has created the light, and fitted our eyes to receive it; He has endowed us with the sense of touch, by which we may confirm and verify the impressions of sight. All physical phenomena are objects of sense, and the senses of sight and touch are the masters of all the other senses.

Can anything, then, be more atrociously perverse, more utterly idiotic, and I may even say impious, than these dark séance investigations? Is it possible to conceive a more melancholy spectacle of intellectual degradation than that presented by a group of human victims assembled for the purpose of "investigating physical manifestations," and submitting, as a primary condition, to be blinded and handcuffed, the room in which they sit being made quite dark, and both hands of each investigator being firmly held by those of his neighbors. That is to say, the primary conditions of making these physical investigations is that each investigator shall be deprived of his natural faculties for doing so.

When we couple this with the fact that these meetings are got up—publicly advertised by adventurers who make their livelihood by the fees paid by their hoodwinked and handcuffed customers—is it at all surprising that those who submit to such conditions should finish their researches in a lunatic asylum?

The gloom, the mystery, the unearthly objects of search, the mysterious noises, and other phenomena so easily manipulated in the presence of those who can see nothing and feel only the sympathetic twitching of another pair of trembling hands, naturally excites very powerfully the poor creatures who pay their half-crowns and half-guineas with any degree of faith; and this unnatural excitement, if frequently repeated, goes on increasing till the brain becomes incurably diseased.

Present space will not permit me to enter upon another branch of this subject, viz.: the moral degradation and the perversion of natural, unsophisticated, and wholesome theology, which these spiritual delusions are generating.

I am no advocate for rectifying moral and intellectual evils by police interference, or I should certainly recommend the bracing air of Dartmoor for the mediums who publicly proclaim that their familiar spirit "Katey" has lately translated a lady through a space of three miles, and through the walls, doors, and ceiling of the house in which a dark séance was being held, and placed her upon the table in the midst of the circle so rapidly that the word "onions" she had just written in her domestic inventory was not yet dried when the lights were brought and she was found there.

This "lady," which her name is Guppy, is, of course, another professional medium, and yet there are people in London who gravely believe this story, and also the appendix, viz.: that another member of the mediumistic firm, finding that Mrs. G. was very incompletely dressed, and much abashed thereby, was translated by the same spirit, Katey, to her house and back again through the door-panel to fetch proper garments. If I could justify the apprehension and imprisonment of poor gipsy fortune-tellers, I certainly should advocate the close confinement of Mrs. Guppy and her male associates, and thus afford the potent spirit, Katey, an opportunity of further manifestation by translating them through the prison walls and back to Lamb's Conduit Street.

(The above letter appeared in the "Birmingham Morning News" of July 18, 1871; the following on November 15. It refers to an article in the "Quarterly Review" of October, 1871.)

The interest excited by Mr. Crookes's investigations on Psychic Force is increasing; the demand for the "Quarterly Review" and the "Quarterly Journal of Science" is so great that Mudie and other proprietors of lending libraries have largely increased their customary supplies, and are still besieged with further excess of demand. Not only borrowers, but purchasers also are supplied with difficulty. I yesterday received a post-card from a bookseller, inscribed as follows: "Cannot get a 'Quarterly Review' in the City, so shall be unable to send it to you until to-morrow." I have waited three days, and am now obliged to go to the reading-room to make my quotations.

There is good and sufficient reason for this, independently of the absence of Parliamentary and war news, and the dearth of political revolutions. Either a new and most extraordinary natural force has been discovered, or some very eminent men specially trained in rigid physical investigation have been the victims of a marvelous, unprecedented, and inexplicable physical delusion. I say unprecedented, because, although we have records of many popular delusions of similar kind and equal magnitude, and speculative delusions among the learned, I can cite no instance of skillful experimental experts being utterly and repeatedly deceived by the mechanical action of experimental test apparatus carefully constructed and used by themselves.

As the interest in the subject is rapidly growing, my readers will probably welcome a somewhat longer gossip on this than I usually devote to a single subject.

Such an extension is the more demanded as the newspaper and magazine articles which have hitherto appeared have, for the most part, by following the lead of the "Quarterly Review," strangely muddled the whole subject, and misstated the position of Mr. Crookes and others. In the first place, all the writers who follow the "Quarterly" omit any mention or allusion to Mr. Crookes's preliminary paper published in July, 1870, which has a most important bearing on the whole subject, as it expounds the object of all the subsequent researches.

Mr. Crookes there states that "Some weeks ago the fact that I was engaged in investigating Spiritualism, so-called, was announced in a contemporary (the "Athenæum"), and in consequence of the many communications I have since received, I think it desirable to say a little concerning the investigations which I have commenced. Views or opinions I cannot be said to possess on a subject which I do not profess to understand. I consider it the duty of scientific men, who have learned exact modes of working, to examine phenomena which attract the attention of the public, in order to confirm their genuineness, or to explain, if possible, the delusions of the honest, and to expose the tricks of the deceivers."

He then proceeds to state the case of Science *versus* Spiritualism thus:—"The Spiritualist tells of bodies weighing 50 or 100 lbs. being lifted up into the air without the intervention of any known force; but the scientific chemist is accustomed to use a balance which will render sensible a weight so small that it would take ten thousand of them to weigh one grain; he is, therefore, justified in asking that a power, professing to be guided by intelligence, which will toss a heavy body to the ceiling, shall also cause his delicately-poised balance to move under test conditions." "The Spiritualist tells of rooms and houses being shaken, even to injury, by superhuman power. The man of science merely asks for a pendulum to be sent vibrating when it is in a glass-case, and supported on solid masonry." "The Spiritualist tells of heavy articles of furniture moving from one room to another without human agency. But the man of science has made instruments which will divide an inch into a million parts, and he is justified in doubting the accuracy of the former observations, if the same force is powerless to move the index of his instrument one poor degree." "The Spiritualist tells of flowers with the fresh dew on them, of fruit, and living objects being carried through closed windows, and even solid brick walls. The scientific investigator naturally asks that an additional weight (if it be only the 1000th part of a grain) be deposited on one pan of his balance when the case is locked. And the chemist asks for the 1000th part of a grain of arsenic to be carried through the sides of a gas tube in which pure water is hermetically sealed."

These and other requirements are stated by Mr. Crookes, together with further exposition of the principles of strict inductive investigation, as it should be applied to such an inquiry. A year after this he published an account of the experiments, which I described in a former letter, and added to his own testimony that of the eminent physicist and astronomer, Dr. Huggins and Serjeant Cox. Subsequently, that is, in the last number of the "Quarterly Journal of Science," he has published the particulars of another series of experiments.

I will not now enter upon the details of these, but merely state that the conclusions of Mr. Crookes are directly opposed to those of the Spiritualists. He positively, distinctly, and repeatedly repudiates all belief in the operations of the supposed spirits, or of any other supernatural agency whatever, and attributes the phenomena he witnessed to an entirely different organ, viz.: to the direct agency of the medium. He supposes that a force analogous to that which the nerves convey from their ganglionic centres to the muscles, in producing muscular contraction, may by an effort of the will be transmitted to external inanimate matter, in such a manner as to influence, in some degree, its gravitating power, and produce vibratory motion. He calls this the *psychic force*.

Now, this is direct and unequivocal *anti*-spiritualism. It is a theory set up in opposition to the supernatural hypotheses of the Spiritualists, and Mr. Crookes's position in reference to Spiritualism is precisely analogous to that of Faraday in reference to table-turning. For the same reasons as those above-quoted, the great master of experimental investigation examined the phenomena called table-turning, and he concluded that they were due to muscular force, just as Mr. Crookes concludes that the more complex phenomena he has examined are due to psychic force.

Speaking of the theories of the Spiritualists, Mr. Crookes, in his first paper (July, 1870). says: "The pseudo-scientific Spiritualist professes to know everything. No calculations trouble his serenity; no hard experiments, no laborious readings; no weary attempts to make clear in words that which has rejoiced the heart and elevated the mind. He talks glibly of all sciences and arts, overwhelming the inquirer with terms like 'electro-biologise,' 'psychologise,' 'animal magnetism,' etc., a mere play upon words, showing ignorance rather than understanding." And further on he says: "I confess that the reasoning of some Spiritualists would almost seem to justify Faraday's severe statement—that many dogs have the power of coming to more logical conclusions."

I have already referred to the muddled misstatement of Mr. Crookes's position by the newspaper writers, who almost unanimously describe him and Dr. Huggins as two distinguished scientific men who have recently been converted to Spiritualism. The above quotations, to which, if space permitted, I might add a dozen others from either the first, the second, or the third of Mr. Crookes's papers, in which he as positively and decidedly controverts the dreams of the Spiritualists, will show how egregiously these writers have been deceived. They have relied very naturally on the established respectability of the "Quarterly Review," and have thus deluded both themselves and their readers. Considering the marvelous range of subjects these writers have to treat, and the acres of paper they daily cover, it is not surprising that they should have been thus misled in reference to a subject carrying them considerably out of their usual track; but the offence of the "Quarterly" is not so venial. It assumes, in fact, a very serious complexion when further investigated.

The title of the article is "Spiritualism and its Recent Converts," and the "recent converts" most specially and prominently named are Mr. Crookes and Dr. Huggins. Serjeant Cox is also named, but not as a *recent* convert; for the reviewer describes him as an old and hopelessly infatuated Spiritualist. Knowing nothing of Serjeant Cox, I am unable to say whether the reviewer's very strong personal statements respecting him are true or false— whether he really is "one of the most gullible of the gullible," etc., though I must protest against the bad taste which is displayed in the attack which is made upon this gentleman. The head and front of his offending consists in having certified to the accuracy of certain experiments; and for having simply done this, the reviewer proceeds, in accordance with the lowest tactics of Old Bailey advocacy, to bully the witness, and to publish disparaging personal details of what he did twenty-five years ago.

Dr. Huggins, who has had nothing further to do with the subject than simply to state that he witnessed what Mr. Crookes described, and who has not ventured upon one word of explanation of the phenomena, is similarly treated.

The reviewer goes out of his way to inform the public that Dr. Huggins is, after all, only a brewer, by artfully stating that, "like Mr. Whitbread, Mr. Lassell, and other brewers we could name, Dr. Huggins attached himself in the first place to the study of astronomy." He then proceeds to sneer at "such scientific amateurs," by informing the public that they "labor, as a rule, under a grave disadvantage, in the want of that broad basis of scientific culture which alone can keep them from the narrowing and pervertive influence of a limited *specialism*."

The reviewer proceeds to say that he has "no reason to believe that Dr. Huggins constitutes an exception" to this rule, and further asserts that he is justified in concluding that Dr. Huggins is ignorant of "every other department of science than *the small subdivision of a branch* to which he has so meritoriously devoted himself." Mark the words, "small subdivision of a branch." Merely a twig of the tree of science is, according to this most unveracious writer, all that Dr. Huggins has ever studied.

If a personal vindication were the business of this letter I could easily show that these statements respecting the avocations, the scientific training, and actual attainments of Dr. Huggins are gross and atrocious misrepresentations; but Dr. Huggins has no need of my championship; his high scientific position, the breadth and depth of his general attainments, and the fact that he is not Huggins the brewer, are sufficiently known to all in the scientific world, with the exception of the "Quarterly" reviewer.

My object is not to discuss the personal question whether book-making and dredging afford better or worse training for experimental inquiry than the marvelously exact and exquisitely delicate manipulations of the modern observatory and laboratory, but to protest against this attempt to stop the progress of investigation, to damage the true interests of science and the cause of truth, by throwing low libellous mud upon any and everybody who steps at all aside from the beaten paths of ordinary investigation.

The true business of science is the discovery of truth; to seek it wherever it may be found, to pursue it through bye-ways as well as highways, and, having found it, to proclaim it plainly and fearlessly, without regard to authority, fashion, or prejudice. If, however, such influential magazines as the "Quarterly Review" are to be converted into the vehicles of artful and elaborate efforts to undermine the scientific reputation of any man who thus does his scientific duty, the time for plain speaking and vigorous protest has arrived.

My readers will be glad to learn that this is the general feeling of the leading scientific men of the metropolis; whatever they may think of the particular investigations of Mr. Crookes, they are unanimous in expressing their denunciations of this article.

The attack upon Mr. Crookes is still more malignant than that upon Dr. Huggins. Speaking of Mr. Crookes's fellowship of the Royal Society, the reviewer says: "We speak advisedly when we say that this distinction *was conferred on him with considerable hesitation*;" and further that "We are assured, on the highest authority, that he is regarded among chemists as a specialist of specialists, *being totally destitute of any knowledge of chemical philosophy, and utterly untrustworthy as to any inquiry which requires more than technical knowledge for its successful conduct*."

The italics in these quotations are my own, placed there to mark certain statements to which no milder term than that of falsehood is applicable. The history of Mr. Crookes's admission to the Royal Society will shortly be published, when the impudence of the above statement respecting it will be unmasked; and the other quotations I have emphasized are sufficiently and abundantly refuted by Mr. Crookes's published works, and his long and able conduct of the *Chemical News*, which is the only and the recognized British periodical representative of chemical science.

If space permitted, I could go on quoting a long series of misstatements of matters of fact from this singularly unveracious essay. The writer seems conscious of its general character, for, in the midst of one of his narratives, he breaks out into a foot-note, stating that "*This* is not an invention of our own, but a fact communicated to us by a highly intelligent witness, who was admitted to one of Mr. Crookes's *séances*." I have taken the liberty to emphasize the proper word in this very explanatory note.

The full measure of the injustice of prominently thrusting forward Dr. Huggins and Mr. Crookes as "recent converts" to Spiritualism will be seen by comparing the reviewer's own definition of Spiritualism with Mr. Crookes's remarks above quoted. The reviewer says that "The fundamental tenet of the Spiritualist is the old doctrine of communication between the spirits of the departed and souls of the living."

This is the definition of the reviewer, and his logical conclusion is that Mr. Crookes is a Spiritualist because he explicitly denies the fundamental tenet of Spiritualism, and Dr. Huggins is a Spiritualist because he says nothing whatever about it.

If examining the phenomena upon which the Spiritualist builds his "fundamental tenet," and explaining them in some other manner, constitutes conversion to Spiritualism, then the reviewer is a far more thoroughgoing convert than Mr. Crookes, who only attempts to explain the mild phenomena of his own experiments, while the reviewer goes in for everything, including even the apotheosis of Mrs. Guppy and her translation through the ceiling, a story which is laughed at by Mr. Crookes and everybody else, excepting a few of the utterly crazed disciples of the "Lamb's Conduit Mediums" and the "Quarterly" reviewer, who actually attempts to explain it by his infallible and ever applicable physiological nostrum of *"unconscious cerebration."*

No marvelous story either of ancient or modern date is too strong for this universal solvent, which according to the reviewer, is the sole and glorious invention of Dr. Carpenter. Space will not now permit me to further describe "unconscious cerebration" and its vast achievements, but I hope to find a corner for it hereafter.

I may add that the name of the reviewer is kept a profound secret, and yet is perfectly well-known, as everybody who reads the article finds it out when he reaches those parts which describe Dr. Carpenter's important physiological researches and discoveries.

Chapter 24

MATHEMATICAL FICTIONS

(BRITISH ASSOCIATION, 1871)

The President's inaugural address, which was going through the press in London while being spoken in Edinburgh, has already been subject to an unusual amount of sharp criticism. For my own part I cannot help regarding it as one of the least satisfactory of all the inaugural addresses that have yet been delivered at these annual meetings. They have been of two types, the historical and the controversial; the former prevailing. In the historical addresses the President has usually made a comprehensive and instructive survey of the progress of the whole range of science during the past year, and has dwelt more particularly on some branch which from its own intrinsic merits has claimed special attention, or which his own special attainments have enabled him to treat with the greatest ability and authority. A few Presidents have, like Dr. Huxley last year, taken up a particular subject only, and have discussed it more thoroughly than they could have done had they also attempted a general historical survey.

Every President until 1871 has scrupulously kept in view his judicial position, and the fact that he is addressing, not merely a few learned men, but the whole of England, if not the whole civilized world. They have therefore clearly distinguished between the established and the debatable conclusions of science, between ascertained facts and mere hypotheses, have kept this distinction so plainly before their auditors that even the most uninitiated could scarcely confound the one with the other.

In Sir William Thomson's address this desirable rule is recklessly violated. He tells his unsophisticated audience that Joule was able "to estimate the average velocity of the ultimate molecules or atoms" of gases, and thus determined the atomic velocity of hydrogen "at 6225 feet per second at temperature 60 degs. Fahr., and 6055 feet at the freezing point;" that "Clausius took fully into account the impacts of molecules upon one another, and the kinetic energy of *relative* motion of the matter constituting an individual atom;" and that "he investigated the relation between their diameters, the number in a given space, and the mean length of path from impact to impact, and so gave the foundation for estimates of the absolute dimensions of atoms." Also that "Loschmidt, in Vienna, had shown, and not much later Stoney, independently, in England, showed how to reduce from Clausius and Maxwell's kinetic theory of gases a superior limit to the number of atoms in a given measurable space."

The confiding auditor follows the President through further disquisitions on the "superlatively grand question, what is the inner mechanism of an atom?" and a minute and most definite description of the "regular elastic vibrations" of "the ultimate atom of sodium," of the manner in which "any atom of gas, when struck and left to itself, vibrates with perfect purity its fundamental note or notes," and how, "in a highly attenuated gas, each atom is very rarely in collision with other atoms, and therefore is nearly at all times in a state of true vibration," while "in denser gases each atom is frequently in collision;" besides, a great deal more, in all of which the existence of these atoms is coolly taken for granted, and treated as a fundamental established scientific fact.

After hearing all these oracular utterances concerning atoms, the unsophisticated listener before mentioned will be surprised to learn that no human being has ever seen an atom of any substance whatever; that there exists absolutely no direct evidence of the existence of any such atoms; that all these atoms of which Sir W. Thomson speaks so confidently and familiarly, and dogmatically, are pure fragments of the imagination.

He will be still further surprised to learn that the bare belief in the existence of ultimate atoms as a merely hypothetical probability is rejected by many of the most eminent of scientific men, and that among those who have disputed the idea of the atomic constitution of matter, is the great Faraday himself; that the question of the existence or non-existence of atoms has recently been rather keenly discussed; and that even on the question of the permissibility of admitting their *hypothetical* existence, scientific opinion is divided; and that such a confident assumption of their existence as forms the basis of this part of the President's address is limited to only a small section of mutually admiring transcendental mathematicians, Sir W. Thomson being the most admired among them, as shown by the address of Professor Tait to Section A.

It would have been perfectly legitimate and most desirable that Sir W. Thomson should give the fullest and most favorable possible statement of the particular hypotheses upon which he and his friends have exercised their unquestionably great mathematical skill; but he should have stated them as what they are, and for what they are worth, and have clearly distinguished between such hypotheses and the established facts of universally admitted science. Instead of doing this, he has so mixed up the actual discoveries of indisputable facts with these mere mathematical fancies as to give them both the semblance of equally authoritative scientific acceptance, and thus, without any intention to deceive anybody, must have misled nearly all the outside public who have heard or read his address.

As these letters are mainly intended for those who are too much engaged in other pursuits to study science systematically, and as most of the readers of such letters will, as a matter of course, read the inaugural address of the President of the British Association, I have accepted the duty of correcting among my own readers the false impression which this address may create.

As a set-off to the authoritative utterances of Sir W. Thomson on the subject of atoms, I quote the following from an Italian philosopher, who, during the present year, is holding in Italy a position very similar to that of the annual President of our British Association.

Professor Cannizzaro has been elected by a society of Italian chemists to act as this year's director of a Chronicle of the Progress of Chemical Science in Italy and abroad. In this capacity he has published an inaugural treatise on the history of modern chemical theory, in the course of which he thus speaks of the over-confident atomic theorists: "They often speak on molecular subjects with as much dogmatic assurance as though they had actually realized the ingenious fiction of Laplace—had constructed a microscope by which they could detect the molecules, and observe the number, forms, and arrangements of their constituent atoms, and even determine the direction and intensity of their mutual actions. Many of these things, offered at what they are worth—that is, as hypotheses more or less probable, or as simple artifices of the intellect—may serve, and really have served, to collocate facts and incite to further investigations which, one day or other, may lead to a true chemical theory; but, when perverted by being stated as truths already demonstrated, they falsify the intellectual education of the students of inductive science, and bring reproach on the modern progress of chemistry."

I translate the above from the first page of the first number of the "Gazetta Chimica Italiana," published at Palermo in January last. Had these words been written in Edinburgh on the evening of the 2d of August, in direct application to Sir William Thomson's address, they could not have described more pointedly and truly the prevailing vice of this production. If space permitted, I could go further back and quote the words of Lord Bacon, from the great text-book of inductive philosophy, wherein he denounces the worship of all such intellectual idols as our modern mathematical dreamers have created, and which they so fervently adore.

An able writer in the *Daily News* of last Friday is very severe upon the biological portion of the President's address, which contains a really original hypothesis. Sir W. Thomson having stated that he is "ready to adopt as an article of scientific faith, true through all space and through all time, that life proceeds from life, and from nothing but life," asks the question, "How then did life originate on the earth?" and tells us that "if a probable solution consistent with the ordinary course of nature can be found, we must not invoke an abnormal act of creative power."

He assumes, with that perfect confidence in mathematical hypotheses which is characteristic of the school of theorists which he leads, that "tracing the physical history of the earth backwards, on strictly dynamical principles, we are brought to a red-hot melted globe, on which no life could exist;" and then, to account for the beginning of life on our earth as it cooled down, he creates another imaginary world, which he brings in collision with a second similar creation, and thereby shatters it to fragments. He further imagines that one of these imaginary broken-up worlds was already stocked with the sort of life which he says can only proceed from life, and that from such a world thus stocked and thus smashed "many great and small fragments carrying seed and living plants and animals would undoubtedly be scattered through space;" and that, "if at the present instant no such life existed upon this earth, one such stone falling upon it might, by what we blindly call *natural* causes, lead to its becoming covered with vegetation."

The conclusion of this paragraph is instructively characteristic of the philosophy of Sir William Thomson and his admirers. He says that "the hypothesis that life originated on this earth through moss-grown fragments of another world may seem *wild and visionary*; all I maintain is that it is *not unscientific*."

I have italicized the phrases which, put together, express the philosophy of this school of modern manufacturers of mathematical hypotheses. It matters not to them how "wild and visionary," how utterly gratuitous any assumption may be, it is not unscientific provided it can be invested in formulæ, and worked out mathematically. These transcendental mathematicians are struggling to carry philosophy back to the era of Duns Scotus, when the greatest triumph of learning was to sophisticate so profoundly an obvious absurdity that no ordinary intellect could refute it.

Fortunately for the progress of humanity, there are other learned men who firmly maintain that the business of science is the discovery and teaching of simple sober truth.

The writer of the *Daily News* article above referred to very charitably suggests that Sir W. Thomson may be "poking fun at some of his colleagues," and compares the moss-grown meteorite hypothesis with the Hindoo parable which explains the stability of the earth by stating that it stands on the back of a monster tortoise, that the tortoise rests upon the back of a gigantic elephant, which stands upon the shell of a still bigger tortoise, resting on the back of another still more gigantic elephant, and so on. Sir W. Thomson, of course, requires to smash two more worlds in order to provide a moss-grown fragment for starting the life upon the world which was broken up for our benefit, and so on backwards *ad infinitum*.

Chapter 25

WORLD-SMASHING

Sir W. Thomson's moss-grown fragment of a shattered world is not yet forgotten. In the current number of the *Cornhill Magazine* (January, 1872) it is very severely handled; the more severely, because the writer, though treating the subject quite popularly, shows the fallacy of the hypothesis, even when regarded from the point of view of Sir W. Thomson's own special department of study. That an eminent mathematician should make a great slip when he ventures upon geological or physiological ground is not at all surprising; it is, in fact, quite to be expected, as there can be no doubt that the close study of *pure* mathematics, by directing the mind to processes of calculation rather than to phenomena, induces that sublime indifference to facts which has characterized the purely mathematical intellect of all ages.

It is not surprising that a philosopher who has been engaged in measuring the imaginary diameter, describing the imaginary oscillations and gyrations of imaginary atoms, and the still more complex imaginary behavior of the imaginary constituents of the imaginary atmospheres by which the mathematical imagination has surrounded these imaginary atoms, should overlook the vulgar fact that neither mosses nor other vegetables, nor even their seeds, can possibly retain their vitality when alternately exposed to the temperature of a blast furnace, and that of two or three hundred degrees below the freezing point; but it is rather surprising that the purely mathematical basis of this very original hypothesis of so great a mathematician should be mathematically fallacious—in plain language, a mathematical blunder.

In order to supply the seed-bearing meteoric fragment by which each planet is to be stocked with life, it is necessary, according to Sir W. Thomson, that two worlds—one at least flourishing with life—shall be smashed; and, in order to get them smashed with a sufficient amount of frequency to supply the materials for his hypothesis, the learned President of the British Association has, in accordance with the customary ingenuity of mathematical theorists, worked out the necessary mathematical conditions, and states with unhesitating mathematical assurance that—"It is as sure that collisions must occur between great masses moving through space, as it is that ships, steered without intelligence directed to prevent collision, could not cross and recross the Atlantic for thousands of years with immunity from collision."

The author of the paper in the *Cornhill* denies this very positively, and without going into the mathematical details, points out the basis upon which it may be mathematically refuted—

viz., that all such worlds are traveling in fixed or regular orbits around their primaries or suns, while each of these primaries travels in its own necessary path, carrying with it all its attendants, which still move about him, just as though he had no motion of his own.

These are the conclusions of Newtonian dynamics, the sublime simplicity of which contrasts so curiously with the complex dreams of the modern atom-splitters, and which make a further and still more striking contrast by their exact and perfect accordance with actual and visible phenomena.

Newton has taught us that there can be no planets traveling at random like the Sir W. Thomson's imaginary ships with blind pilots, and by following up his reasoning, we reach the conclusion, that among all the countless millions of worlds that people the infinity of space, there is no more risk of collision than there is between any two of the bodies that constitute our own solar system.

All the observations of astronomers, both before and since the discovery of the telescope, confirm this conclusion. The long nightly watching of the Chaldean shepherds, the star-counting, star-gauging, star-mapping, and other laborious gazing of mediæval and modern astronomers, have failed to discover any collision, or any motion tending to collision, among the myriads of heavenly bodies whose positions and movements have been so faithfully and diligently studied. Thus, the hypothesis of creation which demands the destruction of two worlds in order to effect the sowing of a seed, is as inconsistent with sound dynamics as it is repugnant to common sense.

This subject suggests a similar one, which was discussed a few months since at the Acadamy of Sciences of Paris. On January 30th last M. St. Meunier read a paper on "The mode of rupture of a star, from which meteors are derived." The author starts with the assumption that meteors have been produced by the rupture of a world, basing this assumption upon the arguments he has stated in previous papers. He discards altogether Sir W. Thomson's idea of a collision between two worlds, but works out a conclusion quite as melancholy.

He begins, like most other builders of cosmical theories, with the hypothesis that this and all the other worlds of space began their existence in a condition of nebulous infancy; that they gradually condensed into molten liquids, and then cooled down till they obtained a thin outside crust of solid matter, resting upon a molten globe within; that this crust then gradually thickened as the world grew older and cooled down by radiation. I will not stop to discuss this nebular and cooling-down hypothesis at present, though it is but fair to state that "I don't believe a bit of it."

Taking all this for granted—a considerable assumption—M. St. Meunier reasons very ably upon what must follow, if we further assume that each world is somehow supplied with air and water, and that the atmosphere and the ocean of each world are limited and unconnected with those of any other world, or with any general interstellar medium.

What, then, will happen as worlds grow old? As they cool down, they must contract; the liquid inside can manage this without any inconvenience to itself, but not so with the outer spherical shell of solid matter. As the inner, or hotter part of this contracts, the cool outside must crumple up in order to follow it, and thus mountain chains and great valleys, lesser hills and dales, besides faults and slips, dykes, earthquakes, volcanoes, etc., are explained.

According to M. St. Meunier, the moon has reached a more advanced period of cosmical existence than the earth. She is our senior; and like the old man who shows his gray hairs and tottering limbs to inconsiderate youth, she shines a warning upon our gay young world, telling her that—

> Let her paint an inch thick, to this favor she must come

—that the air and ocean must pass away, that all the living creatures of the earth must perish, and the desolation shall come about in this wise.

At present, the interior of our planet is described as a molten fluid, with a solid crust outside. As the world cools down with age, this crust will thicken and crack, and crack again, as the lower part contracts. This will form *rainures*, i.e., long narrow chasms, of vast depth, which, like those on the moon, will traverse, without deviation, the mountains, valleys, plains, and ocean-beds; the waters will fall into these, and, after violent catastrophes, arising from their boiling by contact with the hot interior, they will finally disappear from the surface, and become absorbed in the pores of the vastly-thickened earth-crust, and in the caverns, cracks, and chasms which the rending contraction will open in the interior. These cavities will continue to increase, will become of huge magnitude when the outside crust grows thick enough to form its own supporting arch, for then the fused interior will recede, and form mighty vaults that will engulf not the waters merely, but all the atmosphere likewise.

At this stage the earth, according to M. St. Meunier, will be a middle-aged world like the moon; but as old age advances the contraction of the fluid, or viscous interior beneath the outside solid crust will continue, and the *rainures* will extend in length and depth and width, as he maintains they are now growing in the moon. This, he says, must continue till the centre solidifies, and then these cracks will reach that centre, and the world will be split through in fragments corresponding to the different *rainures*.

Thus we shall have a planet composed of several solid fragments held together only by their mutual attractions, but the rotary movement of these will, according to the French philosopher, become unequal, as "the fragments present different densities, and are situated at unequal distances from the centre; some will be accelerated, others retarded; they will rub against each other, and grind away those portions which have the weakest cohesion." The fragments thus worn off will, "at the end of sufficient time, girdle with a complete ring the central star." At this stage the fragments become real meteors, and then perform all the meteoric functions excepting the seed-carrying of Sir W. Thomson.

It would be an easy task to demolish these speculations, though not within the space of one of my letters. A glance at the date of this paper, and the state of Paris and the French mind at the time, may, to some extent, explain the melancholy relish with which the Parisian philosopher works out his doleful speculations. Had the French army marched vigorously to Berlin, I doubt whether this paper would ever have found its way into the "Comptes Rendus." After the fall of Paris, and the wholesale capitulation of the French armies, it was but natural that a patriotic Frenchman, howsoever strong his philosophy, should speculate on the collapse of all the stars, and the general winding-up of the universe.

Chapter 26

THE DYING TREES IN KENSINGTON GARDENS

A great many trees have lately been cut down in Kensington Gardens, and the subject was brought before the House of Commons at the latter part of its last session. In reply to Mr. Ritchie's question, Mr. Adam, the then First Commissioner of Works, made explanations which, so far as they go, are satisfactory—but the distance is very small. He states that all who have watched the trees must have seen that their decay "has become rapid and decided in the last two years," that when the vote for the parks came on many "were either dead or hopelessly dying," that in the more thickly planted portions of the gardens the trees were dead and dying by hundreds, owing to the impoverished soil and the terrible neglect of timely thinning fifty or sixty years ago.

Knowing the sensitiveness of the public regarding tree-cutting, Mr. Adam obtained the co-operation of a committee of experts, consisting of Sir Joseph Hooker, Mr. Clutton, and Mr. Thomas, "so distinguished as a landscape gardener," and the late First Commissioner of Works. They had several meetings, and, as Mr. Adam informs us, "the result has been a unanimous resolution that we ought to proceed at once to clear away the dead and dying trees." This is being done to the extent of "an absolute clearance" in some places, and the removal of numerous trees all over the gardens. We are further told that "the spaces cleared will either be trenched, drained, and replanted, or will be left open, as may appear best." Mr. Adam adds that "the utmost care is being used in the work; that not a tree is being cut that can properly be spared; and that every effort will be made to restore life to the distinguished trees that are dying."

I have watched the proceedings in Kensington Gardens and also in Bushey Park, and have considerable difficulty in describing the agricultural vandalism there witnessed, and expressing my opinion on it, without transgressing the bounds of conventional courtesy towards those who are responsible. I do not refer to the cutting down of the dead and dying trees, but to the proceedings by which they have been officially and artificially killed by those who ought to possess sufficient knowledge of agricultural chemistry to understand the necessary consequences of their conduct.

About forty years have elapsed since Liebig taught to all who were able and willing to learn that trees and other vegetables are composed of two classes of material: 1st, the carbon and elements of water derived from air and rain; and 2d, the nitrogenous and incombustible saline compounds derived from the soil. The possible atmospheric origin of some of the nitrogen is still under debate, but there is no doubt that all which remains behind as incombustible ash, when we burn a leaf, is so much matter taken out of the soil. Every scientific agriculturist knows that certain crops take away certain constituents from the soil, and that if this particular cropping continues without a replacing of those particular constituents of fertility, the soil must become barren in reference to the crop in question, though other crops demanding different food may still grow upon it.

The agricultural vandalism that I have watched with so much vexation is the practice of annually raking and sweeping together the fallen leaves, collecting them in barrows and carts, and then carrying them quite away from the soil in which the trees are growing, or should grow. I have inquired of the men thus employed whether they put anything on the ground to replace these leaves, and they have not merely replied in the negative, but have been evidently surprised at such a question being asked. What is finally done with the leaves I do not know; they may be used for the flower-beds or sold to outside florists. I have seen a large heap accumulated near to the Round Pond.

Now, the leaves of forest trees are just those portions containing the largest proportion of ash; or, otherwise stated, they do the most in exhausting the soil. In Epping Forest, in the New Forest, and other forests where there has been still more "terrible neglect of timely thinning," the trees continue to grow vigorously, and have thus grown for centuries; the leaves fall on the soil wherein the trees grow, and thus continually return to it all they have taken away.

They do something besides this. During the winter they gradually decay. This decay is a process of slow combustion, giving out just as much heat as though all the leaves were gathered together and used as fuel for a bonfire; but the heat in the course of natural decay is gradually given out just when and where it is wanted, and the coating of leaves, moreover, forms a protecting winter jacket to the soil.

I am aware that the plea for this sweeping-up of leaves is the demand for tidiness; that people with thin shoes might wet their feet if they walked through a stratum of fallen leaves. The reply to this is that all reasonable demands of this class would be satisfied by clearing the footpaths, from which nobody should deviate *in the winter time*. Before the season for strolling in the grass returns, Nature will have disposed of the fallen leaves. A partial remedy may be applied by burning the leaves, then carefully distributing their ashes; but this is after all a clumsy imitation of the natural slow combustion above described, and is wasteful of the ammoniacal salts as well as of the heat. The avenues of Bushey Park are not going so rapidly as the old sylvan glories of Kensington Gardens, though the same robbery of the soil is practiced in both places. I have a theory of my own in explanation of the difference, viz., that the cloud of dust that may be seen blowing from the roadway as the vehicles drive along the Chestnut Avenue of Bushey Park, settles down on one side or the other, and supplies material which to some extent, but not sufficiently, compensates for the leaf-robbery.

The First Commissioner speaks of efforts being made to restore life to the distinguished trees that are dying. Let us hope that these include a restoration to the soil of those particular salts that have for some years past been annually carted away from it in the form of dead leaves, and that this is being done not only around the "distinguished" trees, but throughout the gardens.

Any competent analytical chemist may supply Mr. Adam with a statement of what are these particular salts. This information is obtainable by simply burning an average sample of the leaves and analyzing their ashes.

While on this subject I may add a few words on another that is closely connected with it. In some parts of the parks gardeners may be seen more or less energetically occupied in pushing and pulling mowing-machines; and carrying away the grass which is thus cut. This produces the justly admired result of a beautiful velvet lawn; but unless the continuous exhaustion of the soil is compensated, a few years of such cropping will starve it. This subject is now so well understood by all educated gardeners that it should be impossible to suppose it to be overlooked in our parks, as it is so frequently in domestic gardening. Many a lawn that a few years ago was the pride of its owner is now becoming as bald as the head of the faithful, "practical," and obstinate old gardener who so heartily despises the "fads" of scientific theorists.

When natural mowing-machines are used, i.e., cattle and sheep, their droppings restore all that they take away from the soil, minus the salts contained in their own flesh, or the milk that may be removed. An interesting problem has been for some time past under the consideration of the more scientific of the Swiss agriculturists. From the mountain pasturages only milk is taken away, but this milk contains a certain quantity of phosphates, the restoration of which must be effected sooner or later, or the produce will be cut off, especially now that so much condensed milk is exported.

The wondrously rich soil of some parts of Virginia has been exhausted by unrequited tobacco crops. The quantity of ash displayed on the burnt end of a cigar demonstrates the exhausting character of tobacco crops. That which the air and water supplied to the plant is returned as invisible gases during combustion, but all the ash that remains represents what the leaves have taken from the soil, and what should be restored in order to sustain its pristine fertility.

The West India Islands have similarly suffered to a very serious extent on account of the former ignorance of the sugar planters, who used the canes as fuel in boiling down the syrup, and allowed the ashes of those canes to be washed into the sea. They were ignorant of the fact that pure sugar maybe taken away in unlimited quantities without any impoverishment of the land, seeing that it is composed merely of carbon and the elements of water, all derivable from air and rain. All that is needed to maintain the perennial fertility of a sugar plantation is to restore the stems and leaves of the cane, or carefully to distribute their ashes.

The relation of these to the soil of the sugar plantations is precisely the same as that of the leaves of the trees to the soil of Kensington Gardens, and the reckless removal of either must produce the same disastrous consequences.

Chapter 27

THE OLEAGINOUS PRODUCTS OF THAMES MUD: WHERE THEY COME FROM AND WHERE THEY GO

Once upon a time—and not a very long time since—a French chemist left the land of super excellence, and crossed to the shores of foggy Albion. He proceeded to Yorkshire, his object being to make his fortune. He was so presumptuous as to believe that he might do this by picking up something which Yorkshiremen threw away. That something was soapsuds. His chemistry taught him that soap is a compound of fat and alkali, and that if a stronger acid than that belonging to the fat is added to soapsuds, the stronger acid will combine with the alkali and release the fat, the which fat thus liberated will float upon the surface of the liquid, and may then be easily skimmed off, melted together, and sold at a handsome profit.

But why leave the beautiful France and desolate himself in dreary Yorkshire merely to do this? His reason was, that the cloth workers of Yorkshire use tons and tons of soap for scouring their materials, and throw away millions of gallons of soapsuds. Besides this, there are manufactories of sulphuric acid near at hand, and a large demand for machinery grease just thereabouts. He accordingly bought iron tanks, and erected works in the midst of the busiest centre of the woolen manufacture. But he did not make his fortune all at once. On the contrary, he failed to pay expenses, for in his calculations he had omitted to allow for the fact that the soap liquor is much diluted, and therefore he must carry much water in order to obtain a little fat. This cost of carriage ruined his enterprise, and his works were offered for sale.

The purchaser was a shrewd Yorkshireman, who then was a dealer in second-hand boilers, tanks, and other iron wares. When he was about to demolish the works, the Frenchman took him into confidence, and told the story of his failure. The Yorkshireman said little, but thought much; and having finally assured himself that the carriage was the only difficulty, he concluded, after the manner of Mahomet, that if the mountain would not come to him, he might go to the mountain; and then made an offer of partnership on the basis that the Frenchman should do the chemistry of the work, and that he (the Yorkshireman) should do the rest.

Accordingly, he went to the works around, and offered to contract for the purchase of all their soapsuds, if they would allow him to put up a tank or two on their premises. This he did; the acid was added, the fat rose to the surface, was skimmed off, and carried, *without the water*, to the central works, where it was melted down, and, with very little preparation, was converted into "cold-neck grease," and "hot-neck grease," and used, besides, for other lubricating purposes. The Frenchman's science and skill, united with the Yorkshireman's practical sagacity, built up a flourishing business, and the grease thus made is still in great demand and high repute for lubricating the rolling-mills of iron works, and for many other kinds of machinery.

My readers need not be told that there are soapsuds in London as well as in Yorkshire, and they also know that the London soapsuds pass down the drains into the sewers. I may tell them that besides this there are many kinds of acids also passed into London sewers, and that others are generated by the decompositions there abounding. These acids do the Frenchman's work upon the London soapsuds, but the separated fat, instead of rising slowly and undisturbed to form a film upon the surface of the water, is rolled and tumbled amongst its multifarious companion filth, and it sticks to whatever it may find congenial to itself. Hairs, rags, wool, ravellings of cotton, and fibres of all kinds are especially fraternal to such films of fat: they lick it up and stick it about and amid themselves; and as they and the fat roll and tumble along the sewers together, they become compounded and shaped into unsavory balls that are finally deposited on the banks of the Thames, and quietly repose in its hospitable mud.

But there is no peace even there, and the gentle rest of the fat nodules is of short duration. The mud-larks are down upon them, in spite of all their burrowing; they are gathered up and melted down. The filthiest of their associated filth is thus removed, and then, and with a very little further preparation, they appear as cakes of dark-colored hard fat, very well suited for lubricating machinery, and indifferently fit for again becoming soap, and once more repeating their former adventures.

Those gentlemen of the British press whose brilliant imagination supplies the public with their intersessional harvests of sensational adulteration panics, have obtained a fertile source of paragraphs by co-operating with the mud-larks in the manufacture of butter from Thames mud.

The origin of these stories is traceable to certain officers of the Thames police, who, having on board some of these gentlemen of the press engaged in hunting up information respecting a body found in the river, supplied their guests with a little supplementary chaff by showing them a mud-lark's gatherings, and telling them that it was raw material from which "fine Dorset" is produced. A communication from "Our Special Correspondent" on the manufacture of butter from Thames mud accordingly appeared in the atrocity column on the following morning, and presently "went the round of the papers."

Although it is perfectly possible by the aid of modern chemical skill to refine even such filth as this, and to churn it into a close resemblance to butter, the cost of doing so would exceed the highest price obtainable for the finest butter that comes to the London market. A skillful chemist can convert all the cotton fibres that are associated with this sewage fat into pure sugar or sugar-candy, but the manufacture of sweetmeats from Thames mud would not pay any better than the production of butter from the same source, and for the same reason.

Mutton-suet, chop-parings, and other clean, wholesome fat can be bought wholesale for less than fivepence per pound. It would cost above three times as much as this to bring the fat nodules of the Thames mud to as near an approach to butter as this sort of fat. Therefore the Thames mud-butter material would be three times as costly as that obtainable from the butcher. While the supply of mutton-suet is so far in excess of the butter-making demand that tons of it are annually used in the North for lubricating machinery, we need not fear that anything less objectionable—i.e., more costly to purify—will be used as a butter substitute.

Chapter 28

LUMINOUS PAINT

The sun is evidently going out of fashion, and is more and more excluded from "good society" as our modern substitute for civilization advances. "Serve him right!" many will say, for behaving so badly during the last two summers. The old saw, which says something about "early to bed and early to rise" is forgotten: we take "luncheon" at dinner-time, dine at supper-time, make "morning" calls and go to "morning" concerts, etc., late in the afternoon, say "Good morning" until 6 or 7 P.M.; and thus, by sleeping through the bright hours of the morning, and waking up fully only a little before sunset, the demand for artificial light becomes almost overwhelming. Not only do we require this during a longer period each day, but we insist upon more and more, and still more yet, during that period.

The rushlight of our forefathers was superseded by an exotic luxury, the big-flame candle made of Russian tallow, with a wick of Transatlantic cotton. Presently this luxurious innovation was superseded by the "mould candle;" the dip was consigned to the kitchen, and the bloated aristocrats of the period indulged in a *pair* of candlesticks, alarming their grandmothers by the extravagance of burning two candles on one table. Presently the mould candle was snuffed out by the composite; then came the translucent pearly paraffin candle, gas light, solar lamps, moderator lamps, and paraffin lamps. Even these, with their brilliant white flame from a single wick, are now insufficient, and we have duplex and even triplex wicks to satisfy our demand for glaring mockeries of the departed sun.

Some are still living who remember the oil lamps in Cheapside and Piccadilly, and the excitement caused by the brilliancy of the new gas lamps; but now we are dissatisfied with these, and demand electric lights for common thoroughfares, or some extravagant combination of concentric or multiplex gas-jets to rival it.

The latest novelty is a device to render darkness visible by capturing the sunbeams during the day, holding them as prisoners until after sunset, and then setting them free in the night. The principle is not a new discovery; the novelty lies in the application and some improvements of detail. In the "Boy's Own Book," or "Endless Amusement," of thirty or forty years ago, are descriptions of "Canton's phosphorus," or "solar phosphori," and recipes for making them. Burnt oyster-shells or oyster-shells burnt with sulphur, was one of these.

Various other methods of effecting combination between lime or baryta with sulphur are described in old books, the result being the formation of more or less of what modern chemists call calcium sulphide and barium sulphide (or otherwise sulphide of calcium or sulphide of barium). These compounds, when exposed to the sun, are mysteriously acted upon by the solar rays, and put into such a condition that their atoms or molecules, or whatever else constitutes their substance, are set in motion—in that sort of motion which communicates to the surrounding medium the wavy tremor which agitates our optic nerve and produces the sensation of light.

Until lately, this property has served no other purpose than puzzling philosophers, and amusing that class of boys who burn their fingers, spoil their clothes, and make holes in their mothers' table-covers, with sulphuric acid, nitric acid, and other noxious chemicals. The first idea of turning it to practical account was that of making a sort of enamel of one or the other of these sulphides, and using it as a coating for clock-faces. A surface thus coated and exposed to the light during the day becomes faintly luminous at night.

Anybody desirous of seeing the sort of light which it emits, may do so very easily by purchasing an unwashed smelt from the fishmonger, and allowing it to dry with its natural slime upon it, then looking at it in the dark. A sole or almost any other fish will answer the purpose, but I name the smelt from having found it the most reliable in the course of my own experiments. It emits a dull, ghostly light, with very little penetrating power, which shows the shape of the fish, but casts no perceptible light on objects around.

Thus the phosphorescent parish-clock face, with non-phosphorescent figures and hands, would look like a pale ghost of the moon with dark figures round it, and dark hands stretching across, by which the time of the night might possibly be discovered there or thereabouts. This invention has already appeared in a great many paragraphs, but, hitherto, upon very few clock-faces.

Recently it has assumed a more ambitions form—patented, of course. The patentees claim an improved phosphorescent powder, which is capable of being worked up with the medium of paints and varnishes, and thus applied, not merely to clock-faces, but to the whole of the walls and ceilings of any apartment. In this case the faintness of the light will be in some degree compensated by the extent of phosphorescent surface, and it is just possible that the sum total of the light emitted from walls and ceiling may be nearly equal to that of one mould candle. If so, it will have some value as a means of lighting powder magazines and places for storage of inflammable compounds. It is stated that one of the London Dock companies is about to use it for its spirit vaults; also that the Admiralty has already tried the paint at Whitehall, and has ordered two compartments of the *Comus* to be painted with it, in order to test its capability of lighting the dark regions of ironclad ships.

This application can, however, only be limited to those parts which receive a fair amount of light during the day, for unless the composition first receives light, it is not able afterwards to emit it, and this emission or phosphorescence only continues a few hours after the daylight has passed away; five or six hours is the time stated.

A theatrical manager is said to be negotiating for the exclusive right to employ this weird illumination for scenic purposes. The sepulchre scene in "Robert le Diable," or the incantation in "Der Freischutz," or "The Sorcerer," might be made especially effective by its ghostly aid. The name-plates of streets, and buoys at sea might be advantageously coated with such a composition; and many other uses suggest themselves.

There are rival inventors, as a matter of course. The French patentees claim the use of cuttle-fish bones, various sea-shells, etc., mixed with pure lime, sulphur, and calcined sea-salt, besides sulphides of calcium, barium, strontium, uranium, magnesium, or aluminium. They also add phosphorus itself, though for what purpose is questionable, seeing that this substance is only luminous during the course of its oxidation or slow combustion, and after this has ended the resultant phosphoric acid is no more luminous than linseed oil or turpentine. An admixture of phosphorus might temporarily increase the luminosity of a *sample*, but any conclusions based upon this would be quite delusive. They also assert that electrical discharges passed through the paint increase its luminosity. According to some enthusiasts, electricity is to do everything; but these ladies and gentlemen omit to calculate the cost of rousing and feeding this omnipotent giant. In this case electrical machinery for stimulating the paint for anything outside of lecture-table experiments or theatrical and other sensational displays, would be a commercial absurdity.

The Americans, of course, are re-inventing in this direction, but Mr. Edison has not yet appeared on the luminous-paint scene. If he does we shall doubtless hear of something very brilliant, even though we never see it. In the meantime we may safely hope that this application of an old scientific plaything to useful purposes may become of considerable utility, as it evidently opens a wide field for further investigation and progressive improvement, by the application of the enlarged powers which modern science places at the disposal of ingenious inventors. We hope, for the sake of all concerned, that it will not fall into the hands of professional prospectus manufacturers and joint-stock-company mongers, and that the story of its triumphs will be told without any newspaper exaggerations.

Since the above was written—in February, 1880—I have tested this luminous paint (Balmain's patent). Practically, I find it unsatisfactory. In the first place, its endurance is far shorter than is stated. It begins to fade almost immediately the light is withdrawn, and in the course of an hour or two it is, for all practical use—though not absolutely—extinguished. Besides this it emits a very unpleasant odor painfully resembling sewage and sulphureted hydrogen. This is doubtless due to the sulphur compound, but is, I have no doubt, quite harmless in spite of its suggestions.

Chapter 29

THE ORIGIN AND PROBABLE DURATION OF PETROLEUM

In spite of the enormous quantities of mineral oil that are continuously drawn from the earth, and the many places from which it may thus be drawn, geologists are still puzzled to account for it. If it were commonly associated with coal the problem of its origin would be solved at once. We should then be satisfied that natural mineral oil is produced in the same manner as the artificial product, i.e., by the heating and consequent distillation of certain kinds of coal or of bituminous shales; but, as a matter of fact, it is but rarely that petroleum is found in the midst of coal seams, though it is sometimes so found.

I visited, some years ago, a coal-mine in Shropshire, known as "the tarry pit," thus named on account of the large quantity of crude mineral oil of a rather coarse quality that exuded from the strata pierced by the shaft. It ran down the sides of the shaft, filled the "sumph" (i.e., the well at the bottom of the shaft in which the water draining from the mine should accumulate for pumping), and annoyed the colliers so seriously that they refused to work in the mine unless the nuisance were abolished. It was abolished by "tubbing" the shaft with an oil-proof lining built round that part from which the oil issued. The "tar" as the crude oil was called, was then pumped out of the sumph, and formed a pool which has since been filled up by the *débris* of the ordinary mine workings.

A publican in the Black Country of South Staffordshire discovered an issue of inflammable vapor in his cellar, collected it by thrusting a pipe into the ground, and used it for lighting and warming purposes, as well as an attraction to customers.

These and other cases that might be cited, although exceptional, are of some value in helping us to form a simple and rational theory of the origin of this important natural product. They prove that mineral oil *may* be produced in connection with coal seams and apparently from the coal itself. A sound theory of the origin of petroleum is of practical as well as theoretical value, inasmuch as the very practical question of the probable permanency of supply depends entirely on the nature of the origin of that supply. Some very odd theories have been put forth, especially in America.

Seeing that petroleum is commonly found associated with sandstone and limestone, especially in cavities of the latter, it has been supposed that these minerals somehow produce it. Turning back to the *Grocer* for April 18, 1872, I find some speculations of this kind quoted from the *Petroleum Monthly*. The writer sets aside altogether, as an antiquated and exploded fallacy, the idea that petroleum is produced from coal, and maintains "that petroleum is mainly produced from, or generated through, limestone," and argues that the generation of petroleum by such rocks is a continuous process, from the fact that exhausted wells have recovered after being abandoned, his explanation being "that the formerly abandoned territory was given up because the machinery for extracting petroleum from the earth exceeded in its power of exhausting the fluid the generative powers by which it is produced;" these generative powers somehow residing in the limestone and sandstone, but how is not specified.

Some writers have, however, gone a little further toward answering the question of how limestone may generate petroleum. They have pointed to the fossilized remains of animals, their shells, etc., existing in the limestone, and have supposed that the animal matter has been distilled, and has thus formed the oil.

If such a process could be imitated artificially by distilling some of the later deposits of similar fossil character this theory would have a better basis, or even if a collection of oysters, mussels, or any other animal matters could by distillation be shown to produce an oil similar to petroleum.

The contrary is the case. We may obtain oil from such material, but it is utterly different from any kind of mineral oil, while, on the other hand, by distilling natural bituminous shales, or cannel coal, or peat, we obtain a crude oil almost identical with natural petroleum, and the little difference between the two is perfectly accounted for by the greater rapidity of our methods of distillation as compared with the slow natural process. We may go on approximating more and more nearly to the natural petroleum by distilling more and more slowly. As it is, the refined products of the natural and artificial oil which is commercially distilled in Scotland, are scarcely distinguishable—some of them are not at all distinguishable—the solid paraffin, for example. I now offer my own theory of the origin of oil springs.

To render this the more intelligible, let us first consider the origin of ordinary water springs. St. Winifred's Well, at Holywell, in Flintshire, maybe taken as an example, not merely on account of its magnitude, but because it is quite typical, and is connected with limestone and sandstone in about the same manner as are the petroleum wells of Pennsylvania.

Here we have a wondrous uprush of water just between the sandstone and mountain limestone rocks, which amounts to above twenty tons per minute, and flows down to the Dee, a small river turning several water-mills. It is certain that all this water is not generated either by the limestone or the sandstone from which it issues, nor can it be all "generated" on the spot. The true explanation of its origin is simple enough.

The mountain limestone underlies the coal measures and crops up obliquely at Holywell; against this oblique subterranean wall of compact rock impermeable to water, abuts a great face of down-sloping strata of porous sandstone and porous shales. These porous rocks receive the rain which falls on the slopes of the Hope Mountain and other hills which they form; this water sinks into the millstone grit of these hills and percolates downwards until it reaches the limestone barrier, into which it cannot penetrate.

It here accumulates as a subterranean reservoir which finds an outlet at a convenient natural fissure, and, as the percolation is continuous, the spring is a constant one. Some of the water travels many miles underground before it thus escapes. Hundreds of other smaller instances might be quoted, the above being the common history of springs which start up whenever the underground waters that flow through porous rocks or soil meet with compact rocks or impermeable clay, and thus, being able to proceed no further downwards, accumulate and produce an overflow which we call a "spring."

If water can thus travel underground, why not oil?

Although the oil springs or oil wells are not immediately above or below coal seams, they are all within "measurable distance" of great coal formations—the oil territory of Pennsylvania is, in fact, surrounded by coal, some of it anthracite, which is really a coke, such as would be produced if we artificially distilled the hydrocarbons from coal, and then compressed the residue, as the anthracite has certainly been pressed by the strata resting upon it.

The rocks in immediate contact and proximity to coal seams—"the coal measures," as they are called—are mostly porous, some of them very porous, and thus if at any period of the earth's long history a seam of coal became heated, as we know so many strata are, and have been heated, a mineral oil would certainly be formed, would first permeate the porous rocks as vapor, then be condensed and make its way through them, following their "dip" or inclination until it reached a barrier such as the limestone forms.

It would thus in after-ages be found, not among the coal where it was formed, but at the limestone or other impermeable rock by which its further percolation was arrested.

This is just where it actually is found.

Limestone, although not porous like shales and sandstones, is specially well adapted for storing large subterranean accumulations, on account of the great cavities to which it is liable. Nearly all the caverns in this country, in Ireland where they abound, in America, and other parts of the world, are in limestone rocks; they are especially abundant in the "carboniferous limestone" which underlies the coal measures, and this is explained by the fact that limestone may be dissolved by rain-water that has oozed through vegetable soil or has soaked fallen leaves or other vegetable matter, and thereby become saturated with carbonic acid.

Where the petroleum finds a crevice leading to such cavities it must creep through it and fill the space, thereby forming one of the underground reservoirs supplying those pumping wells that have yielded such abundance for a while and then become dry. But if this theory is correct it does not follow that the drying of such a well proves a final stoppage of the supply, for if the cavity and crevice are left, more oil may ooze into the crevice and flow into the cavity, and this may continue again and again throughout the whole oil district so long as the surrounding feeders of permeable strata continue saturated, or nearly so. The magnitude of these feeding grounds may far exceed that of the district wherein the springs occur, or where profitable wells may be sunk, seeing that the localizing of profitable supply depends mainly on the stoppage of further oozing away by the action of the impermeable barrier.

A well sunk into the oozing strata itself would receive a very small quantity, only that which, in the course of its passage came upon the well sides, while at the junction between the permeable and the impermeable rocks the accumulation may include all that reached the whole surface of such junction or contact—many square miles.

To test this theory thoroughly it would be necessary to make borings, not merely at the wells, but in their neighborhood, where the porous rocks dip towards the limestone, and to bring up sample cores of these porous rocks, and carefully examine them. Dr. Sterry Hunt has done this in the oil-yielding limestone rocks of Chicago, but not in those of the nearest coal-measures.

As the oil industry of America is of such great national importance, an investigation of this kind is worthy of the energies of the American Government geologists. It would throw much light on the whole subject, and supply data from which the probable duration of the oil supply might be approximately calculated.

Such an investigation might even do more than this. By proving the geological conditions upon which depend the production of petroleum springs, new sources may be discovered, just as new coal-seams have been discovered, in accordance with geological prediction, or as the practical discovery of the Austrian gold-fields was so long preceded by Sir Roderick Murchison's theoretical announcement of their probable existence.

When the "kerosene wells" were first struck, the speculations concerning their probable permanency were wild and various. Some maintained that it was but a spurt, a freak of nature limited to a narrow locality, and would soon be over; others asserted forthwith that American oil, like everything else American, was boundless. Neither had any grounds for their assertions, and therefore made them with the usual boldness of mere dogmatism.

Then came a period of scare, started by the fact that wells which at first spouted an inflammable mixture of oil and vapor high into the air soon became quiescent, and from "spouting wells" became "flowing wells," merely pouring out on the surface a small stream at first, which gradually declined to a dribble, and finally ceased to flow at all. Even those that started modestly as flowing wells did the latter, and thus appeared to become exhausted.

This exhaustion, however, was only apparent, as was proved by the application of pumps, which drew up from wells that had ceased either to spout or flow, large and apparently undiminishing quantities of crude oil.

Further observation and thought revealed the cause of these changes. It became understood that the spouting was due to the tapping of a rock-cavity containing oil of such varying densities and volatility that some of it flew out as a vapor, or boiled at the mean temperature of the air of the country or that of the surrounding rocks. Such being the case, the cavity was filled with high-pressure oil-vapor straining to escape. If the bore-hole tapped the crown or highest curve of the roof of such an oil-cavern, it opened directly into the vapor there accumulated, and the vapor itself rushed out with such force that a pillar of fire was raised in the air if a light came within some yards of the orifice. We are told of heavy iron boring-rods that were shot up to wondrous heights—and we may believe these stories if we please.

If the bore-hole struck lower down, somewhere on the sloping sides or in the shallow lower branches of the oil-cavern, it dipped at once into liquid oil, and this oil, being pressed by the elastic vapor of the upper part, was forced up as a jet of spouting oil.

In either case these violent proceedings soon came to an end, for as the vapor or oil poured out, the space above the oil-level where the vapor had been confined was increased, and its pressure diminished, till at last it barely sufficed to raise the oil to the surface, and afterwards failed to do that.

It is quite clear from this that the supplies are not "inexhaustible." The quantity of vapor having been limited, there must also be a limit to the quantity of oil giving off this vapor; the space in the oil-cavern occupied by this vapor having been limited, there must be a limit to the space occupied by the oil. The quantity of oil may be ten times, a hundred times, a thousand times, or ten thousand times, greater than that of the vapor, but in either or any case it must come to an end at last, sooner later.

If there were but a few wells here and there, as at other similar places, such as Rangoon, the Persian oil-wells, etc., the pumping might continue for centuries and centuries; but this is not the case in America. The final boundaries of the oil-bearing strata may not yet have been reached; but so far as they are known they are riddled through and through, and pumped in every direction, so that the end must come at last, though with our present knowledge we cannot say *when*.

We can, however, say *how* it must come. It will not be a sudden stoppage, but a gradual exhaustion indicated by progressive diminution of supply. We shall not be suddenly deprived of this important source of light and cheerfulness; but we may at any time begin to feel the pinch of scarcity and consequent rise of price. This rise of price will check the demand, and bring forth other supplies from sources that now cannot be profitably worked on account of the cheapness of American petroleum.

Many of the countries now largely supplied from America have oil-springs of their own, which a rise of price will speedily bring into paying operation.

We have nothing to fear. The fact that in spite of the ruinous prices that have recently prevailed the Scotch oil-makers continue to exist at all, shows us what they may do with a rise of even a few pence per gallon. The thickness and area of the dark shales from which their oil is distilled are so great that their exhaustion is very far remote indeed. The Americans have similar shales to fall back upon when the spontaneous product ceases to flow, but they are quite incapable of competing with us at home on equal terms—that is, when both have to obtain the oil as a manufactured product of artificial distillation.

If anything like moderation were possible in America, the first indications of scarcity would be followed by some economy in working; but this is not to be anticipated. It is more likely that the first rise of prices will attract additional speculation, and the sinking of more wells in the hope of large profits, and this of course will shorten the period of gradual exhaustion, the commencement of which may, for aught we know, be very near at hand, especially if the new projects for using petroleum as furnace fuel under steam boilers, and for the smelting, puddling, and founding of iron and other metals, are carried out as they may be so easily at present prices, and with the aid of pipe-lines to carry the crude or refined oil from the wells to any part of the great American continent where it may be required in large quantities.

The old story of the goose that laid the golden eggs seems to be in course of repetition in Transatlantic Petrolia.

* * *

Since the above was written I have received from Dr. Sterry Hunt a copy of his interesting "Chemical and Geological Essays," in one of which he expounds a theory of the origin of petroleum. He states that it appears to him "that the petroleum, or rather the materials from which it has been formed, existed in the limestone rocks from the time of their first deposition," and "that petroleum and similar bitumens have resulted from a peculiar transformation of vegetable matters, or in some cases of animal tissues analogous to these in composition."

The objections on page 275 apply to the animal tissues of this theory, and as regards the vegetable matter I think it fails from the want of anything like an adequate supply in these limestone rocks.

Chapter 30

THE ORIGIN OF SOAP

A history of soap would be very interesting. Who invented it? When and where did it first come into common use? How did our remote ancestors wash themselves before soap was invented? These are historical questions that naturally arise at first contemplation of the subject; but, as far as we are aware, historians have failed to answer them. We read a great deal in ancient histories about anointing with oil and the use of various cosmetics for the skin, but nothing about soap.

These ancients must have been very greasy people, and I suspect that they washed themselves pretty nearly in the same way as modern engine-drivers clean their fingers, by wiping off the oil with a bit of cotton-waste.

We are taught to believe that the ancient Romans wrapped themselves round with togas of ample dimensions, and that these togas were white. Now, such togas, after encasing such anointed oily skins, must have become very greasy. How did the Roman laundresses or launders—historians do not indicate their sex—remove this grease? Historians are also silent on this subject.

A great many curious things were found buried under the cinders of Vesuvius in Pompeii, and sealed up in the lava that flowed over Herculaneum. Bread, wine, fruits, and other domestic articles, including several luxuries of the toilet, such as pomades or pomade-pots, and rouge for painting ladies' faces, but no soap for washing them. In the British Museum is a large variety of household requirements found in the pyramids of Egypt, but there is no soap, and we have not heard of any having been discovered there.

Finding no traces of soap among the Romans, Greeks, or Egyptians, we need not go back to the pre-historic "cave men," whose flint and bone implements were found embedded side by side with the remains of the mammoth bear and hyena in such caverns as that at Torquay, where Mr. Pengelly has, during the last eighteen years, so industriously explored.

All our knowledge, and that still larger quantity, our ignorance, of the habits of antique savages, indicate that solid soap, such as we commonly use, is a comparatively modern luxury; but it does not follow that they had no substitute. To learn what that substitute may probably have been we may observe the habits of modern savages, or primitive people at home and abroad.

This will teach us that clay, especially where it is found having some of the unctuous properties of fuller's-earth, is freely used for lavatory purposes, and was probably used by the Romans, who were by no means remarkable for anything approaching to true refinement.

They were essentially a nasty people, the habits of the poor being "cheap and nasty;" of the rich, luxurious and nasty. The Roman nobleman did not sit down to dinner, but sprawled with his face downwards, and took his food as modern swine take theirs. At grand banquets, after gorging to repletion, he tickled his throat in order to vomit and make room for more. He took baths occasionally, and was probably scoured and shampooed as well as oiled, but it is doubtful whether he performed any intermediate domestic ablutions worth naming.

A refinement upon washing with clay is to be found in the practice once common in England, and still largely used where wood fires prevail. It is the old-fashioned practice of pouring water on the wood ashes, and using the "lees" thus obtained. These lees are a solution of alkaline carbonate of potash the modern name of potash being derived from the fact that it was originally obtained from the ashes under the pot. In like manner soda was obtained from the ashes of seaweeds and of the plants that grow near the seashore, such as the *salsover soda*, etc.

The pot-ashes or pearl-ashes being so universal as a domestic bi-product, it was but natural that they should be commonly used, especially for the washing of greasy clothes, as they are to the present day. Upon these facts we may build up a theory of the origin of soap.

It is a compound of oil or fat with soda or potash, and would be formed accidentally if the fat on the surface of the pot should boil over and fall into the ashes under the pot. The solution of such a mixture if boiled down would give us soft soap.

If oil or fat became mixed with the ashes of soda plants, it would produce hard soap. Such a mixture would most easily be formed accidentally in regions where the olive flourishes near the coast, as in Italy and Spain for example, and this mixture would be Castile soap, which is still largely made by combining refuse or inferior olive oil with the soda obtained from the ashes of seaweed.

The primitive soap-maker would, however, encounter one difficulty—that arising from the fact that the potash or soda obtained by simple burning of the wood or seaweed is more or less combined with carbonic acid, instead of being all in the caustic state which is required for effective soap-making. The modern soap-maker removes this carbonic acid by means of caustic lime, which takes it away from the carbonate of soda or carbonate of potash by simple exchange—i.e., caustic lime *plus* carbonate of soda becoming caustic soda *plus* carbonate of lime, or carbonate of potash *plus* caustic lime becoming caustic potash *plus* carbonate of lime.

How the possibility of making this exchange became known to the primitive soap-maker, or whether he knew it at all, remains a mystery, but certain it is that it was practically used long before the chemistry of the action was at all understood. It is very probable that the old alchemists had a hand in this.

In their search for the philosopher's stone, the elixir of life, or drinkable gold, and for the universal solvent, they mixed together everything that came to hand, they boiled everything that was boilable, distilled everything that was volatile, burnt everything that was combustible, and tortured all their "simples" and their mixtures by every conceivable device, thereby stumbling upon many curious, many wonderful, and many useful results. Some of them were not altogether visionary—were, in fact, very practical, quite capable of understanding the action of caustic lime on carbonate of soda, and of turning it to profitable account.

It is not, however, absolutely necessary to use the lime, as the soda plants when carefully burned in pits dug in the sand of the sea-shore may contain but little carbonic acid if the ash is fluxed into a hard cake like that now commonly produced, and sold as "soda ash." This contains from three to thirty per cent of carbonate, and thus some samples are nearly caustic, without the aid of lime.

As cleanliness is the fundamental basis of all true physical refinement, it has been proposed to estimate the progress of civilization by the consumption of soap, the relative civilization of given communities being numerically measured by the following operation in simple arithmetic:—Divide the total quantity of soap consumed in a given time by the total population consuming it, and the quotient expresses the civilization of that community.[28]

The allusion made by Lord Beaconsfield, at the Lord Mayor's dinner in 1879, to the prosperity of our chemical manufactures was a subject of merriment to some critics, who are probably ignorant of the fact that soap-making is a chemical manufacture, and that it involves many other chemical manufactures, some of them, in their present state, the results of the highest refinements of modern chemical science.

While the fishers of the Hebrides and the peasants on the shores of the Mediterranean are still obtaining soda by burning seaweed as they did of old, our chemical manufacturers are importing sulphur from Sicily and Iceland, pyrites from all quarters, nitrate of soda from Peru and the East Indies, for the manufacture of sulphuric acid, by the aid of which they now make enormous quantities of caustic soda from the material extracted from the salt mines of Cheshire and Droitwich. These sulphuric acid works and these soda works are among the most prosperous and rapidly growing of our manufacturing industries, and their chief function is that of ministering to soap-making, in which Britain is now competing triumphantly with all the world.

By simply considering how much is expended annually for soap in every decent household, and adding to this the quantity consumed in laundries and by our woolen and cotton manufacturers, a large sum total is displayed. Formerly, we imported much of the soap we used at home; now, in spite of our greatly magnified consumption, we supply ourselves with all but a few special kinds, and export very large and continually increasing quantities to all parts of the world; and if the arithmetical rule given above is sound, the demand must steadily increase as civilization advances.

---

[28] The scientific pedant of the Middle Ages displayed his profundity by continually quoting Aristotle and other "ancients." His modern successor does the like by decorating his pages with displays of algebraical formulæ. In order to secure the proper respect of *my* readers I here repeat the equation that I enunciated many years ago, "$c = s/p$" where c stands for civilization, s for the quantity of soap consumed per annum, and p the population of a given community.

Chapter 31

OILING THE WAVES

The recent gales have shown that if "Britannia rules the waves" her subjects are very turbulent and costly. Our shipping interests are now of enormous magnitude, and they are growing year by year. We are, in fact, becoming the world's carriers on the ocean, and are thus ruling the waves in a far better sense than in the old one. Our present mercantile rule adds to the wealth of our neighbors instead of destroying it, as under the old warlike rule.

Everything concerning these waves is thus of great national interest, the loss of life and sacrifice of wealth by marine casualties being so great. Some curious old stories are extant, describing the exploits of ancient mariners in stilling the waves by pouring oil upon them. Both Plutarch and Pliny speak of it as a regular practice. Much later than this, in a letter dated Batavia, January 5, 1770, written by M. Tengragel, and addressed to Count Bentinck, the following passage occurs:—"Near the islands Paul and Amsterdam we met with a storm, which had nothing particular in it worthy of being communicated to you, except that the captain found himself obliged, for greater safety in wearing the ship, to pour oil into the sea to prevent the waves breaking over her, which had an excellent effect, and succeeded in preserving us. As he poured out but a little at a time, the East India Company owes, perhaps, its ship to only six demi-aumes of olive oil. I was present on deck when this was done, and should not have mentioned this circumstance to you, but that we have found people here so prejudiced against the experiment as to make it necessary for the officers on board and myself to give a certificate of the truth on this head, of which we made no difficulty."

The idea was regarded with similar prejudice by scientific men until Benjamin Franklin had his attention called to it, as he thus narrates:—"In 1757, being at sea in a fleet of ninety-six sail, bound for Louisbourg, I observed the wakes of two of the ships to be remarkably smooth, while all the others were ruffled by the wind, which blew fresh. Being puzzled with the differing appearance, I at last pointed it out to the captain, and asked him the meaning of it. 'The cooks,' said he, 'have, I suppose, been just emptying their greasy water through the scuppers, which has greased the sides of the ships a little.' And this answer he gave me with an air of some little contempt, as to a person ignorant of what everybody else knew. In my own mind, I first slighted the solution, though I was not able to think of another."

Franklin was not a man to remain prejudiced; he accordingly investigated the subject, and the results of his experiments, made upon a pond on Clapham Common, were communicated to the Royal Society. He states that after dropping a little oil on the water, "I saw it spread itself with surprising swiftness upon the surface, but the effect of smoothing the waves was not produced; for I had applied it first upon the leeward side of the pond, where the waves were largest, and the wind drove my oil back upon the shore. I then went to the windward side, where they began to form; and there the oil, though not more than a teaspoonful, produced an instant calm over a space several yards square, which spread amazingly, and extended itself gradually till it reached the lee side, making all that quarter of the pond (perhaps half an acre) as smooth as a looking-glass."

Franklin made further experiments at the entrance of Portsmouth Harbor, opposite the Haslar Hospital, in company with Sir Joseph Banks, Dr. Blagden, and Dr. Solander. In these experiments the waves were not destroyed, but were converted into gentle swelling undulations with smooth surfaces. Thus it appeared that the oil destroys small waves, but not large billows.

Franklin's explanation is, "that the wind blowing over water covered with a film of oil cannot easily *catch* upon it, so as to raise the first wrinkles, but slides over it and leaves it smooth as it finds it."

Further investigations have since been made which confirm this theory. The first action of the wind in blowing up what the sailors call "a sea," is the production of a ripple on the surface of the water. This ripple gives the wind a strong hold, and thus larger waves are formed, but on these larger there are smaller waves, and on these smaller waves still smaller ripples. All this roughness of surface goes on helping the wind, till at last the mightiest billows are formed, which then have an oscillation independent of the wind that formed them. Hence the oil cannot at once subdue the great waves that are already formed, but may prevent their formation if applied in time. Even the great waves are moderated by the oil stopping the action of the wind which sustains and augments them.

Quite recently, Captain David Gray made some experiments at the north bar of Peterhead, where a very heavy surf breaks over in rough weather. On a rough day he dropped a bottle full of oil into the sea. The oil floating out of the bottle, converted the choppy waves over a large area "into an expanse of long undulating rollers, smooth and glassy, and so robbed of all violence that a small open boat could ride on them in safety."

This result is quite in accordance with what we are told respecting the ancient practice of the fishermen of Lisbon, who were accustomed to empty a bottle of oil into the sea when they found on their return to the river that there was a dangerous surf on the bar, which might fill their boats in crossing it.

As regards Peterhead, it is proposed to lay perforated pipes across the mouth of the harbor, and to erect tanks from which these pipes may be supplied with oil, and thus pour a continuous and widely distributed stream into the sea in bad weather. The scheme was mooted some time ago, but I am not aware whether it has yet been carried out. Its success or failure must mainly be determined by the cost, and this will largely depend upon the kind of oil that is used. A series of well-conducted experiments upon the comparative areas protected by different kinds of oil would be very interesting and practically useful, for, until this has been ascertained, a proper selection cannot be made. How long will it last? is another question.

I have frequently seen such tracks as Franklin observed out at sea, and have climbed to the masthead in order to sight the ship that produced them, without seeing any. Several of such smooth shining tracks have been observed at the same time, but no ship visible, and this in places where no sail has been seen for days before or after. The poet's description of "the trackless ocean" is by no means "founded on fact."

The Plymouth Breakwater contains 3,369,261 tons of stone, and cost the British Government a million and a half. The interest on this at 4 per cent amounts to 60,000*l.* per annum. If the above statements are reliable, some of the wholesale oil merchants who read this might contract to becalm a considerable area of the Channel for a smaller amount.

Further experiments have been made at Peterhead since the above was written. The following account, from the *Times* of those made on February 27, 1882, is interesting:

"On Monday the long-wished-for easterly gale to test the experiment of throwing oil on the troubled waters reached Peterhead. It may be mentioned that the harbor of Peterhead is singularly exposed, and with an east or north-east gale is very dangerous of approach. Mr. Shields, of Perth, has laid the oil apparatus to be used in quelling the troubled waters. It consists of an iron pipe which conveys oil and extends from a wooden house behind the seawall at Roanhead down through a natural gullet in the rocks about 150 yards long and about 50 yards beyond the mouth of the gullet into about seven fathoms of water; at this point the iron pipe is joined to a guttapercha pipe, which extends across the harbor entrance outside the bar and is perforated at distances 12½ yards apart. Through the guttapercha pipe the oil reaches the sea. On Monday the wind was not so strong as to make the experiment so complete as could have been wished; still, there was a heavy swell. Early in the forenoon the pumps were put in motion and the leakage space in the pipe filled; but unfortunately it was found, soon after the oil began to rise to the surface of the bay, that the supply in the cask had become exhausted, and those who were conducting the experiment did not consider themselves at liberty to order a fresh cask of oil without Mr. Shield's sanction. But while the experiment was only partial it was highly satisfactory. At the same time, the film did not extend sufficiently far to prevent the waves forming and curving to broken water. As soon, however, as they reached the oil-covered neck the observers from the pier-head could easily discern the influence at work. Waves which came in crested gradually assumed the shape of undulating bodies of water, and, once formed, they rolled unbroken towards the breakwater. On Wednesday morning there was a heavy sea at the north breakwater. The oil valves were opened, and immediately the effect was manifest. The waves, which had before clashed with fury against the breakwater, assumed a rolling motion and were quite crestless. Indeed, it was admitted that the oil had rendered the entrance comparatively safe, *but the effect was not so abiding as could have been wished.*"

As regards the want of duration there noted, I venture to make a suggestion.

Oils vary so greatly in their rate of outspreading over water and the character of the film they form, that some years ago Mr. Moffatt, of Glasgow, proposed to use these differences as a test for the adulterations of one kind of oil with other and cheaper kinds.

I made a number of experiments verifying some of his results.

From these it is evident that the duration of the becalming effect will vary with different oils, and therefore further experiments upon these difference should be made, in order to select that kind which is the most effective, with due regard, of course, to cost.

The oil indicated by my experiments as combining permanency and cheapness, and altogether the most suitable and attainable is the "*dead oil*" refuse of the gas-works. This may be used in its crude and cheapest condition.

Chapter 32

ON THE SO-CALLED "CRATER NECKS" AND "VOLCANIC BOMBS" OF IRELAND

A PAPER READ AT THE GEOLOGISTS' ASSOCIATION, DECEMBER 6, 1878

Mr. Hull, "Physical Geography and Geology of Ireland," p. 68, under the head of "Volcanic Necks and Basaltic Dykes," says that "although the actual craters and cones of eruption have been swept from the surface of the country by the ruthless hand of time, yet the old "necks" by which the volcanic mouths were connected with the sources of eruption can occasionally be recognized; they sometimes appear as masses of hard trap, columnar or otherwise, projecting in knolls or hills above the upper surface of the sheets through which they pierce."

In other cases, the "neck" consists of a great pipe choked up by bombs and blocks of trap, more or less consolidated, bombs which have been shot into the air and have fallen back again. He then refers to one of these near Portrush, and proceeds to state that the rock on which stands the ruined Castle of Dunluce, "is formed of bombs of all sizes up to six feet in diameter, of various kinds of basalt, dolerite, and amygdaloid firmly cemented, and presenting a precipitous face to the sea."

In a note dated September, 1877, Mr. Hull states that subsequent examination, since the above was written, of the rock of Dunluce Castle and the cliffs adjoining, has led him "to suspect that we have here, instead of old volcanic necks, simply pipes, formed by the filtration out of the chalk into which the basaltic masses have fallen and slipped down, thus giving rise to their fragmental appearance."

Further on (page 146) he describes without any sceptical comment, "the remarkable mass of agglomerate made up (as on the southern flanks of Slieve Gullion) of bombs of granite, which have been torn up from the granite mass of the hills below, and blown through the throat of an old crater." Other geologists still adhere firmly to the bomb theory, some ascribing the bombs to subaqueous rather than subaerial ejection.

Immediately under Dunluce Castle is a sea-worn cavern or tunnel, which is about 40 or 50 feet high at its mouth, affording a fine section of this curious conglomerate. The floor of the cavern which slopes upwards from the sea is strewn with a beach of boulders. The resemblance of this beach to those I had recently examined at the foot of the boulder-clay cliffs of Galway Bay (and described in a paper read to the British Association), suggested the explanation of the origin of the rock I am about to offer.

In shape and size they are exactly like the Galway shore boulders, those nearest the sea being the most rounded; higher up the slope, where less exposed to wave action, they are subangular. They differ from the Galway boulders in being chiefly basaltic instead of being mainly composed of carboniferous limestone. Some of these at Dunluce are granitic, and a few, if I am not greatly mistaken, are of carboniferous limestone. I had not at hand the means of positively deciding this.

Neither could I find any unquestionable examples of glacial striation among them, though at the upper part I saw some lines on boulders that were very suggestive of partially obliterated scratches.

On looking at the cavern walls surrounding me the theory so obviously suggested by the boulders on the floor was strikingly confirmed by their structure and general appearance. The imbedded "bombs" are subangular, and of irregular shape and varying composition, and the matrix of the rock is a brick-like material just such as would be formed by the baking of boulder clay; the inference that I was looking upon a bank or deposit of glacier drift that had been baked by volcanic agency was irresistible.

I was unable to see on any part of the extensive section, or among the fragments below, a single specimen of an unequivocal volcanic bomb; no approach to anything like those described by Sir Samuel Baker in his "Nile Tributaries of Abyssinia," the miniature representatives of which, ejected from the Bessemer converter, I have figured and described in *Nature*, vol. 3, pp. 389 and 410, where Sir Samuel Baker's description is quoted.

I have witnessed the fall of masses of lava during a minor eruption of an inner crater of Mount Vesuvius. These as they fell upon the ground around me were flattened out into thin cakes. There was no approach to the formation of subangular masses, like those displayed upon the Dunluce cavern walls.

Some years ago a project for melting the basaltic rock known as "Rowley Rag," and casting it into moulds for architectural purposes was carried out near Oldbury, and I had an opportunity of watching the experiment, which was conducted on a large scale at great expense by Messrs. Chance.

It was found that if the basalt cooled rapidly it became a black obsidian, and to prevent the formation of such brittle material, the castings, and the moulds, which enclosed them, had to be kept at a red-heat for some days, and very gradually cooled.[29]

It is physically impossible that lava ejected under water, in lumps no larger than these boulders, could have the granular structure which they display.

The fundamental idea upon which this bomb theory is based will not bear examination. Such bombs could not have been shot into either air or water and have fallen back again into the volcanic neck at any other time than during an actual eruption; and at such time they could not have remained where they fell, and have become embedded in any such matrix as now contains them. True volcanic bombs and ordinary spattering lumps of lava, are, as we know, flung obliquely out of active craters, and distributed around, while those which are

---

[29] Geologists who may be interested in seeing the results of this experiment, will find on the Edgbaston Vestry Hall, in Enville Road, near the Five Ways, Birmingham, some columns, massive window pieces, doorways, and ornamental steps cast from the fused Rowley Rag and slowly cooled.

ejected perpendicularly into the air and return are re-ejected, and finally pulverized into volcanic dust if this perpendicular ejection and return are continued long enough.

In the course of a rapid drive round the Antrim coast I observed other examples of this peculiar conglomerate, and have reason to believe that it is far more common than is generally supposed. I found it remarkably well displayed at a place almost as largely visited as the Giant's Causeway, and where it nevertheless appears to have been hitherto unnoticed, viz., Carrick-a-Rede, where the public car stops to afford visitors an opportunity of examining or crossing the rope bridge, etc.

Here the whole formation is displayed in a manner that strikingly illustrates my theory.

There is an overlying stream of basalt forming the surface of the isolated rock, and this basalt rests directly upon a base of conglomerate, having exactly the appearance that would result from the slow baking of a mass of boulder clay.

The sea gully that separates the insular rock from the mainland displays a fine section above eighty feet in thickness, and has the advantage of full daylight as compared with Dunluce Cave. That this is no mere neck or pipe is evident from its extent. Its position below the basalt cap refutes the above quoted subsequent explanation, which Mr. Hull and others have recently adopted.

The heterogeneous bomb-like character of the boulders is not so strongly marked as in the Dunluce rock, and this may arise from the closer proximity of the basalt, which, coming here in direct contact, would be likely to heat the clay matrix (itself formed mainly of ice-ground basalt) to incipient fusion, and thereby render it more like the basalt boulders it contains than the other clay that had been less intensely heated on account of greater distance from the lava-flow.

The path leading to the ladder by which the bridge is approached passes over such conglomerate, and further extensions are seen in sections around. I saw sufficient in the course of my hurried visit to indicate the existence of a large area of this particular formation.

At a short distance from Carrick-a-Rede, on the way to Ballycastle, the car passes in sight of considerable deposits of ordinary boulder clay uncovered and unaltered.

The blocks of basalt, etc., embedded in this correspond in general size and shape with the "bombs," excepting that some of the latter have a laminated, or shaly, character near their surfaces.

I regret my inability to do justice to this subject in consequence of the fact that the above explanation of the origin of this curious formation only suggested itself when hurrying homeward after a somewhat protracted visit to Ireland. As I may not have an opportunity of further investigation for some time to come, I offer the hypothesis in this crude form in order that it may be discussed, and either confirmed or refuted by the geologists of the Ordnance Survey, or others who have better opportunities of observation than I can possibly command.

Should this conglomerate prove to be, as I suppose, a drift deposit altered by a subsequent flow of lava, it will supply exceedingly interesting data for the determination of the chronological relations of the glacial epoch to that period of volcanic activity to which the lavas of the N.E. of Ireland are due. Though it will nowise disturb the general conclusion that the great eruptions that overspread the cretaceous rocks of this region, and supplied the boulders of my supposed metamorphosed drift, occurred during the Miocene period, it will show that this volcanic epoch was of vastly greater duration than is usually supposed; or that there must have been two or more volcanic epochs—pre-glacial, as usually understood, and post-glacial, in order to supply the lava overflowing the drift.

This post-glacial extension of the volcanic period has an especial interest in Ireland, as the "Annals of the Four Masters," and other records of ancient Irish history and tradition, abound in accounts of physical changes, many of which correspond remarkably with those of recent occurrence in the neighborhood of active and extinct volcanoes.

In a paper read before the Royal Irish Academy, June 23, 1873, and published in its "Proceedings," Dr. Sigerson has collected some of the best authenticated of these accounts, and compares them with similar phenomena recently observed in Naples, Sicily, South America, Siberia, etc. etc. The "great sobriety of diction, and circumstantial precision of statement," of names, dates, etc., which characterize these accounts render them well worthy of the sort of comparison with strictly scientific data which Dr. Sigerson has made.

As we now know that man existed in Britain during the inter-glacial, if not the pre-glacial period, and as so violent a volcanic disturbance as that which poured out the lavas of Antrim and the Mourne district could scarcely have subsided suddenly, but was probably followed by ages of declining activity, it is not at all surprising that this period of minor activity should have extended into that of tradition and the earliest of historical records.

Chapter 33

TRAVERTINE

The old exclamation about Augustus finding Rome of brick and leaving it of marble, deceives many. Ancient Rome was by no means a marble city, although the quarries of Massa and Carrara are not far distant. The staple-building materials of the Imperial City, even in its palmiest days, were brick and travertine. The brick, however, was very different from the porous cakes of crudely burnt clay of which the modern metropolis of the world is built. I have examined on the spot a great many specimens, and found them all to be of remarkably compact structure, somewhere between the material of modern terra-cotta and that of common flower-pots, and similarly intermediate in color. The Roman builders appear to have had no standard size; the bricks vary even in the same building—the Coliseum for example; all that I have seen are much thinner than our bricks—we should call them tiles.

But the most characteristic material is the travertine. The walls of the Coliseum are made up of a mixture of this and the tiles above-mentioned. The same is the case with most of the other very massive ruins, as the baths, etc. Many of the temples with columns and facing of marble have inner walls built of this mixture, while others are entirely of travertine.

I was greatly surprised at the wondrous imperishability of this remarkable material. In buildings of which the smooth crystalline marble had lost all its sharpness and original surface, this dirty, yellow, spongy-looking limestone remained without the slightest indication of weathering. A most remarkable instance of this is afforded by the temple of Neptune at Paestum, in Calabria. This is the most perfect ruin of a pure classic temple that now remains in existence, and in my opinion is the finest. I prefer it even to the Parthenon.

We have a little sample of it in London. The Doric columns at the entrance of the Euston station are copies of those of its peristyle. The originals are of travertine, the blocks forming them are laid upon each other without mortar or cement, and so truly flattened that in walking round the building and carefully prying, I could find no crevice into which a slip of ordinary writing paper, or the blade of a pen-knife could be inserted. Yet this temple was an antiquarian monument in the days of the Roman emperors.

The rough natural surface of the stone is exposed, and at first sight appears as though weathered, but this appearance is simply due to its natural sponge-like structure. It appears to have been coated with some sort of stucco or smoothing film, which, either by forming a thin layer, or possibly by only filling up the pores of the travertine, gave a smooth surface upon which the coloring was applied. This is now only indistinctly visible here and there, and if I remember rightly, some have disputed its existence.

But this travertine, though so familiar to the Italian, is such a rarity here that some further description of its structure and composition may be demanded. It is a limestone formed by *chemical* precipitation. Most limestones are more or less of organic origin, are agglomerations of shells, corals, etc., but this is formed by the same kind of action as that which produces the stalactites in limestone caverns. It has some resemblance to the incrustation formed on boilers by calcareous water. Although the material of so many ancient edifices, it is, geologically speaking, the youngest of all the hard rocks. Its formation is now in progress at some of the very quarries that supplied Imperial Rome.

On the Campagna, between Rome and Tivoli, is a small circular lake, from which a stream of tepid water, that wells up from below, is continually flowing. Its local name is the "The Lake of Tartarus." The water, like that of Zoedone, or soda-water or champagne, is supersaturated with carbonic acid that was forced into it while under pressure down below. This carbonic acid has dissolved some of the limestones through which the subterranean water passes, and when it comes to the surface, the carbonic acid flies away like that which escapes when we uncork a bottle of soda-water, though less suddenly, and the lime losing its solvent is precipitated, and forms a crust on whatever is covered by the water.

When I visited this lake in the month of February it was surrounded by a *chevaux de frise* of an extraordinary character; thousands of tubes of about half an inch to one inch in diameter outside, with calcareous walls about one eighth of an inch in thickness. These were standing up from two to three feet high, and so close together that we had to break our way through the dense palisade they formed in order to reach the margin of the lake. After some consideration and inquiry, their origin was discovered. They are the encrusted remains of bullrushes that had flourished in the summer and died down since. During the time of their growth the water had risen, and thus they became coated with a crust of compact travertine. This deposition takes place so rapidly that a piece of lace left in the lake for a few hours comes out quite stiff, every thread being coated with limestone. Such specimens, and twigs similarly covered, are sold to tourists or prepared by them if they have time to stop. Sir Humphry Davy drove a stick into the bottom of the lake and left it standing upright in the water from May to the following April, and then had some difficulty in breaking with a sharp pointed hammer the crust formed round the stick. This crust was several inches in thickness. That which I saw round the ex-bullrushes may have all been formed in a few days or weeks. The rivulet that flows from the lake deposits travertine throughout its course, and when it overflows leaves every blade of grass that it covers encrusted with this limestone.

Near to the Lake of Tartarus is the *Solfatara* lake which contains similar calcareous water, but strongly impregnated with sulphureted hydrogen; it consequently deposits a mixture of carbonate and sulphide of calcium, a sort of porous tufa, some of it so porous that it floats like a stony scum, forming what the cicerone call "floating islands." Lyell, in his "Principles of Geology," confounds these lakes, and describes Tartarus under the name of Solfatara.

The travertine used as a building stone is chiefly derived from the quarries of Ponte Lucano, and is the deposit that was formed on the bed of a lake like that of Tartarus. The celebrated cascade of the Anio at Tivoli forms calcareous stalactites, and all the country round has rivulets, caverns, and deposits, where this formation may be seen in progress or completed.

It varies considerably in structure, some specimens are compact and smooth, others have the appearance of a petrified moss, and great varieties may be found among the materials of a single building. It is, however, usually rough and more or less spongy-looking, as above stated, but this structure does not seem to affect its stability, at least, not in the climate of Italy. Whether it would stand long frosts is an open question. The night frosts at and about Rome are rather severe, but usually followed by a warm sunny day; thus there is no great penetration of ice.

Every specimen I have examined shows a remarkable compactness of *molecular* structure in spite of visible porosity. All give out a clear metallic ring when struck, and the intimate surface, if I may so describe the surface of the warm-like structure it sometimes displays, is always clear and smooth as though varnished. To this I attribute its durability. Lest the above description should appear self-contradictory, I will explain a little further. If melted glass were run into threads, and those threads while soft were allowed to agglomerate loosely into a convoluted mass, it would, as regarded in mass, have a porous or spongy-looking structure, but nevertheless its *molecular* structure would be compact and vitreous; there would be mechanical but not molecular, porosity. Travertine is similar.

Have we any travertine in England? This is a practical question of some importance, and one to which I have no hesitation in replying, Yes. There is plenty formed and forming in the neighborhood of Matlock, but that which I have seen on the face of caverns, etc., is not so compact and metal-like as the Italian. This, however, does not prove the entire absence of the useful travertine. Not having any commercial interest in the search, I have only looked at what has come in my way, but have little doubt that there are other kinds besides those I saw. I have also seen travertine in course of formation in Ireland, where I think there is a fine field for exploration in the mountain limestone regions, which have been disturbed by volcanic action of the Miocene period. The travertines of Italy are found in the neighborhood of extinct volcanoes.

The classic associations of this material, its remarkable stability, and the faculty with which it may be worked, render it worthy of more attention than it has yet received from British builders.

Chapter 34

THE ACTION OF FROST IN WATER-PIPES AND ON BUILDING MATERIALS

Popular science has penetrated too deeply now to render necessary any refutation of the old popular fallacy which attributed the bursting of water-pipes to the thaw following a frost; everybody now understands that the thaw merely renders the work of the previous freezing so disastrously evident. Nevertheless, the general subject of the action of freezing water upon our dwellings is not so fully understood by all concerned as it should be. Builders and house-owners should understand it thoroughly, as most of the domestic miseries resulting from severe winters may be greatly mitigated, if not entirely prevented, by scientific adaptation in the course of building construction. Now-a-days tenants know something about this and select accordingly. Thus the market value of a building may be increased by such adaptation.

Solids, liquids, and gases expand as they are heated. This great general law is, however, subject to a few exceptions, the most remarkable of which is that presented by water. Let us suppose a simple experiment. Imagine a thermometer tube with its bulb and stem so filled with water that when the water is heated nearly to its boiling point it will rise to nearly the top of the long stem. Now let us cool it. As the cooling proceeds the water will descend, and this descending will continue until it attains the temperature marked on our ordinary thermometer as 39°, or more strictly 39-2/10; then a strange inversion occurs. As the temperature falls below this, the water rises gradually in the stem until the freezing point is reached.

This expansion amounts to 1/7692 part of the whole bulk of the water, or 100,000 parts become 100,013. So far the amount of expansion is very small, but this is only a foretaste of what is coming. Lower the temperature still further, the water begins to freeze, and at the moment of freezing it expands suddenly to an extent equalling 1/15 of its bulk, i.e., of the bulk of so much water as becomes solidified. The temperature remains at 32° until the whole of the water is frozen.

Fortunately for us, the freezing of water is always a slow process, for if this conversion of every 15 gallons into 16 took place suddenly, all our pipes would rip open with something like explosive violence. But such sudden freezing of any considerable quantity of water is practically impossible, on account of the "latent heat" of liquid water, which amounts to 142½°. All this is given out in the act of freezing. It is this giving out of so much heat that keeps the temperature of freezing water always at 32°, even though the air around may be much colder. No part of the water can fall below 32° without becoming solid, and that portion which solidifies gives out enough heat to raise 142½ times its own quantity from 31° to 32°.

The slowness of thawing is due to the same general fact. An instructive experiment may be made by simply filling a saucepan with snow or broken ice, and placing it over a common fire. The slowness of the thawing will surprise most people who have not previously tried the experiment. It takes about as long to melt this snow as it would to raise an equal weight of water from 32° to 174°. Or, if a pound of water at 174° be mixed with a pound of snow at 32°, the result will be two pounds of water at 32°; 142° will have disappeared without making the snow any warmer, it will all have been used up in doing the work of melting.

The force with which the great expansion due to freezing takes place is practically irresistible. Strong pieces of ordnance have been filled with water, and plugged at muzzle and touch-hole. They have burst in spite of their great thickness and tenacity. Such being the case, it is at first sight a matter of surprise that frozen water-pipes, whether of lead or iron, ever stand at all. They would not stand but for another property of ice, which is but very little understood, viz., its *viscosity*.

This requires some explanation. Though ice is what we call a solid, it is not truly solid. Like other apparent solids it is not perfect rigid, but still retains some degree of the possibility of flowing which is the characteristic of liquids. This has been shown by filling a bombshell with water, leaving the fuse-hole open and freezing it. A shell of ice is first formed on the outside, which of course plugs up the fuse-hole. Then the interior gradually freezes, but the expansion due to this forces the ice out of the fuse-hole as a cylindrical stick, just as putty might be squeezed out, only that the force required to mould and eject the ice is much greater.

I have constructed an apparatus which illustrates this very strikingly. It is an iron syringe with cylindrical interior of about half an inch in diameter, and a terminal orifice of less than 1/20 of an inch in diameter. Its piston of metal is driven down by a screw. Into this syringe I place small fragments of ice, or a cylinder of ice fitted to the syringe, and then screw down the piston. Presently a thin wire of ice is squirted forth like vermicelli when the dough from which it is made is similarly treated, showing that the ice is plastic like the dough, provided it is squeezed with sufficient force.

This viscosity of ice is displayed on a grand scale in glaciers, the ice of which actually flows like a river down the glacier valley, contracting as the valley narrows and spreading out as it widens, just as a river would; but moving only a few inches daily according to the steepness of the slope and the season, slower in winter than in summer.

Upon this, and the slowness of the act of freezing, depends the possibility of water in freezing in iron pipes without bursting them. Even iron yields a little before bursting, but ordinary qualities not sufficiently to bear the expansion of 1/15 of their contents. What happens then? The cylinder of ice contained in the tube elongates as it freezes, provided always the pipe is open at one or both ends. But there is a limit to this, seeing that the friction of such a tight-fitting core, even of slippery ice, is considerable, and if the pipe be too long, the resistance of this friction may exceed the resistance of tenacity of the pipe. I am unable to give any figures for such length; the subject does not appear to have been investigated as it should be, and as it might well be by our wealthy water companies.

We all know that lead pipes frequently succumb, but a little observation shows that they do so only after a struggle. The tenacity of lead is much less than that of iron (about 1/20 of that of ordinary wrought iron), but it yields considerably before breaking. It has, in fact, the property of viscosity similar to that of ice. At Woolwich the lead used for elongated rifle bullets is squirted like the ice in my syringe above described, powerful hydraulic pressure being used.

This yielding saves many pipes. It would save all *new* pipes if the lead were pure and uniform; but as this is not the case, they may burst at a weak place, the yielding being shown by the bulge that commonly appears at the broken part.

From the above it will be easily understood that a pipe which is perfectly cylindrical—other conditions equal—will be less likely to burst than one which is of varying diameter, as the sliding from a larger to a smaller portion of the pipe must be attended with great resistance, or a certain degree of block, beyond what would be due to the mere friction along a pipe of uniform diameter.

Let us now consider the relative merits of lead and iron as material for water-pipes in places where exposure to frost is inevitable. Lead yields more than iron, and so far has an advantage; this, however is but limited. As lead is practically inelastic, every stretch remains, and every stretch diminishes the capacity for further stretching; the lead thus stretched at one frost is less able to stretch again, and has lost some of its original tenacity. Hence the superiority of new leaden pipes. Iron is elastic within certain limits, and thus the iron pipe may yield a little without permanent strain or "distress," and if its power of elastic resistance is not exceeded, it regains its original size without becoming sensibly weaker. Add to this its great tenacity, its nonliability to be indented, or otherwise to vary in diameter, and we have a far superior material.

But this conclusion demands some qualification. There is iron and iron, cast-iron and wrought-iron, and very variable qualities of each of these. I need scarcely add that common brittle cast-iron is quite out of the question for such purposes, though there is a new kind of cast-iron or semi-steel coming forward that may possibly supersede all other kinds; but this opens too wide a subject for discussion in the present paper, the main object of which has been a popular exposition of the general physical laws which must be obeyed by the builder, or engineer, who desires to construct domestic or other buildings that will satisfy the wants of intelligent people.

The mischievous action of freezing water is not confined to the pipes that are constructed to receive or convey it. Wherever water may be, if that water freezes, it must expand in the degree and with the force already described. If it penetrates stone or brick, or mortar or stucco, and freezes therein, one of two things must occur—either the superfluous ice must exude at the surface or to neighboring cavities, or the saturated material must give way, and split or crumble according to the manner and degree of penetration. To understand this, the reader must remember what I stated about the little-understood *viscosity* of ice, as well as its expansion at the moment of freezing.

Bricks are punished, but not so severely as might be anticipated, seeing how porous are some of the common qualities, especially those used in London. They are so amply porous that the water not only finds its way into them, but the pores are big enough and many enough for the ice to demonstrate its viscosity by squeezing out and displaying its crystalline structure in the form of snow-like efflorescence on the surface. This may have been observed by some of my readers during a severe frost. It is commonly confounded with the hoar-frost that whitens the roofs of houses, but which is very rarely deposited on perpendicular wall faces.

The mortar most liable to suffer is that which is porous and pulverulent within, but has been cleverly faced or pointed with a crust of more compact material. This outer film prevents the exuding of the expanding ice crystals, is thrust forth bodily, and retained by ice-cement during the frost, but it falls in scales when this temporary binding material thaws. Mortar that is compact throughout does not suffer to any appreciable extent. This is proved by the condition of the remains of Roman brickwork that still exist in Britain and other parts of Europe. Some of the old shingle walls at Brighton and other parts of the south coast, where the chalk for lime-burning was at the builder's feet, and where his mortar is so thickly laid between the irregular masses of flint, also show the possible duration of good mortar. The jerry builder's mortar, made of the riddlings of burnt clay ballast and dust-hole refuse just flavored with lime, crumbles immediately, because these materials do not combine with the lime as fine siliceous sand gradually does, to form an impermeable glassy silicate.

Stucco is punished by two distinct modes of action. The first is where the surface is porous, and the water permeates accordingly and freezes. This, of course, produces superficial crumbling, which should not occur at all upon good material protected by suitable paint. The other case, very deplorable in many instances, is where the water finds a space between the inner surface of the stucco and the outer surface of the material upon which it is laid. This water, when frozen, of course, expands, and wedges away the stucco bodily, causing it to come down in masses at the thaw. This, however, only occurs after severe frosts, as the ordinary mild frosts of our favored climate seldom endure long enough to penetrate to any notable depth of so bad a conductor as stone or stucco. It is worthy of note that water is a still worse conductor than stone.

Building stones are so various both in chemical composition and mechanical structure that the action of freezing water is necessarily as varied as the nature of the material. The highly siliceous granites (or, rather, porphyries that commonly bear the name of granite) are practically impermeable to water so long as they are free from any chemical decomposition of their feldspathic constituents; but when we come to sandstones and limestones, or intermediate material, very wide differences prevail.

The possible width of this difference is shown in the behavior of the unselected material in its natural home. Certain cliffs and mountains have stood for countless ages almost unchanged by the action of frost; others are breaking up with astonishing rapidity in spite of apparent solidity of structure. The Matterhorn, or Mont Cervin, one of the most gigantic of the giant Alps, 15,200 feet high, is rendered especially dangerous to ambitious climbers by the continual crashing down of fragments that are loosened when the summer sun melts the ice that first separated and then for a while held them in their original places. All the glaciers of the Alps are more or less streaked with "moraines," which are fragments of the mountains that freezing water has detached.

Our stone buildings would suffer proportionally if some selection of material were not made. Generally speaking, this selection is based upon the experience of previous practical trials. Certain quarries are known to have supplied good material of a certain character, and this quarry has, therefore, a reputation which is usually of no small value to its fortunate owner. Other quarries are opened in the neighborhood wherever the rock resembles that of the tested quarry.

Sometimes, however, materials are open for selection that have not been so well tested, and a method of testing which is more expeditious and less expensive than constructing a building and watching the result, is very desirable. The subject of testing building materials in special reference to their resistance of frost was brought before the Academy of Science of Paris by M. Brard some years since.

In his preliminary experiments he used small cubes of the stone to be tested, soaked them in water, and then exposed them to the air in frosty weather, or subjected them to the action of freezing mixtures. Afterwards he found that by availing himself of the expansive force which certain saline solutions exert at the moment of crystallization, he could conveniently imitate the action of freezing without the aid of natural or artificial frost. Epsom salts, nitre, alum, sulphate of iron, Glauber's salts, etc., were tried. The last named, Glauber's salt (or sulphate of soda), which is very cheap, was found to be the best for the purpose.

His method of applying the test is as follows: Cut the specimens into two-inch cubes, with flat sides and sharp edges and corners, mark each specimen with a number, either by ink or scratching, and enter in a book all particulars concerning it. Make a saturated solution of the sulphate of soda in rain or distilled water, by adding the salt until no more will dissolve; perfect saturation being shown by finding, after repeated stirring, that a little of the salt remains at the bottom an hour or two after the solution was made. Heat this solution in a suitable vessel, and when it boils put in the marked specimens one by one, and keep them immersed in the boiling solution for half an hour. Take out the specimens separately and suspend them by threads, each over a separate vessel containing some of the liquid in which they were boiled, but which has been carefully strained to free it from any solid particles. In the course of a day or two, as the cubes dry, they will become covered with an efflorescence of snow-like crystals; wash these away by simply plunging the specimen into the vessel below, and repeat this two or three times daily for four or five days or longer. The most suitable vessel for the purpose is a glass "beaker," sold by vendors of chemical apparatus.

In comparing competing samples, be careful to treat all alike, i.e., boil them together in the same solution, and dip them an equal number of times at equal intervals.

Having done this, the result is now to be examined. If the stone is completely resistant the cube will remain smooth on its surfaces and sharp at its edges and corners, and there will be no particles at the bottom of the vessel. Otherwise, the inability of the stone to resist the test will be shown by the disfigurement of the cube or the small particles wedged off and lying at the bottom of the liquid. Care must be taken not to confound these with crystals of the salt which may also be deposited. These crystals are easily removed by adding a little more water or warming the solution.

For strict comparison the fragments thus separated should be weighed in a delicate balance, such as is used in chemical analysis.

Chapter 35

THE CORROSION OF BUILDING STONES

About fifty years ago two eminent French chemists visited London, and rather "astonished the natives" by a curious feature of their dress. They wore on their hats large patches of colored paper. Coming, as they did, from Paris, many supposed that this was one of the latest Paris fashions, and the dandies of the period narrowly escaped the compulsion to follow it. They probably would have done so had the Frenchmen shown any attempt at decorative shaping of the paper. They neglected this because it was litmus paper, and their object in attaching it to their hats was to test the impurities of the London atmosphere.

Blue litmus paper, as everybody knows now-a-days, turns red when exposed to an acid. The French chemists found that their hat-decorations changed color, and indicated the presence of acid in the air of London; but when they left the metropolis and wandered in the open fields their blue litmus paper retained its original color. By using alkaline paper they contrived to collect enough of the acid to test its composition. They found it to be the acid which is formed by the burning of sulphur, and attributed its existence to the sulphur of our coal. At this time the domestic use of coal was scarcely known in Paris.

Subsequent experiments have proved that they were right; that the air of London contains a very practical quantity of sulphurous and sulphuric acids, which are due to the combustion of that yellow shining material more or less visible in most kinds of coal, and has been occasionally supposed to be gold. It is iron pyrites, a compound of iron and sulphur. When heated the sulphur is separated and burns, producing sulphurous acid, which, exposed to moist air, gradually takes up more oxygen and becomes sulphuric acid, which in concentrated solution is oil of vitriol. In the air it is very much diluted by diffusion, but is still strong enough to do mischief to some kinds of building materials.

In manufacturing towns, such as Birmingham and Sheffield, the quantity of this acid in the air is much greater than in London, and there its mischief is consequently more distinctly visible. The church of St. Philip, which stands nearly in the middle of Birmingham, and is surrounded by an old churchyard, was so corroded by this acid that the stone peeled away on all sides, and its condition was most deplorable. The tombstones were similarly disintegrated on their surfaces, and inscriptions quite obliterated. It became so bad that a few years ago restoration was necessary, and it was newly faced accordingly.

Some of the old tombstones that are preserved may still be seen against the church wall, and their peculiar structure is well worthy of study. They display a lamination or peeling away due to unequal corrosion, certain layers of the material of the stone having been evidently eaten away more rapidly than others. Anybody visiting Birmingham may easily examine these, as St. Philip's churchyard is situated between the two railway stations of New Street and Snow Hill, and is but two minutes' walk from either.

Other stone buildings in the town have suffered, but in very different degrees, and some have quite escaped, proving the necessity of careful selection of material wherever coal fires abound. In Birmingham the action of coal fires is assisted by other sources of acid vapor. The process of "pickling" brass castings, i.e., brightening their surface, by dipping first in common nitric acid ("pickle acky") and then in water, is attended with considerable evolution of acid fumes. Besides this very widespread use of acid, there are several chemical manufactories that throw still more acid into the air immediately surrounding them.

As an example of the action of the atmospheric acids of London upon building stones, I have but to name the Houses of Parliament, which have only been rescued from superficial ruin by the patchwork replacing of certain blocks of stone, and various devices of siliceous and other washings that have been carried out at great cost to the nation. That such an unsuitable material should have been used is disgraceful to all concerned. The ruin commenced before the building was finished. At the time when its erection commenced there were abundant evidences of the ruinous action of London atmosphere on some kinds of stone and the capability of others to resist it, for while many modern buildings are peeling and crumbling, some of the oldest in the midst of the city show scarcely any signs of corrosion.

The Birmingham and Midland Institute was established and in practical operation a few years before the present noble building was erected. I was the first teacher there and conducted the Science classes in the temporary premises in Cannon street. Having observed with some interest the disintegration of St. Philip's Church and other buildings, I was anxious for the safety of the new Institute buildings, and accordingly made some experiments upon the material proposed to be used by the architect. My method of testing was very simple, and as the practical result has verified my anticipations I think it might be adopted by others.

First, I immersed some lumps of the stone in moderately strong solutions of sulphuric and hydrochloric acids successively, and observed whether any visible action occurred after some days. There was none. I then roughly tested the crushing pressure of small samples in their natural state, and subjected similar sized pieces to the same test after they had been immersed in the acids. I found thus that there were no evidences of internal disintegration even after several days' immersion, and therefore inferred that the stone would stand the acid vapors of the Birmingham atmosphere. This has been the case with that portion of the building that was built of the material I tested. As I know nothing of the stone which is used for the extension of the building under the present architect, Mr. Chamberlain, I am unable to make any forecast of its probable durability.

The experiments I made at the time named with this and other building materials justified the conclusion that the worst of all material for exposure to acid atmospheres is a sandstone, the particles of which are held together by limestone, or are otherwise surrounded by or intermingled with limestone; and that the best of *ordinary* material is a pure sandstone quite free from lime. I do not here consider such luxurious material as granite or porphyries.

Compact limestone, such as good homogeneous marble, stands fairly well, although it is slowly corroded. The corrosion, however, in this case, is purely superficial and tolerably uniform. It is a very slow washing away of the surface, without any disintegration such as occurs where a small quantity of limestone acts as binding material to hold together a large quantity of siliceous or sandy material, and where the agglomeration is porous, and the stone is so laid that a downward infiltration of water can take place; for it must be remembered that although the acid originally exists as vapor in the air, it is taken up by the falling rain, and the mischief is directly done to the stone by the acidified water. This, of course, is very weak acid indeed. That which I used for testing the stone was many thousand times stronger, but then I exposed the stone for only a few days instead of many thousand days.

As above stated, my experiments were but rude, but I think it would be quite worth while to construct crushing apparatus capable of registering accurately the pressure used, and to operate with standard solutions of acid upon carefully squared blocks of standard size, and thus to make comparative tests of various samples of stone when competitions for building materials are offered. In the case of the Birmingham and Midland Institute building there was no such competition, the choice was left entirely to the architect, and my examination was unofficially conducted upon the material already chosen with the intent of protesting if it failed. As it stood the test I merely reported the results informally to the architect, the late Sir Edward Barry, no further action being demanded.

Chapter 36

FIRE-CLAY AND ANTHRACITE

For household fire-places, whether open or closed, these may be regarded as the material and the fuel of the future, and should be more generally and better understood than they are.

The merits of fire-clay were fully appreciated and described nearly a hundred years ago by that very remarkable man, Benjamin Thompson, Count of Rumford. Any sound scientific exposition of the relative value of fire-clay and iron as fire-place materials can be little more or less than a repetition of what he struggled to teach at the beginning of the present century.

It is impossible to fairly understand this subject unless we start with a firm grasp of first principles. The business before us is to get as much heat as possible from fuel burning in a certain fashion, and to do this with the smallest possible emission of smoke.

Substances that are hotter than their surroundings communicate their excess of temperature in three different ways; 1st, by *Conduction*; 2d, by *Convection*; 3d, by *Radiation*. All of these are operating in every form of fire-place, but in very different proportions according to certain variations of construction.

To demonstrate the conduction of heat, hold one end of a pin between the finger and thumb, and the other end in the flame of a candle. The experiment will terminate very speedily. Then take a piece of a lucifer match of the same length as the pin, and hold that in the candle. This may become red-hot and flaming without burning the fingers, as the pin did at a much lower temperature. It matters not whether the pin be held upwards, downwards, or sideways, the heat will travel throughout its substance, and this sort of traveling is called "conduction," and the pin a "conductor" of heat. The conducting power of different substances varies greatly, as the above experiment shows. Metals generally are the best conductors, but they differ among themselves; silver is the best of all, copper the next. Calling (for comparison sake) the conductivity of silver 1000, that of copper is 736, gold 532, brass 236, iron 119, marble and other building stones 6 to 12, porcelain 5, ordinary brick earth only 4, and fire-brick earth less than this. Thus we may at once start upon our subject, with the practical fact that iron conducts heat thirty times more readily than does fire-brick.

Convection is different from conduction, inasmuch as it is effected by the movements of the something which has been heated by contact with something else. Water is a very bad conductor of heat, much worse than fire-brick, and yet, as we all know, heat is freely transmitted by it, as when we boil water in a kettle. If, however, we placed the water in a fire-clay kettle, and applied the heat at the top we should have to wait for our tea until to-morrow or the next day. When the heat is applied below, the hot metal of the kettle heats the bottom film of water by *direct contact*; this film expands, and thus, being lighter, rises through the rest of the water, heating other portions by contact as it meets them, and so on throughout. The heat is thus conveyed, and the term "convection" is based on the view that each particle is a carrier of heat as it proceeds. Air conveys heat in the same manner; so may all gases and liquids, but no such convection is possible in solids. The common notion that "heat ascends" is based on the well-known facts of convection. It is the heated gas or liquid that really ascends. No such preference is given to an upward direction, when heat is conducted or radiated.

Radiation is a flinging off of heat in all directions by the heated body. Radiation from solids is mainly superficial, and it depends on the nature of the heated surface. The rougher and the more porous the surface of a given substance the better it radiates. Bright metals are the worst radiators; lampblack the best, and fire-brick nearly equal to it. To show the effect of surface, take three tin canisters of equal size, one bright outside, the second scratched and roughened, the third painted over with a thin coat of lampblack. Fill each with hot water of the same temperature, and leave them equally exposed. Their rates of radiation will then be measurable by their rates of cooling. The black will cool the most rapidly, the rough canister next, and the bright one the slowest.

Radiant heat may be reflected like light from bright surfaces, the reflecting substance itself becoming heated in a proportion which diminishes just as its reflecting powers increase. Good reflectors are bad radiators and bad absorbers of heat, and the power of *absorbing* heat, or becoming superficially hot when exposed to radiant heat, is exactly proportionate to radiating efficiency.

Fire-clay is a good absorber of radiant heat, i.e., it becomes readily heated when near to hot coals or flames, without requiring actual contact with them. It is an equally good radiator.

Let us now apply these facts to fire-clay in fireplaces, beginning with ordinary open grates used for the warming of apartments; first supposing that we have an ordinary old-fashioned grate all made of iron—front, sides, and back, as well as bars, and next that we have another of similar form and position, but all the fire-box and the back and cheeks of the grate made of fire-clay.

It is evident that the fire-clay not in actual contact with the coals, but near to them, will absorb more heat than the iron, and thus become hotter. Even at the same temperature it will radiate much more heat than iron, but being so much hotter this advantage will be proportionately increased. An open fireplace lined throughout with fire-clay thus throws into the room a considerable amount of its own radiation in addition to that thrown out from the coal.

But what becomes of this portion of the heat when the fireplace is all of metal? It is carried up the chimney by convection, for the metal, while it parts with less heat by radiation, gives up more to the air by direct contact. Therefore, if we must burn our coals inside the chimney, we lose less by burning them in a fire-clay box than in a metal box.

Count Rumford demonstrates this, and described the best form of open firegrate that can be placed in an ordinary English hole-in-the-wall fireplace. The first thing to be done, according to his instructions, is to brick up your large square fireplace recess, so that the back of it shall come forward to about 4 inches from the front inside face of the chimney, thus contracting the *throat* of the chimney, just behind the mantel, to this small depth (Rumford's device for sweeping need not be here described). The sides or "covings" of this shallowed recess are now to be sloped inwards so that each one shall horizontally be at an angle of 135 deg. to the plane of this new back, and meet it at a distance of six or more inches apart, according to the size of grate required. The covings will thus spread out at right angles with each other, and leave an annular opening to be lined with fire-brick, and run straight up to the chimney. The fire-bars and grate-bottom to be simply let into this as far forward as possible.

By this simple arrangement we get a fire-grate with a narrow flat back and out-sloping sides; all these three walls are of fire-brick; the back radiates perpendicularly across the room; and the sloping sides radiate outwards, instead of merely across the fire from one to the other, as when they are square to the walls.

At Rumford's time our ordinary fireplaces were square recesses; now we have adopted something like his suggestion in the sloping sides of our register grates, and we bring our fireplaces forward. We have gone backwards in material, by using iron, but this, after all, may be merely due to the ironmongery interest overpowering that of the bricklayers. The preponderance of this interest in the South Kensington Exhibition may account for the fact that Rumford's simple device was not to be seen in action there. It could not pay anybody to exhibit such a thing, as nobody can patent it, and nobody can sell it. I have seen the Rumford arrangement carried out in office fireplaces with remarkable success. To apply it anywhere requires only an intelligent bricklayer, a few bricks, and some iron bars.

Although nobody exhibited this, a very near approach to it was described in an admirable lecture delivered at South Kensington, by Mr. Fletcher, of Warrington. In one respect Mr. Fletcher goes further than Count Rumford in the application of fire-clay. He makes the bottom of the fire-box of a slab of fire-clay instead of ordinary iron fire-bars. This demands a little more trouble and care in lighting the fire, owing to the absence of bottom-draught, but when the fire is well started the advantages of this further encasing in fire-clay are considerable. They depend upon another effect of the superior radiant and absorbent properties of fire-clay that I will now explain.

So far, I have only described the beneficial effect of its radiation on the room to be heated, but it performs a further duty inside the fireplace itself. Being a bad conductor, it does not readily carry away the heat of the burning coal that rests upon it, and being also an excellent absorber, it soon becomes very hot—i.e., superficially hot, or hot where its heat is effective. This action may be seen in a common register stove with fire-clay back and iron sides. When the fire is brisk the back is visibly red-hot, while the sides are still dull. If, after such a fire has burnt itself out, we carefully examine the ashes, there will be found more fine dust in contact with the fire-brick than with the iron—i.e., evidence of more complete combustion there; and one of the advantages justly claimed by Mr. Fletcher is, that with his solid fire-clay bottom there will be no unburnt cinders—nothing left but the incombustible mineral ash of the coal. Economy and abatement of smoke are the necessary concomitants of such complete combustion.

A valuable "wrinkle" was communicated by Mr. Fletcher. The powdered fire-clay that is ordinarily sold is not easily applied on account of its tendency to crumble and peel off the back and sides of the stove after the first heating. In order to overcome this, and obtain a fine compact lining, Mr. Fletcher recommends the mixing of the fireclay powder with a solution of water-glass (silicate of soda) instead of simple water. It acts by forming a small quantity of glassy silicate of alumina, which binds the whole of the clay together by its fusion when heated.

Londoners, and, in fact, Englishmen generally, have hitherto regarded anthracite as a museum mineral and a curiosity, rather than an everyday coal-scuttle commodity. If it is to be the fuel of the future, it is very desirable that we should all know something about its merits and demerits, as well as the possibilities of supply.

Anthracite is a natural coke. From its position in the earth, and its relations to bituminous coal, as well as from its composition, we are justified in regarding it as a coal that was originally bituminous, but which has been altered by heat, acting under great pressure. In the great coal-field of South Wales, to which we must look for our main supply of anthracite, we are able to trace the action of heat in producing a whole series of different classes of coal in a single seam, which at one part is highly bituminous—soft, flaming coal, like the Wallsend, then it becomes harder and less bituminous, then semi-bituminous "steam coal," then less and less flaming, until at last we have the hard, shiny form of purely carbonaceous coal, that may be handled without soiling the fingers, and which burns without flame, like coke or charcoal. This change proceeds as the seam extends from the east towards the west. In some places the coal at the base of a hill may be anthracite, while that on the outcrop above it may be bituminous.

An artificial anthracite may be made by heating coal in a closed vessel of sufficient strength to resist the expansion of the gases that are formed. It differs from coke in being compact, is not porous, and therefore, of course, much denser, a given weight occupying less space.

That we Englishmen should be about the last of all the coal-using peoples to apply anthracite to domestic purposes is a very curious fact, but so it is. In America it is the ordinary fuel, and this is the case in all other countries where it is obtainable at the price of bituminous coal. Our perversity in this respect shows out the more strikingly when we go a little further into the subject by comparing the two classes of coal in reference to our methods of using them, and when we consider the fact that our South Wales anthracite is far superior to the American.

Our open fires only do their small fraction of useful work by radiation. Their convection is all up the chimney. Such being the case, and we being theoretically regarded as rational beings, it might be supposed that for our national and especially radiating fireplaces we should have selected a coal of especial radiating efficiency, but, instead of this, we do the opposite. The flaming coal is just that which flings the most heat up the chimney, and the least into the room, and, as though we were all struggling to destroy as speedily as possible the supposed physical basis of our prosperity, we select that coal which in our particular fireplaces burns the most wastefully. If we had closed iron stoves with long stove-pipes in the room, giving to the air the heat they had obtained by the convective action of the flame and smoke, there might be some reason for using the flaming coal, as the flame would thereby do useful work, but, as it is, we stubbornly persist in using only the radiated heat, and at the same time select just the coal which supplies the smallest quantity of what we require.

No scientific dissertation is necessary to prove the superior radiating power of an anthracite fire to anybody who has ever stood in the front of one. This is most strikingly demonstrated by those grates that stand well forward, and are kept automatically filled with the radiant-carbon.

Let us now see *why* anthracite is a better radiator than bituminous coal. This is due to its chemical composition. Of all the substances that we have upon the earth carbon in its ordinary black form is the best radiator. Anthracite contains from 90 to 94 per cent of pure carbon, bituminous coal from 70 to 85, and much of this being combined with hydrogen burns away as flame. On a rough average we may say that the fixed or solid carbon capable of burning with a smokeless flameless glow, amounts to 65 per cent in ordinary British bituminous coal, against an average of 92 per cent in British anthracite. The advantages of anthracite as a fuel for open radiating grates are nearly in the proportion of these figures. Besides this it contains about half the quantity of ash. Thus we see that from a purely selfish point of view, and quite irrespective of our duty to our fellow-citizens as regards polluting the atmosphere, anthracite is preferable to ordinary coal on economical grounds, supposing we can obtain it at the same price as bituminous coal, which is now the case.

Another great advantage of anthracite is its cleanliness, It may be picked up in the fingers without soiling them, and it is similarly cleanly throughout the house. It produces no "blacks," no grimy dust, and if it were generally in use throughout London one half of the house-cleaning would be saved. White curtains, blinds, etc., might hang quite four times as long, and then come down not half so dirty as now. The saving in soap alone, without counting labor, would at once return a handsome percentage on the capital outlay required for reconstructing all our fireplaces.

Let us now look on the other side, and ask what are the disadvantages of anthracite, and why is it not at once adopted by everybody? There is really only one disadvantage, viz., the greater difficulty of starting an anthracite fire. Practically this is considerable, seeing that laziness is universal and ever ready to find excuses when an innovation is proposed that stands in its way. To light an anthracite fire in an ordinary fireplace the bellows are required unless a specially suitable draught or fire-lighter is used. Some recommend that an admixture of bituminous coal should be used to start it, but this is a feeble device calculated to lead to total failure, seeing that the sole originator and sustainer of our ordinary use of bituminous coal is domestic ignorance and indolence, and if both kinds of coal are kept in a house a common English servant will stubbornly use the easy-lighting kind, and solemnly assert that the other cannot be used at all. The only way to deal with this obstacle, the human impediment, is to say, "This you must use, or go." This is strictly just, as a simple enforcement of duty.

At the same time some help should be supplied in the way of artificial modes of creating a draught in starting an anthracite fire. This may be done by temporarily closing the front of the fire by a "blower," or better still by selecting one of the grates specially devised for burning anthracite, of which so many now are made. Another and rather important matter is to obtain the anthracite in suitable condition. It is a very hard coal, too hard to be broken by the means usually at hand in ordinary houses. For domestic purposes it should always be delivered broken up of suitable size, from that of an egg to a cocoa-nut. For furnaces, of course, large lumps are preferable.

Then, again, anthracite must not be stirred and poked about; once fairly started it burns steadily and brightly, demanding only a steady feeding. The best of the special grates are more or less automatic in the matter of feeding, and thus the trouble of lighting is fully compensated by the absence of any further trouble.

As regards the supply. This for London and the greater part of England will doubtless be derived from the great coal-field of South Wales. The total quantity of available coal in this region after deducting the waste in getting, was estimated by the Government Commissioners at 32,456 millions of tons. It is very difficult or impossible to correctly estimate the proportion of anthracite in this, but supposing it to be one tenth of true anthracite it gives us 3245 millions of tons, or about enough for the domestic supply of the whole country during 100 years, assuming that it shall be used less wastefully than we are now using bituminous coal, which would certainly be the case. But, including the imperfect anthracite, the quantity must be far larger than this, and we have to add the other sources of anthracite.

We need not, therefore, have any present fear of insufficient supply; probably before the 100 years are ended we shall find other sources of anthracite, or even have become sufficiently civilized to abolish altogether our present dirty devices, and to adopt rational methods of warming and ventilating our houses. When we do this any sort of coal may be used.

Chapter 37

COUNT RUMFORD'S COOKING-STOVES

In the preceding chapter I described Count Rumford's modification of the English open firegrate which eighty years ago was offered to the British nation without any patent or other restrictions. Its non-adoption I believe to be mainly due to this—it was nobody's monopoly, nobody's business to advertise it, and, therefore, nobody took any further notice of it; especially as it cannot be made and sold as a separate portable article.

An ironmonger or stove-maker who should go to the expense of exhibiting Rumford's simple structure of fire-bricks and a few bars, described in the last chapter, would be superseding himself by teaching his customers how they may advantageously do without him.

The same remarks apply to his stoves for cooking purposes. They are not iron boxes like our modern kitcheners, but are brick structures, matters of masonry in all but certain adjuncts, such as bars, fire-doors, covers, oven-boxes, etc., which are very simple and inexpensive. Even some of Rumford's kitchen utensils, such as the steamers, were cheaply covered with wood, because it is a bad conductor, and therefore wastes less heat than an iron saucepan lid.

Rumford was no mere theorist, although he contributed largely to pure science. His greatest scientific discoveries were made in the course of his persevering efforts to solve practical problems. I must not be tempted from my immediate subject by citing any examples of these, but may tell a fragment of the story of his work so far as it bears upon the subject of cooking-ranges.

He began life as a poor schoolmaster in New Hampshire, when it was a British colony. He next became a soldier; then a diplomatist; then in strange adventurous fashion he traveled on the Continent of Europe, entered the Bavarian service and began his searching reform of the Bavarian army by improving the feeding and the clothing of the men. He became a practical working cook in order that they should be supplied with good, nutritious, and cheap food.

But this was not all. He found Munich in a most deplorable condition as regards mendicity; and took in hand the gigantic task of feeding, clothing, and employing the overwhelming horde of paupers, doing this so effectually that he made his "House of Industry" a true workhouse; it paid all its own expenses, and at the end of six years left a net profit of 100,000 florins.

I mention these facts in conformation of what I said above concerning his practical character. Economical cookery was at the root of his success in this maintenance of a workhouse without any poor-rates.

After doing all this he came to England, visited many of our public institutions, reconstructed their fireplaces, and then cooked dinners in presence of distinguished witnesses, in order to show how little need be expended on fuel, when it is properly used.

At the Foundling Institution in London he roasted 112 lbs. of beef with 22 lbs. of coal, or at a cost of less than threepence. The following copy of certificate, signed by the Councillor of War, etc., shows what he did at Munich: "We whose names are underwritten certify that we have been present frequently when experiments have been made to determine the expense of fuel in cooking for the poor in the public kitchen of the military workhouse at Munich, and that when the ordinary dinner has been prepared for 1000 persons, the expense for fuel has not amounted to quite 12 kreutzers." Twelve kreutzers is about 4½d. of our money. Thus only 1-50th of a farthing was expended on cooking each person's dinner, although the peas which formed the substantial part of the soup required five hours, boiling. The whole average daily fuel expenses of the kitchen of the establishment amounted to 1-20th of a farthing for each person, using wood, which is much dearer than coal. At this rate, *one ton of wood should do the cooking for ten persons during two years and six days, or one ton of coal would supply the kitchen of such a family three and a half years.*

The following is an abstract of the general principles which he expounds for the guidance of all concerned in the construction of cooking stoves.

1) All cooking fires should be enclosed.
2) Air only to be admitted from below and under complete control. All air beyond what is required for the supply of oxygen "is a thief."
3) All fireplaces to be surrounded by non-conductors, *brickwork, not iron.*
4) The residual heat from the fireplace to be utilized by long journeys in returning flues, and by *doing the hottest work first.*
5) Different fires should be used for different work.

The first of these requirements encounters one of our dogged insular prejudices. The slaves to these firmly believe that meat can only be roasted by hanging it up to dry in front of an open fire; their savage ancestors having held their meat on a skewer or spit over or before an open fire, modern science must not dare to demonstrate the wasteful folly of the holy sacrifice. Their grandmothers having sent joints to a bakehouse, where other people did the same, and having found that by thus cooking beef, mutton, pork, geese, etc., some fresh, and some stale, in the same oven, the flavors became somewhat mixed, and all influenced by sage and onions, these people persist in believing that meat cannot be roasted in any kind of closed chamber.

Rumford proved the contrary, and everybody who has fairly tried the experiment knows that a properly ventilated and properly heated roasting oven produces an incomparably better result than the old desiccating process.

Rumford's roaster was a very remarkable contrivance, that seems to have been forgotten. It probably demands more intelligence in using it than is obtainable in a present-day kitchen. When the School Boards have supplied a better generation of domestic servants we may be able to restore its use.

It is a cylindrical oven with a double door to prevent loss of heat. In this the meat rests on a grating over a specially constructed gravy and water dish. Under the oven are two "blowpipes," i.e., stout tubes standing just above the fire so as to be made red hot, and opening into the oven at the back, and above the fireplace in front, where there is a plug to be closed or open as required. Over the front part of the top of the oven is another pipe for carrying away the vapor. It is thus used: The meat is first cooked in an atmosphere of steam formed by the boiling of water placed in the bottom of the double dish, over which the meat rests. When by this means the meat has been raised throughout its whole thickness to the temperature at which its albumen coagulates, the plugs are removed from the blow-pipes, and *then* the special action of roasting commences by the action of a current of superheated air which enters below and at the back of the oven, travels along and finds exit above and in front of the steam-pipe before named.

The result is a practical attainment of theoretical perfection. Instead of the joint being dried and corticated outside, made tough, leathery, and flavorless to about an inch of depth, then fairly cooked an inch further, and finally left raw, disgusting, and bloody in the middle, as it is in the orthodox roasting by British cooks, the whole is uniformly cooked throughout without the soddening action of mere boiling or steaming, as the excess of moisture is removed by the final current of hot dry air thrown in by the blow-pipes, which at the same time give the whole surface an uniform browning that can be regulated at will without burning any portion or wasting the external fat.

Rumford's second rule, that air be admitted only from below, and be limited to the requirements, is so simple that no comment upon it is needed. Although we have done so little in the improvement of domestic fireplaces, great progress has been made in engine furnaces, blast furnaces, and all other fireplaces for engineering and manufacturing purposes. Every furnace engineer now fully appreciates Rumford's assertion that excess of cold air is a thief.

The third rule is one which, as I have already stated, stands seriously in the way of any commercial "pushing" of Rumford's kitchen ranges. Those which he figures and describes are all of them masonic structures, not ironmongery; the builder must erect them, they cannot be bought ready-made; but, now that public attention is roused, I believe that any builder who will study Rumford's plans and drawings, which are very practically made, may do good service to himself and his customers by fitting up a few houses with true Rumford kitcheners, and offering to reconstruct existing kitchen ranges, especially in large houses.

The fourth rule is one that is sorely violated in the majority of kitcheners, and without any good reason. The heat from the fire of any kitchener, whether it be of brick or iron, should first do the work demanding the highest temperature, viz., roasting and baking, then proceed to the boiler or boilers, and after this be used for supplying the bed-rooms and bathroom, and the housemaid, etc., with hot water for general use, as Rumford did in his house at Brompton Row, where his chimney terminated in metal pipes that passed through a water-tank at the top of the house.

Linen-closets may also be warmed by this residual heat.

The fifth rule is also violated to an extent that renders the words uttered by Rumford nearly a century ago as applicable now as then. He said, "Nothing is so ill-judged as most of those attempts that are frequently made by ignorant projectors *to force the same fire to perform different services at the same time.*"

Note the last words, "same time." In the uses above mentioned the heat does different work successively, which is quite different from the common practice of having flues to turn the flame of one fire in opposite directions, to split its heat and make one fireplace appear to do the work of two.

Every householder knows that the kitchen fire, whether it be an old-fashioned open fireplace, or a modern kitchener of any improved construction, is a very costly affair. He knows that its wasteful work produces the chief item of his coal bill, but somehow or other he is helpless under its infliction. If he has given any special attention to the subject he has probably tried three or four different kinds without finding any notable relief. Why is this? I venture to make a reply that will cover 90 per cent, or probably 99 per cent of these cases, viz., that he has never considered the main source of waste, which Rumford so clearly defines as above, and which was eliminated in all the kitchens that he erected.

Let us suppose the case of a household of ten persons, but which in the ordinary course of English hospitality *sometimes* entertains twice that number. What do we find in the kitchen arrangements? Simply that there is one fireplace suited for the maximum requirements, i.e., sufficient for twenty, even though that number may not be entertained more than half a dozen times in the course of a year. To cook a few rashers of bacon, boil a few eggs, and boil a kettle of water for breakfast, a fire sufficient to cook for a dinner party of twenty is at work. This is kept on all day long, because it is just possible that the master of the house may require a glass of grog at bedtime. There may be dampers and other devices for regulating this fire, but such regulation, even if applied, does very little so long as the capacity of the grate remains, and as a matter of ordinary fact the dampers and other regulating devices are neglected altogether; the kitchen fire is blazing and roaring to waste from 6 or 7 A.M. to about midnight, in order to do about three hours and a half work, i.e., the dinner for ten, and a nominal trifle for the other meals.

In Rumford's kitchens, such as those he built for the Baron de Lerchenfeld and for the House of Industry at Munich, the kitchener is a solid block of masonry of work-bench height at top, and with a deep bay in the middle, wherein the cook stands surrounded by his boilers, steamers, roasters, ovens, etc., all within easy reach, each one supplied by its own separate fire of very small dimensions, and carefully closed with non-conducting doors. Each fire is lighted when required, charged with only the quantity of fuel necessary for the work to be done, and then extinguished or allowed to die out.

It is true that Rumford used wood, which is more easily managed in this way than coal. If we worked as he did, we might use wood likewise, and in spite of its very much higher price do our cooking at half its present cost. This would effect not merely "smoke abatement" but "smoke extinction" so far as cooking is concerned. But the lighting of fires is no longer a troublesome and costly process as in the days of halfpenny bundles of firewood. To say nothing of the improved fire-lighters, we have gas everywhere, and nothing is easier than to fix or place a suitable Bunsen or solid flame burner under each of the fireplaces (an iron gaspipe, perforated *below* to avoid clogging, will do), and in two or three minutes the coals are in full blaze; then the gas may be turned off. The writer has used such an arrangement in his study for some years past, and starts his fire in full blaze in three minutes quite independent of all female interference.

I have no doubt that ultimately gas will altogether supersede coal for cooking; but this and all other scientific improvements in domestic comfort and economy must be impossible with the present generation of uneducated domestics, whose brains (with few exceptions) have become torpid and wooden from lack of systematic exercise during their period of growth.

Chapter 38

THE "CONSUMPTION OF SMOKE"

A great deal has been spoken and written on this subject, but practically nothing has been *done*. At one time I shared the general belief in its possibility, and accordingly examined a multitude of devices for smoke-consuming, and tried several of the most promising, chiefly in furnaces for metallurgical work, for steam boilers and stills. None of them proved satisfactory, and I was driven to the conclusion that smoke-consumption is a delusion, and further, that *economical* consumption of smoke is practically impossible. When smoke is once formed, the cost of burning it far exceeds the value of the heat that is produced by the combustion of its very flimsy flocculi of carbon. It is a fiend that once raised cannot be exorcised, a Frankenstein that haunts its maker, and will not be appeased.

To describe in detail the many ingenious devices that have been proposed and expensively patented and advertised for this object, would carry me far beyond the intended limits of this paper. I must not even attempt this for a selected few, as even among them there is none that can be pronounced satisfactory.

The common idea is that if the smoke be carried back to the fire that produced it, and made to pass through it again, a recombustion or consumption of the smoke will take place. This is a mistake, as a little reflection will show. First, let us ask why did this particular fire produce such smoke? Everybody now-a-days can answer this question, as we all know that smoke is a result of imperfect combustion, and, knowing this, it can easily be understood that to return the carbonic acid and excess of carbon to the already suffocated fire can only add smother to smotheration, and make the smoky fire more smoky still.

There is, however, one case in which a fire *appears* to thus consume its own smoke, but the appearance is delusive. I refer to fires lighted from above. These, if properly managed, are practically smokeless, and it is commonly supposed that smoke passes from the raw coal below through the burning coal above, and is thereby consumed. The fact is, however, that no such smoke is formed. That which under these conditions comes from the coal beneath, when gradually heated by the fire above, is combustible *gas*, and this gas is burned as it passes through the fire. In this case the formation or non-formation of smoke depends mainly on how this gas is burned, whether completely or incompletely. If the air supplied for its combustion is insufficient, smoke will be formed as it is when we turn up an Argand gas-flame so high that the gas is too great in proportion to the quantity of air that can enter the glass chimney.

Herein lies the fundamental principle. We may *prevent* smoke, though we cannot *cure* it, and this prevention depends upon how we supply air to the gas which the coal gives off when heated, and upon the condition of this gas when we bring it in contact with the air by which its combustion is to be effected. We must always remember that coal when its temperature is sufficiently heated, whether in a gas retort or fireplace, gives off a series of combustible hydrocarbon gases and vapors, and all we have to do in order to obtain smokeless fires is to secure the complete combustion of these.

Now we know that to burn a given quantity of gas we must supply it with a sufficient quantity of oxygen, i.e., of the active principle of the air; but this is not all: we all know well enough that if cold coal-gas and cold air be brought together in any proportion whatever no combustion occurs. A certain amount of heat is necessary to start the chemical combination of oxygen with hydrogen and carbon, which combination is the combustion, or burning.

Therefore, when the coal gas and the air are brought together one or the other, or both, must be heated up to a certain point in order that the combustion be complete. If cold there is no combustion; if insufficiently heated, there is imperfect combustion, however well the supplies may be regulated.

A very simple experiment that anybody may make illustrates this. When an ordinary open fire is burning brightly and clearly without flame, throw a few small pieces of raw coal into the midst of the glowing coals. They will flame fiercely, but without smoking. Then throw a heap of coal or one large lump on a similar fire. Now you will have dense volumes of smoke, and little or no flame, simply because the cooling action of the large bulk of coal in the course of distillation brings the temperature of its gases below that required for their complete combustion.

This simple experiment supplies a most important practical lesson, as well as a philosophical example. The best of all smoke-abatement machines is an intelligent and conscientious stoker, and every contrivance for smoke abatement must, in order to be efficient, either be fed by such a stoker or provided with some automatic arrangement by which the apparatus itself does the work of such a stoker by supplying the fresh fuel just when and where it is wanted.

Cornish experience is very instructive in this respect. The engines that pump the water from the mines do a definitely measurable amount of work, and are made to register this. The stoker is a skilled workman, and prizes are given to those who obtain the largest amount of "duty" from given engines per ton of coal consumed. Instead of pitching his coal in anyhow, cramming his fire-hole, and then sitting down to sleep or smoke in company with his chimney, the Cornish, or other good fireman, feeds little and often, and deftly sprinkles the contents of his shovel just where the fire is the brightest and the hottest, and the bars are the least thickly covered. The result is remarkable. A colliery proprietor of South Staffordshire was visiting Cornwall, and went with a friend to see his works. On approaching the engine-house and seeing a whitewashed shaft with no smoke issuing from its mouth, he expressed his disappointment at finding that the engine was not at work. To all who have been accustomed to the "Black Country," where coal is so shamefully wasted because it is cheap, the tall clean whitewashed shafts of Cornwall, all so smokeless, present quite an astonishing appearance.

This is not a result of "smoke-consuming" apparatus, but mainly of careful firing. It was in the first place promoted by the high price of coal due to the cost of carriage before the Cornish railways were constructed, and it brought about a curious result. Horse-power for horse-power the cost of fuel for working Cornish pumping engines has been brought below that of pumping engines in the places where the price of coal per ton was less than one-half. Another coal famine that should raise the price of coal in London to 60*s.* per ton, and keep it there for two or three years, would effect more smoke abatement than we can hope to result from the present and many future South Kensington efforts. I need scarcely dwell upon the necessity for a due supply of air. This is well understood by everybody. An over supply of air does mischief, by carrying away wastefully a proportionate quantity of heat. The waste due to this is sometimes very serious.

After reviewing all that has been done, the conclusion that London cannot become a clean, smokeless, and beautiful city, so long as we are dependent upon open fire-grates of anything like ordinary construction, and fed with bituminous coal, is inevitable. The general use of anthracite would effect the desired change, but there is no hope of its becoming general without legislative compulsion, and Englishmen will not submit to this.

One of the most hopeful schemes is that which was propounded a short time since by Mr. Scott Moncrieff. Instead of receiving our coal in its crude state he proposes that we should have its smoke-producing constituents removed before it is delivered to us; that it should be made into a sort of artificial semi-anthracite at the gas-works by a process of half distillation, which would take away not *all* the flaming gas as at present, but that portion which is by far the richest to the gas-maker and the most unmanageable in common fires. We should thus have a material which, instead of being so difficult to light as coke and anthracite, would light more easily than crude coal, and at the same time our gas would have far greater illuminating power, as it would all be drawn off during the early period of distillation, when it is at its richest. From a given quality of coal the difference would be as twenty-four candles to sixteen. The ammonia which we now throw into the air, the naphtha and coal-tar products, which we waste, are so valuable that they would pay all the expenses at the gas-works and leave a handsome profit. We should thus get gas so much better that two burners would do the work now obtained from three. We should get all we require for lighting purposes and plenty more for heating; the intermediate profits of the coal merchant would be abolished, and our solid fuel of far better quality could be supplied twenty or thirty per cent cheaper than at present, provided always that the gas monopoly were abolished, "a consummation most devoutly to be wished for."

Mr. Moncrieff (who brought forward his scheme without any company-mongering, or claims for patent rights) estimates the saving to London at £2,125,000 per annum, over and above the far greater saving that would result from the abolition of smoke.

In connection with this scheme I may mention a fact that has not been hitherto noted, viz., that we have perforce and unconsciously done a little in this direction already. Formerly London was supplied almost exclusively with "Wallsend" and other sea-borne coals of a highly bituminous composition—soft coals that fused in the grate and caked together. Partly owing to exhaustion of the seams, and partly to the competition of railway transit, we now obtain a large proportion of hard coal from the Midlands. This is less smoky and less sooty, and hence the Metropolitan smoke nuisance has not increased quite as greatly as the population.

But I will now conclude by repeating that whatever scheme be chosen, "smoke abatement" is to be achieved, *not by smoke-consumption, but by smoke-prevention.*

Chapter 39

THE AIR OF STOVE-HEATED ROOMS

Whatever opinions may be formed of the merits of the exhibits at South Kensington, one result is unquestionable—the exhibition itself has done much in directing public attention to the very important subject of economizing fuel and the diminution of smoke. We sorely need some lessons. Our national progress in this direction has been simply contemptible, so far as domestic fireplaces are concerned.

To prove this we need only turn back to the essays of Benjamin Thompson, Count of Rumford, published in London just eighty years ago, and find therein nearly all that the Smoke Abatement Exhibition *ought* to teach us, both in theory and practice—lessons which all our progress since 1802, plus the best exhibits at South Kensington, we have yet to learn.

This small progress in domestic heating is the more remarkable when contrasted with the great strides we have made in the construction and working of engineering and metallurgical furnaces, the most important of which is displayed in the Siemens regenerative furnace. A climax to this contrast is afforded by a speech made by Dr. Siemens himself, in which he defends our domestic barbarisms with all the conservative inconvincibility of a born and bred Englishman, in spite of his German nationality.

The speech to which I refer is reported in the "Journal of the Society of Arts," December 9, 1881, and contains some curious fallacies, probably due to its extemporaneous character; but as they have been quoted and adopted not only in political and literary journals, but also by a magazine of such high scientific standing as *Nature* (see editorial article January 5, 1882, p. 219), they are likely to mislead many.

Having already, in my "History of Modern Invention, etc.," and in other places, expressed my great respect for Dr. Siemens and his benefactions to British industry, the spirit in which the following plain-spoken criticism is made will not, I hope, be misunderstood either by the readers of "Knowledge" or by Dr. Siemens himself.

I may further add that I am animated by a deadly hatred of our barbarous practice of wasting precious coal by burning it in iron fire-baskets half buried in holes within brick walls, and under shafts that carry 80 or 90 per cent of its heat to the clouds; that pollute the atmosphere of our towns, and make all their architecture hideous; that render scientific and efficient ventilation of our houses impossible; that promote rheumatism, neuralgia, chilblains, pulmonary diseases, bronchitis, and all the other "ills that flesh is heir to" when roasted on one side and cold-blasted on the other; that I am so rabid on this subject, that if Dr. Siemens, Sir F. Bramwell, and all others who defend this English abomination, were giant windmills in full rotation, I would emulate the valor of my chivalric predecessor, whatever might be the personal consequences.

Dr. Siemens stated that the open fireplace "communicates absolutely no heat to the air of the room, because air, being a perfectly transparent medium, the rays of heat pass clean through it."

Here is an initial mistake. It is true that air which has been artificially deprived of *all* its aqueous vapor is thus completely permeable by heat rays, but such is far from being the case with the water it contains. This absorbs a notable amount even of bright solar rays, and a far greater proportion of the heat rays from a comparatively obscure source, such as the red-hot coals and flame of a common fire. Tyndall has proved that 8 to 10 per cent of all the heat radiating from such a source as a common fire is absorbed in passing through only 5 feet of air in its ordinary condition, the variation depending upon its degree of saturation with aqueous vapor.

Starting with the erroneous assumption that the rays of heat pass "clean through" the air of the room, Dr. Siemens went on to say that the open fireplace "gives heat only by heating the walls, ceiling, and furniture, and here is the great advantage of the open fire;" and, further, that "if the air in the room were hotter than the walls, condensation would take place on them, and mildew and fermentation of various kinds would be engendered; whereas, if the air were cooler than the walls, the latter must be absolutely dry."

Upon these assumptions, Dr. Siemens condemns steam-pipes and stoves, hot-air pipes, and all other methods of directly heating the *air* of apartments, and thereby making it warmer than were the walls, the ceiling, and furniture when the process of warming commenced. It is quite true that stoves, stove-pipes, hot-air pipes, steam-pipes, etc., do this; they raise the temperature of the air directly by *convection*, i.e., by warming the film of air in contact with their surfaces, which film, thus heated and expanded, rises towards the ceiling, and, on its way, warms the air around it, and then is followed by other similarly-heated ascending films. When we make a hole in the wall, and burn our coals within such cavity, this convection proceeds up the chimney in company with the smoke.

But is Dr. Siemens right in saying that the air of a room, raised by convection above its original temperature, and above that of the walls, deposits any of its moisture on these walls? I have no hesitation in saying very positively that he is clearly and demonstrably wrong; that no such condensation can possibly take place under the circumstances.

Suppose, for illustration sake, that we start with a room of which the air and walls are at the freezing point, 32° F., before artificial heating (any other temperature will do), and, to give Dr. Siemens every advantage, we will further suppose that the air is fully saturated with aqueous vapor, i.e., just in the condition at which some of its water might be condensed. Such condensation, however, can only take place by cooling the air *below* 32°, and unless the walls or ceiling or furniture are capable of doing this they cannot receive any moisture due to such condensation, or, in other words, they must fall below 32° in order to obtain it by cooling the film in contact with them. Of course Dr. Siemens will not assert that the stoves or steam-pipes (enclosing the steam, of course), or the hot-air or hot-water pipes, will lower the *absolute* temperature of the walls by heating the air in the room.

But if the air is heated more rapidly than are the walls, etc., the *relative* temperature of these will be lower. Will condensation of moisture *then* follow, as Dr. Siemens affirms? Let us suppose that the air of the room is raised from 30° to 50° *by convection purely*; reference to tables based on the researches of Regnault, shows that at 32° the quantity of vapor required to saturate the air is sufficient to support a column of 0·182 inch of mercury, while at 50° it amounts to 0·361, or nearly double. Thus the air, instead of being in a condition of giving away its moisture to the walls, has become thirsty, or in a condition to *take moisture away from them* if they are at all damp. This is the case whether the walls remain at 32° or are raised to any higher temperature short of that of the air.

Thus the action of close stoves and of hot surfaces or pipes of any kind is exactly the opposite of that attributed to them by Dr. Siemens. They dry the air, they dry the walls, they dry the ceiling, they dry the furniture and everything else in the house.

In *our* climate, especially in the infamous jerry-built houses of suburban London, this is a great advantage. Dr. Siemens states his American experience, and denounces such heating by convection because the close stoves *there* made him uncomfortable. This was due to the fact that the winter atmosphere of the United States is very dry, even when at zero. But air, when raised from 0° to 60°, acquires about twelve times its original capacity for water. The air thus simply heated is desiccated, and it desiccates everything in contact with it, especially the human body. The lank and shriveled aspect of the typical Yankee is, I believe, due to this. He is a desiccated Englishman, and we should all grow like him if our climate were as dry as his.[30] The great fires that devastate the cities of the United States appear to me to be due to this general desiccation of all building materials, rendering them readily inflammable and the flames difficult of extinction.

When an undesiccated Englishman, or a German endowed with a wholesome John Bull rotundity, is exposed to this superdried air, he is subjected to an amount of bodily evaporation that must be perceptible and unpleasant. The disagreeable sensation experienced by Dr. Siemens in the stove-heated railway cars, etc., were probably due to this.

An English house, enveloped in a foggy atmosphere, and encased in damp surroundings, especially requires stove-heating, and the most inveterate worshipers of our national domestic fetish, the open grate, invariably prefer a stove or hot-pipe-heated room, when they are unconscious of the source of heat, and their prejudice hoodwinked. I have observed this continually, and have often been amused at the inconsistency thus displayed. For example, one evening I had a warm contest with a lady, who repeated the usual praises of a cheerful blaze, etc., etc. On calling afterwards, on a bitter snowy morning, I found her and her daughters sitting at work in the billiard-room, and asked them why. "Because it is so warm and comfortable." This room was heated by an eight-inch steam-pipe, running around and under the table, to prevent the undue cooling of the indiarubber cushions, and thus the room was warmed from the middle, and equally and moderately throughout. The large reception-room, with blazing fire, was scorching on one side, and freezing on the other, at that time in the morning.

The permeability of ill-constructed iron stoves to poisonous carbonic oxide, which riddles through red-hot iron, is a real evil, but easily obviated by proper lining, The frizzling of particles of organic matter, of which we hear so much, is—if it really does occur—highly advantageous, seeing that it must destroy organic poison-germs.

Under some conditions, the warm air of a room *does* deposit moisture on its cooler walls. This happens in churches, concert-rooms, etc., when they are but occasionally used in winter time, and mainly warmed by animal heat, by congregational emanations of breath-vapor, and

---

[30] In each of my three visits to America I lost about thirty pounds in weight, which I recovered within a few months of my return to the "home country" (of English-speaking nations).—Richard A. Proctor.

perspiration—i.e., with warm air supersaturated with vapor. Also, when we have a sudden change from dry, frosty weather to warm and humid. Then our walls may be streaming with condensed water. Such cases were probably in the mind of Dr. Siemens when he spoke; but they are quite different from stove-heating or pipe-heating, which increase the vapor capacity of the heated air, without supplying the demand it creates.

Chapter 40

VENTILATION BY OPEN FIREPLACES

The most stubborn of all errors are those which have been acquired by a sort of inheritance, which have passed dogmatically from father to son, or, still worse, from mother to daughter. They may become superstitions without any theological character. The idea that the weather changes with the moon, that wind "keeps off the rain," are physical superstitions in all cases where they are blindly accepted and promulgated without any examination of evidence.

The idea that our open fireplaces are necessary for ventilation is one of these physical superstitions, which is producing an incalculable amount of physical mischief throughout Britain. A little rational reflection on the natural and necessary movements of our household atmospheres demonstrates at once that this dogma is not only baseless, but actually expresses the opposite of the truth. I think I shall be able to show in what follows, 1st, that they do no useful ventilation; and, 2d, that they render systematic and really effective ventilation practically impossible.

Everybody knows that when air is heated it expands largely, becomes lighter, bulk for bulk, than other air of lower temperature; and therefore, if two portions of air of unequal temperatures, and free to move, are in contact with each other, the colder will flow under the warmer, and push it upwards. The latter postulate must be kept distinctly in view, for the rising of warm air is too commonly regarded as due to some direct uprising activity or skyward affinity of its own, instead of being understood as an indirect result of gravitation. It is the downfalling of the cooler air that causes the uprising of the warmer.

Now, let us see what, in accordance with the above-stated simple laws, must happen in an ordinary English apartment that is fitted, as usual, with one or more windows more or less leaky, and one or more doors in like condition, and a hole in the wall in which coal is burning in an iron cage immediately beneath a shaft that rises to the top of the house, the fire-hole itself having an extreme height of only 24 to 30 inches above the floor, all the chimney above this height being entirely closed. (I find by measurement that 24 inches is the usual height of the upper edge of the chimney opening of an ordinary "register" stove. Old farm-house fireplaces are open to the mantlepiece.)

Now, what happens when a heap of coal is burning in this hole? Some of the heat—from 10 to 20 per cent, according to the construction of the grate—is radiated into the room, the rest is conveyed by an ascending current of air up the chimney. As this ascending current is rendered visible by the smoke entangled with it, no further demonstration of its existence is needed.

But how is it pushed up the chimney? Evidently by cooler air, that flows into the room from somewhere, and which cooler air must get under it in order to lift it. In ordinary rooms this supply of air is entirely dependent upon their defective construction—bad joinery; it enters only by the crevices surrounding the ill-fitting windows and doors, no specially designed opening being made for it. Usually the chief inlet is the space under the door, through which pours a rivulet of cold air, that spreads out as a lake upon the floor. This may easily be proved by holding a lighted taper in front of the bottom door-chink when the window and other door—if any—are closed, and the fire is burning briskly. At the same time more or less of cold air is poured in at the top and the side spaces of the door and through the window-chinks. The proportion of air entering by these depends upon the capacity of the bottom door-chink. If this is large enough it will do nearly all the work, otherwise every other possible leakage, including the key-hole, contributes.

But what is the path of the air which enters by these higher level openings? The answer to this is supplied at once by the fact that such air being colder than that of the room, it must fall immediately it enters. The rivulet under the door is thus supplemented by cascades pouring down from the top and sides of the door and the top and sides of the windows, all being tributaries to the lake of cold air covering the floor.

The next question to be considered is, what is the depth of this lake? In this, as in every other such accumulation of either air or water, the level of the upper surface of the lake is determined by that of its outlet. The outlet in this case is the chimney hole, through which all the overflow pours upwards; and, therefore, the surface of the flowing stratum of cold air corresponds with the upper part of the chimney hole, or of the register, where register stoves are used.

Below this level there is abundant ventilation, above it there is none. The cat that sits on the hearth-rug has an abundant supply of fresh air, and if we had tracheal breathing apertures all down the sides of our bodies, as caterpillars have, those on our lower extremities might enjoy the ventilation. If we squatted on the ground like savages something might be said for the fire-hole ventilator. But as we are addicted to sitting on chairs that raise our breathing apparatus considerably above the level of the top of the register, the maximum efficiency of the flow of cold air in the lake below is expressed by the prevalence of chilblains and rheumatism.[31]

The atmosphere in which our heads are immersed is practically stagnant; the radiations from the fire, plus the animal heat from our bodies, just warm it sufficiently to enable the cool entering air to push it upwards above the chimney outlet and the surface of the lower moving stratum, and to keep it there in a condition of stagnation.

If anybody doubts the correctness of this description, he has only to sit in an ordinary English room where a good fire is burning—the doors and windows closed, as usual—and then to blow a cloud by means of pipe, cigar, or by burning brown paper or otherwise, when the movements below and the stagnation above, which I have described, will be rendered visible. If there is nobody moving about to stir the air, and the experiment is fairly made, the

---

[31] Since the above was written, a correspondent in Paris tells me that a caricature exists, representing a Frenchman enjoying an open fire by standing on his head in the middle of the room.

level of the cool lake below will be distinctly shown by the clearing away of the smoke up to the level of the top of the register opening, towards which it may be seen to sweep.

Above this the smoke-wreaths will remain merely waving about, with slight movements due to the small inequalities of temperature caused by the fraction of heat radiated into the room from the front of the fire. These movements are chiefly developed near the door and windows, where the above-mentioned cascades are falling, and against the walls and furniture, where feeble convection currents are rising, due to the radiant heat absorbed by their surfaces. The stagnation is the most complete about the middle of the room, where there is the greatest bulk of vacant airspace.

When the inlet under the door is of considerable dimensions, there may be some escape of warmer upper air at the top of the windows, if their fitting is correspondingly defective. These, however, are mere accidents; they are not a part of the vaunted chimney-hole ventilation, but interferences with it.

There is another experiment that illustrates the absence of ventilation in such rooms where gas is burning. It is that of suspending a canary in a cage near the roof. But this is cruel; it kills the bird. It would be a more satisfactory experiment to substitute for the canary-bird any wingless biped who, after reading the above, still maintains that our fire-holes are effective ventilators.

Not only are the fire-holes worthless and mischievous ventilators themselves, but they render efficient ventilation by any other means practically impossible. The "Arnott's ventilator" that we sometimes see applied to the upper part of chimneys is marred in its action by the greedy "draught" below.

The tall chimney-shaft, with a fire burning immediately below it, dominates all the atmospheric movement in the house, unless another and more powerful upcast shaft be somewhere else in communication with the apartments. But in this case the original or ordinary chimney would be converted into a downcast shaft pouring air downwards into the room, instead of carrying it away upwards. I need not describe the sort of ventilation thus obtainable while the fire is burning and smoking.

Effective sanitary ventilation should supply gentle and uniformly-diffused currents of air of moderate and equal temperature throughout the house. We talk a great deal about the climate here and the climate there; and when we grow old, and can afford it, we move to Bournemouth, Torquay, Mentone, Nice, Algiers, etc., for better climates, forgetting all the while that the climate in which we practically live is not that out-of-doors, but the indoor climate of our dwellings, the which, in a properly constructed house, may be regulated to correspond to that of any latitude we may choose. I maintain that the very first step towards the best attainable approximation to this in our existing houses is to brick up, cement up, or otherwise completely stop up, all our existing fire-holes, and abolish all our existing fires.

But what next? The reply to this will be found in the next chapter.

Chapter 41

DOMESTIC VENTILATION

A LESSON FROM THE COAL-PITS

We require in our houses an artificial temperate climate which shall be uniform throughout, and at the same time we need a gentle movement of air that shall supply the requirements of respiration without any gusts, or draughts, or alternations of temperature. Everybody will admit that these are fundamental *desiderata*, but whoever does so becomes thereby a denouncer of open-grate fireplaces, and of every system of heating which is dependent on any kind of stoves with fuel burning in the rooms that are to be inhabited. All such devices concentrate the heat in one part of each room, and demand the admission of cold air from some other part or parts, thereby violating the primary condition of uniform temperature. The usual proceeding effects a specially outrageous violation of this, as I showed in the last chapter.

I might have added domestic cleanliness among the *desiderata*; but in the matter of fireplaces, the true-born Briton, in spite of his fastidiousness in respect to shirt-collars, etc., is a devoted worshiper of dirt. No matter how elegant his drawing-room, he must defile it with a coal-scuttle, with dirty coals, poker, shovel, and tongs, dirty ash-pit, dirty cinders, ashes, and dust, and he must amuse himself by doing the dirty work of a stoker towards his "cheerful, companionable, pokeable" open fire.

It is evident that, in order to completely fulfil the first-named requirements, we must, in winter, supply our model residence with fresh artificially-warmed air, and in summer with fresh cool air. How is this to be done? An approach to a practical solution is afforded by examining what is actually done under circumstances where the ventilation problem presents the greatest possible difficulties, and where, nevertheless, these difficulties have been effectually overcome. Such a case is presented by a deep coal mine. Here we have a little working world, inhabited by men and horses, deep in the bowels of the earth, far away from the air that must be supplied in sufficient quantities, not only to overcome the vitiation due to their own breathing, but also to sweep out the deadly gaseous emanations from the coal itself.

Imagine your dwelling-house buried a quarter of a mile of perpendicular depth below the surface of the earth, and its walls giving off suffocating and explosive gases in such quantities that steady and abundant ventilation shall be a matter of life or death, and that in spite of this it is made so far habitable that men who spend half their days there retain robust health and live to green old age, and that horses after remaining there day and night for many months actually improve in condition. Imagine, further, that the house thus ventilated has some hundreds of small, very low-roofed rooms, and a system of passages or corridors with an united length of many miles, and that its inhabitants count by hundreds.

Such dwellings being thus ventilated and rendered habitable for man and beast, it is idle to dispute the practical possibility of supplying fresh air of any given temperature to a mere box of brick or stone, standing in the midst of the atmosphere, and containing but a few passages and apartments.

The problem is solved in the coal-pit by simply and skilfully controlling and directing the natural movements of unequally-heated volumes of air. Complex mechanical devices for forcing the ventilation by means of gigantic fan-wheels, etc., or by steam-jets, have been tried, and are now generally abandoned. An inlet and an outlet are provided, *and no air is allowed to pass inwards or outwards by any other course than that which has been pre-arranged for the purposes of efficient ventilation*. I place especial emphasis on this condition, believing that its systematic violation is the primary cause of the bungling muddle of our domestic ventilation.

Let us suppose that we are going to open a coal-pit to mine the coal on a certain estate. We first ascertain the "dip" of the seam, or its deviation from horizontality, and then start at the *lowest* part, not, as some suppose, at that part nearest to the surface. The reason for this is obvious on a little reflection, for if we began at the shallowest part of an ordinary water-bearing stratum we should have to drive down under water; but, by beginning at the lowest part and driving upwards, we can at once form a "sumpf," or bottom receptacle, to receive the drainage, and from which the accumulated water may be pumped. This, however, is only by the way, and not directly connected with our main subject, the ventilation.

In order to secure this, the modern practice is to sink two pits, "a pair," as they are called, side by side, at any convenient distance from each other. If they are deep, it becomes necessary to commence ventilation of the mere shafts themselves in the course of sinking. This is done by driving an air-way—a horizontal tunnel from one to the other, and then establishing an "upcast" in one of them by simply lighting a fire there. This destroys the balance between the two communicating columns of air; the cooler column in the shaft without a fire, being heavier, falls against the lighter column, and pushes it up just as the air is pushed up one leg of an U tube when we pour water down the other. Even in this preliminary work, if the pits are so deep that more than one air-way is driven, it is necessary to stop the upper ways and leave only the lowest open, in order that the ventilation shall not take a short and useless cut, as it does up our fireplace openings.

Let us now suppose that the pair of pits are sunk down to the seam, with a further extension below to form the water sumpf. There are two chief modes of working a coal-seam: the "pillar and stall" and the "long wall," or more modern system. For present illustration, I select the latter as the simplest in respect to ventilation. This method, as ordinarily worked, consists essentially in first driving roads through the coal, from the pits to the outer boundary of the area to be worked, then cutting a cross road that shall connect these, thereby exposing a "long wall" of coal, which, in working, is gradually cut away towards the pits, the roof remaining behind being allowed to fall in.

Let us begin to do this by driving, first of all, two main roads, one from each pit. It is evident that as we proceed in such burrowing, we shall presently find ourselves in a *cul de sac* so far away from the outer air that suffocation is threatened. This will be equally the case with both roads. Let us now drive a cross-cut from the end of each main road, and thus establish a communication from the downcast shaft through its road, then through the drift to the upcast road and pit. But in order that the air shall take this roundabout course, we must close the direct drift that we previously made between the two shafts, or it will proceed by that shorter and easier course. Now we shall have air throughout both our main roads, and we may drive on further, until we are again stopped by approximate suffocation. When this occurs, we make another cross-cut, but in order that it may act we must stop the first one. So we go on until we reach the working, and then the long wall itself becomes the cross communication, and through this working-gallery the air sweeps freely and effectually.

In the above I have only considered the simplest possible elements of the problem. The practical coal-pit in full working has a multitude of intervening passages and "splits," where the main current from the downcast is divided, in order to proceed through the various streets and lanes of the subterranean town as may be required, and these divided currents are finally reunited ere they reach the upcast shaft which casts them all out into the upper air.

In a colliery worked on the pillar and stall system—i.e., by taking out the coal so as to leave a series of square chambers with pillars of coal in the middle to support the roof—the windings of the air between the multitude of passages is curiously complex, and its absolute obedience to the commands of the mining engineer proves how completely the most difficult problems of ventilation may be solved when ignorance and prejudice are not permitted to bar the progress of the practical applications of simple scientific principles.

Here the necessity of closing all false outlets is strikingly demonstrated by the mechanism and working of the "stoppings" or partitions that close all unrequired openings. The air in many pits has to travel several miles in order to get from the downcast to the upcast shaft, though they may be but a dozen yards apart. (Formerly the same shaft served both for up and down cast, by making a wooden division (a *brattice*) down the middle. This is now prohibited, on account of serious accidents that have been caused by the fracture of the *brattice*.)

But it would not do to carry the coal from the workings to the pit by these sinuous air-courses. What, then, is done? A direct road is made for the coal, but if it were left open, the air would choose it: this is prevented by an arrangement similar to that of canal locks. Valve-doors or "stoppings" are arranged in pairs, and when the "hurrier" arrives with his *corve*, or pit carriage, one door is opened, the other remaining shut; then the *corve* is hurried into the space between the doors, and the entry-door is closed; now the exit-door is opened, and thus no continuous opening is ever permitted.

Only one such opening would derange the ventilation of the whole pit, or of that portion fed by the split thus allowed to escape. It would, in fact, correspond to the action of our open fireplaces in rendering effective ventilation impossible.

The following, from the report of the Lords' Committee on Accidents in Coal Mines, 1849, illustrates the magnitude of the ventilation arrangements then at work. In the Hetton Colliery there were two downcast shafts and one upcast, the former about 12 feet and the latter 14 feet diameter. There were three furnaces at the bottom of the upcast, each about 9 feet wide with about 4 feet length of grate-bars; the depth of the upcast and one downcast 900 feet, and of the other downcast 1056 feet. The quantity of air introduced by the action of these furnaces was 168,560 cubic feet per minute, at a cost of about eight tons of coal per day. The rate of motion of the air was 1097 feet per minute (above 12 miles per hour). This whole current was divided by splitting into 16 currents of about 11,000 cubic feet each per minute, having, on an average, a course of 4¼ miles each. This distance was, however, very irregular—the greatest length of course being 9-1/10 miles; total length 70 miles. Thus 168,560 cubic feet of air were driven through these great distances at the rate of 12 miles per hour, and at a cost of 8 tons of coal per day.

All these magnitudes are greatly increased in coal-mines of the present time. As much as 250,000 cubic feet of air per minute are now passed through the shafts of one mine.

The problem of domestic ventilation as compared with coal-pit ventilation involves an additional requirement, that of warming, but this does not at all increase the difficulty, and I even go so far as to believe that cooling in summer may be added to warming in winter by one and the same ventilating arrangement. As I am not a builder, and claim no patent rights, the following must be regarded as a general indication, not as a working specification, of my scheme for domestic ventilation and the regulation of home climate.

The model house must have an upcast shaft, placed as nearly in the middle of the building as possible, with which every room must communicate either by a direct opening or through a lateral shaft. An ordinary chimney built in the usual manner is all that is required to form such a main shaft.

There must be no stoves nor any fireplaces in any room excepting the kitchen, of which anon. All the windows must be made to fit closely, as nearly air-tight as possible. No downcast shaft is required, the pressure of the surrounding outer atmosphere being sufficient. Outside of the house, or on the ground floor (on the north side, if possible), should be a chamber heated by flues, hot air, steam, a suitable stove, or water-pipes, and with one adjustable opening communicating with the outer fresh air, and another on the opposite side connected by a shaft or air-way with the hall of the ground floor and the general staircase.

Each room to have an opening at its upper part communicating with the chimney, like an Arnott's ventilator, and capable of adjustment as regards area of aperture, and other openings of corresponding or excessive combined area leading from the hall or staircase to the lower part of the room. These may be covered with perforated zinc or wire gauze, so that the air may enter in a gentle, broken stream.

All the outer house-doors must be double, i.e., with a porch or vestibule, and only one of each pair of doors opened at once. These should be well fitted, and the staircase air-tight. The kitchen to communicate with the rest of the house by similar double doors, and the kitchen fire to communicate directly with the upcast shaft or chimney by as small a stove-pipe as practicable. The kitchen fire will thus start the upcast and commence the draught of air from the warm chamber through the house towards the several openings into the shaft. In cold weather, this upcast action will be greatly reinforced and maintained by the general warmth of all the air in the house, which itself will bodily become an upcast shaft immediately the inner temperature exceeds that of the air outside.

But the upcast of warm air can only take place by the admission of fresh air through the heating chamber, thence to hall and staircase, and thence onward through the rooms into the final shaft or chimney.

The openings into and out of the rooms being adjustable, they may be so regulated that each shall receive an equal share of fresh warm air; or, if desired, the bedroom chimney valves may be closed in the daytime, and thus the heat economized by being used only for the day rooms; or, *vice versâ*, the communication between the upcast shaft and the lower rooms may be closed in the evening, and thus all the warm air be turned into the bedrooms at bedtime.

If the area of the entrance apertures of the rooms exceeds that of the outlet, only the latter need be adjusted; the room doors may, in fact, be left wide open without any possibility of "draught," beyond the ventilation current, which is limited by the dimension of the opening from the room into the shaft or chimney.

So far, for winter time, when the ventilation problem is the easiest, because then the excess of inner warmth converts the whole house into an upcast shaft, and the whole outer atmosphere becomes a downcast. In the summer time, the kitchen fire would probably be insufficient to secure a sufficiently active upcast.

To help this there should be in one of the upper rooms—say an attic—an opening into the chimney secured by a small well-fitting door; and altogether enclosed within the chimney a small automatic slow-combustion stove (of which many were exhibited in South Kensington, that require feeding but once in twenty-four hours), or a large gas-burner. The heating-chamber below must now be converted into a cooling chamber by an arrangement of wet cloths, presently to be described, so that all the air entering the house shall be reduced in temperature.

Or the winter course of ventilation may be reversed by building a special shaft connected with the kitchen fire, which, in this case, must not communicate with the house shaft. This special shaft may thus be made an upcast, and the rooms supplied with air from above down the house shaft, through the rooms, and out of the kitchen *viâ* the winter heating-chamber, which now has its communication with the outside air closed.

Reverting to the first-named method, which I think is better than the second, besides being less expensive, I must say a few concluding words on an important supplementary advantage which is obtainable wherever all the air entering the house passes through one opening, completely under control, like that of our heating-chamber. The great evil of our town atmosphere is its dirtiness. In the winter it is polluted with soot particles; in the dry summer weather, the traffic and the wind stir up and mix with it particles of dust, having a composition that is better ignored, when we consider the quantity of horse-dung that is dried and pulverized on our roadways. All the dust that falls on our books and furniture was first suspended in the air we breathe inside our rooms. Can we get rid of any practically important portion of this?

I am able to answer this question, not merely on theoretical grounds, but as a result of practical experiments described in the following chapter, in which is reprinted a paper I read at the Society of Arts, March 19, 1879, recommending the enclosure of London back yards with a roofing of "wall canvas," or "paperhanger's canvas," so as to form cheap conservatories. This canvas, which costs about threepence per square yard, is a kind of coarse, strong, fluffy gauze, admitting light and air, but acting very effectively as an air filter, by catching and stopping the particles of soot and dust that are so fatal to urban vegetation.

I propose, therefore, that this well-tried device should be applied at the entrance aperture of our heating chamber, that the screens shall be well wetted in the summer, in order to obtain the cooling effect of evaporation, and in the winter shall be either wet or dry, as may be found desirable. The Parliament House experiments prove that they are good filters when wetted, and mine that they act similarly when dry.

By thus applying the principles of colliery ventilation to a specially-constructed house, we may, I believe, obtain a perfectly controllable indoor climate, with a range of variation not exceeding four or five degrees between the warmest and the coldest part of the house, or eight or nine degrees between summer and winter, and this may be combined with an abundant supply of fresh air everywhere, all filtered from the grosser portions of its irritant dust, which is positively poisonous to delicate lungs, and damaging to all. The cost of fuel would be far less than with existing arrangements, and the labor of attending to the one or two fires and the valves would also be less than that now required in the carrying of coal-scuttles, the removal of ashes, the cleaning of fireplaces, and the curtains and furniture they befoul by their escaping dust and smoke.

It is obvious that such a system of ventilation may even be applied to existing houses by mending the ill-fitting windows, shutting up the existing fire-holes, and using the chimneys as upcast shafts in the manner above described. This may be done in the winter, when the problem is easiest, and the demand for artificial climate the most urgent; but I question the possibility of summer ventilation and tempering of climate in anything short of a specially-built house or a materially altered existing dwelling. There are doubtless some exceptions to this, where the house happens to be specially suitable and easily adapted, but in ordinary houses we must be content with the ordinary devices of summer ventilation by doors and windows, plus the upper openings of the rooms into the chimneys expanded to their full capacity, and thus doing, even in summer, far better ventilating work than the existing fire-holes opening in the wrong place.

I thus expound my own scheme, not because I believe it to be perfect, but, on the contrary, as a suggestive project to be practically amended and adapted by others better able than myself to carry out the details. The feature that I think is novel and important is that of consciously and avowedly applying to domestic ventilation the principles that have been so successfully carried out in the far more difficult problem of subterranean ventilation.

The dishonesty of the majority of the modern builders of suburban "villa residences" is favorable to this and other similar radical household reforms, as thousands of these wretched tenements must sooner or later be pulled down, or will all come down together without any pulling the next time we experience one of those earthquake tremors which visit England about once in a century.

Chapter 42

HOME GARDENS FOR SMOKY TOWNS

The poetical philanthropists of the shepherd and shepherdess school, if any still remain, may find abundant material for their doleful denunciations of modern civilization on journeying among the house-tops by any of our over-ground metropolitan and suburban railways, and contemplating therefrom the panorama presented by a rapid succession of London back yards. The sandy Sahara, and the saline deserts of Central Asia, are bright and breezy, rural and cheerful, compared with these foul, soot-smeared, lumber-strewn areas of desolation.

The object of this paper is to propose a remedy for these metropolitan measle-spots, by converting them into gardens that shall afford both pleasure and profit to all concerned.

A very obvious mode of doing this would be to cover them with glass, and thus convert them into winter gardens or conservatories. The cost of this at once places it beyond practical reach; but even if the cost were disregarded, as it might be in some instances, such covering in would not be permissible on sanitary grounds; for, doleful and dreary as they are, the back yards of London perform one very important and necessary function; they act as ventilation-shafts between the house-backs of the more densely populated neighborhoods.

At one time I thought of proposing the establishment of horticultural home missions for promoting the dissemination of flower-pot shrubs in the metropolis, and of showing how much the atmosphere of London would be improved if every London family had one little sweetbriar bush, a lavender plant, or a hardy heliotrope to each of its members; so that a couple of million of such ozone generators should breathe their sweetness into the dank and dead atmosphere of the denser central regions of London.

A little practical experience of the difficulty of growing a clean cabbage, or maintaining alive any sort of shrub in the midst of our soot-drizzle, satisfied me that the mission would fail, even though the sweetbriars were given away by the district visitors; for these simple hardy plants perish in a mid-London atmosphere unless their leaves are periodically sponged and syringed, to wash away the soot particles that otherwise close their stomata and suffocate the plant.

It is this deposit that stunts or destroys all our London vegetation, with the exception of those trees which, like the planes have a deciduous bark and cuticle.

Some simple and inexpensive means of protecting vegetation from London soot are, therefore, most desirable.

When the Midland Institute commenced its existence in temporary buildings in Cannon Street, Birmingham, in 1854, I was compelled to ventilate my class-rooms by temporary devices, one of which was to throw open the existing windows, and protect the students from the heavy blast of entering air by straining it through a strong gauze-like fabric stretched over the opening.

After a short time the tammy became useless for its intended purpose; its interstices were choked with a deposit of carbon. On examining this, I found that the black deposit was all on the outside, showing that a filtration of the air had occurred. Even when the tammy was replaced by perforated zinc, puttied into the window frames in the place of glass panes, it was found necessary to frequently wash the zinc, in order to keep the perforations open.

The recollection of this experience suggested that if a gauze-like fabric, cheaper and stronger than the tammy, can be obtained, and a sort of greenhouse made with this in the place of glass, the problem of converting London back-yards into gardens might be solved.

After some inquiries and failures in the trial of various cheap fabrics, I found one that is already to be had, and well adapted to the purpose. It is called "wall canvas," or "scrim," is retailed at 3½d. per yard, and is one yard wide. If I am rightly informed, it may be bought in wholesale quantities at about 2¼d. per square yard, i.e., one farthing per square foot. This fabric is made of coarse unbleached thread yarn, very strong and open in structure. The light passes so freely through it that when hung before a window the loss of light in the room is barely perceptible. When a piece is stretched upon a frame, a printed placard, or even a newspaper, may be read through it.

The yarn being loosely spun, fine fluffy filaments stand out and bar the interstices against the passage of even very minute carbonaceous particles. These filaments may be seen by holding it up to the light.

The fabric being one yard wide, and of any length required, all that is needed for a roof or side walls is a skeleton made of lines or runs of quartering, at 3 feet distance from each other. The cost of such quartering, made of pitch pine, the best material for outside work, is under one penny per foot run; of common white deal, about three farthings. Thus the cost of material for a roof, say a lean-to from a wall-top to the side of a house, which would be the most commonly demanded form of 30 feet by 10 feet, i.e., 300 square feet, would be—

	s.	d.
110 feet of quartering (11 lengths) at 1d.	9	2
300 square feet of canvas, at 1¼	6	3
Nails and tacks, say	1	0
	16	5

The size of the quartering proposed is 2½ by 1¼ inch, which, laid edgewise, would bear the weight of a man on a plank while nailing down the canvas. The canvas has a stout cord-like edge or selvage, that holds the nails well.

I find that what are called "French tacks" are well suited for nailing it down. They are made of wire, well pointed, have good-sized flat clout heads, and are very cheap. They are incomparably superior to the ordinary rubbish sold as "tin tacks" or "cut tacks." The construction of such a conservatory is so simple that any industrious artisan or clerk with any mechanical ingenuity could, with the aid of a boy, do it all himself. No special skill is required for any part of the work, and no other tools than a rule, a saw, and a hammer. Side posts and stronger end rails would in some cases be demanded.

I have not been able to fairly carry out this project, inasmuch as I reside at Twickenham, beyond the reach of the black showers of London soot. I have, however, made some investigations relative to the climate which results from such enclosure.

This was done by covering a small skeleton frame with the canvas, putting it upon the ground over some cabbage plants, etc., and placing registering thermometers on the ground inside, and in similar position outside the frame; also by removing the glass cover of a cucumber frame, and replacing it by a frame on which the canvas is stretched.

I planted 300 cabbages in November last, in rows on the open ground, and placed the canvas-covered frame over 18 of them. At the present date, March 15, only 26 of the 282 outside plants are visible above the ground. All the rest have been cut off by the severe frost. Under the frame *all* are flourishing.

I find that the difference between the maximum and the minimum temperatures varies with the condition of the sky. In cloudy weather, the difference between the inside and the outside rarely exceeds 2° Fahr., and occasionally there is no difference. In clear weather the difference is considerable. During the day the outside thermometer registers from four or five to seven or eight degrees above that within the screen during the sunshine. At night the minimum thermometers show a difference which in one case reached 14°, i.e., between 23d and 24th February, when the lowest temperature I have observed was reached. The outside thermometer then fell to 8° Fahr., the inside to 22°. On the night of the 24th and 25th they registered 15½° outside, 25½° inside. On other, or ordinary clear frosty nights, with E. and N. and N.E. winds, the difference has ranged between 4° and 6°, usually within a fraction of the average, 5°.

The uniformity of this during the recent bright frosty nights, followed by warm sunny days, has been very remarkable, so much so that I think I may venture to state that 5° may be expected as the general protecting effect of a covering of such canvas from the mischievous action of our spring frosts which are due to nocturnal radiation into free space. Thus we obtain a climate, the mean of which would be about the same as outside, but subject to far less variation. How will this affect the growth of plants desirable to cultivate in the proposed canvas conservatories?

In the first place, we must not expect the results obtainable under glass, which by freely transmitting the bright solar rays, and absorbing or resisting the passage of the obscure rays from the heated soil, produces, during sunshine, a tropical climate here in our latitudes. We may therefore at once set aside any expectation of rearing exotic plants of any kind; even our native and acclimatized plants, which require the maximum heat of English sunshine, are not likely to flourish.

On the other hand, all those which demand moderate protection from sudden frosts, especially from spring frosts, and which flourish when we have a long mild spring and summer, are likely to be reared with especial success.

This includes nearly all our table vegetables, our salads, kitchen herbs, and British fruits, all our British and many exotic ferns, and, I believe, most of our out-of-door plants, both wild and cultivated.

As the subject of ornamental flowers is a very large one, and one with the cultivation of which I have very little practical acquaintance, I will pass it over; but must simply indicate that, in respect to ferns, the canvas enclosure offers a combination of most desirable conditions. The slight shade, the comparatively uniform temperature, and the moderated exhalation, are just those of a luxuriant fern dingle.

Respecting the useful or economic products I can speak with more confidence, that being my special department in our family or home gardening, which, as physical discipline, I have always conducted myself, with a minimum of professional aid.

My experience of a small garden leads me to give first place to salads. A yard square of rich soil, well managed, will yield a handsome and delicious weekly dish of salad nearly all the year round; and, at the same rate, seven or eight square yards will supply a daily dish— including lettuces, endives, radishes, spring onions, mustard, and various kinds of cress, and fancy salads, all in a state of freshness otherwise unattainable by the Londoner. My only difficulty has arisen from irregularity of supply. From the small area allowed for salads, I have been over-supplied in July, August, and September, and reduced to in-door or frame-grown mustard and cress during the winter. With the equable insular climate obtainable under the canvas, this difficulty will be greatly diminished; and besides this, most of the salads are improved by partial shade, lettuces and endives more blanched and delicate than when exposed to scorching sun, radishes less fibrous, mustard, cress, etc., milder in flavor and more succulent.

The multitude of savory kitchen herbs that are so sadly neglected in English cookery (especially in the food of the town artisan and clerk), all, with scarcely an exception, demand an equable climate and protection from our destructive spring frosts. These occupy very little space, less even than salads, and are wanted in such small quantities at a time, and so frequently, that the hard-worked housewife commonly neglects them altogether, rather than fetch them from the greengrocer's in their exorbitantly small pennyworths. If she could step into the back yard, and gather her parsley, sage, thyme, winter savory, mint, marjoram, bay leaf, rosemary, etc., the dinner would become far more savory, and the demand for the alcoholic substitutes for relishing food proportionably diminished.

My strongest anticipations, however, lie in the direction of common fruits—apples, pears, cherries, plums of all kinds, peaches, nectarines, gooseberries, currants, raspberries, strawberries, etc.

The most luxuriant growth of cherries, currants, gooseberries, and raspberries I have ever seen in any part of the world that I have visited, is where they might be least expected, viz., Norway; not the South of Norway merely, but more particularly in the valleys that slope from the 500 square miles of the perpetual ice desert of the Justedal down to the Sognefjord, latitude 61° to 61½°, considerably to the north of the northernmost of the Shetland Islands. The cherry and currant trees are marvelous there.

In the garden of one of the farm stations (Sande) I counted 70 fine bunches of red currants growing on six inches of one of the overladen down-hanging stems of a currant bush. Cherries are served for dessert by simply breaking off a small branch of the tree and bringing it to the table—the fruit almost as many as the leaves.

This luxuriance I attribute to two causes. First, that in that part of Norway the winter breaks up suddenly at about the beginning of June, and not until then, when night frosts are no longer possible, do the blossoms appear. It was on the 24th August that I counted the 70 bunches of ripe currants. The second cause is the absence of sparrows and other destructive small birds that devour our currants for the seeds' sake before they ripen, and our cherries immediately on ripening. These are preceded by the bullfinches that feed on the tender hearts of the buds of most of our fruit trees. Those who believe the newspaper myths which represent such thick-billed birds eating caterpillars, should make observations and experiments for themselves as I have done.

In our canvas conservatories neither sparrows nor caterpillars, nor wasps, or other fruit-stealers will penetrate, nor will the spring frosts nip the blossoms that open out in April. All the conditions for full bearing are there fulfilled, and the ripening season, though not so intense, will be prolonged. We shall have an insular Jersey climate in London, where the mean temperature is higher than in the country around, and, if I am not quite deluded, we shall be able to grow the choicest Jersey pears, those that best ripen by hanging on the tree until the end of December, and fine peaches, which are commonly destroyed by putting forth their blossoms so early. All the hundred and one varieties of plums and damsons, greengages, etc., that can grow in temperate climates will be similarly protected from the frosts that kill their early blossoms, and the birds and the wasps that will not give them time to ripen slowly.

I have little doubt that if my project is carried out, any London householder, whether rich or poor, may indulge in delicious desserts of rich fruit all grown on the sites of their own now dirty and desolate back-yards; that if prizes be given for the most prolific branches of cherry and plum trees, gooseberry and currant bushes, the gardens of the Seven-dials and of classic St. Giles's may carry off some of the gold medals; and that, by judicious economy of space and proper pruning of the trees, the canvas conservatories may be made not only to serve as orchard houses, but also to grow the salads, kitchen herbs, and green vegetables for cookery, under the fruit trees or close around their stems.

Among the suitable vegetables, I may name a sort of perennial spinach which yields a wonderful amount of produce on a small area. Four years ago I took the house in which I now reside, and found the garden overgrown with a weed that appeared like beet, the leaves being much larger than ordinary spinach. I tried in vain to eradicate it, then gave some leaves to my fowls. They ate them greedily. After this I had some boiled, and found that the supposed weed is an excellent spinach, which may be sown broadcast in thick patches, without any interspaces, and cut down again and again all the year round, fresh leaves springing up from the roots until the autumn, when it throws up tall flowering stems, and yields an abundant crop of seeds. I have some now, self-sown, that have survived the whole of the late severe winter, while turnip-tops, cabbages, and everything else have perished. I have sown the ordinary spinach seed in the usual manner in rows, and comparing it with the self-sown dense patches of this intruder, find the latter produces, square yard against square yard, six or eight times as much of available eatable crop.

None of my friends who are amateur gardeners know this variety; but a few days since, I called on Messrs. James Carter and Co., the wholesale seedsmen of Holborn, and described it. They gave me a packet of what they call "Perpetual spinach beet," which, as may be seen by comparison with the seeds of those I have here of my own growing, is probably the same. Messrs. Carter and Co. tell me that the plant is very little known, and the seed scarce from want of cultivation and demand. I therefore step so far aside to describe and recommend it as specially suited for obtaining large crops on small areas.[32]

I also recommend a mode of growing cabbages that I have found very profitable, viz., to sow the seed broadcast in richly manured beds or patches and leave the plants crowding

---

[32] I tried the seeds given to me by Messrs. Carter, and find them to produce the same plant as my own, which I still cultivate very successfully. I now sow it in the spring as a kitchen garden border.

together; cut them down while very young, without destroying the centre bud; let them sprout again and again. They thus yield a succession of crops, every leaf of which is eatable. This, instead of transplanting and growing large plants, which, however desirable for sale in the market, are far less profitable for home use. Celery may be grown in like manner, and cut down young and green for boiling.

Some collateral advantages may be fairly anticipated in cases where the back-yard is fully enclosed by the canvas.

In the first place, the air coming into the house from the back will be more or less filtered from the grimy irritant particles with which our London atmosphere is loaded, besides obtaining the oxygen given off by the growing plants, and the ozone which recent investigations have shown to be produced where aromatic plants—such as kitchen herbs—are growing. Lavender, which is very hardy, and spreads spontaneously, might be grown for this purpose.

Back-doors might be left open for ventilation, without danger of intrusion or of slamming by gusts of wind. The air thus admitted would be tempered both in summer and winter. By wetting the canvas, which may easily be done by means of a small garden engine, or hand syringe, the exceptionally hot summer days that are so severely felt in London might be moderated to a considerable extent. The air under the canvas being cooler than that in front would enter from below, while the warmer air would be pushed upwards and outwards to the front.

Although such conservatories may be erected, as already stated, by artisans or other tenants of small houses, I do not advocate dependence on this; but, on the contrary, regard them as more properly constituting landlord's fixtures, and recommend their erection by owners of small house property in London and other large towns. A workman who will pay a trifle extra for such a garden, is likely to be a better and more permanent tenant than one who is content with the slovenly squalor of ordinary back premises.

I base this opinion on some experience of holding small houses in the outskirts of Birmingham (Talbot Street, Winson Green.) These have small gardens, while most of those around have none. They are held by weekly tenure, and, during eighteen years, I have not lost a week's rent from voids; the men who would otherwise shift their dwelling when they change workshops, prefer to remain and walk some distance rather than lose their little garden crops; and when obliged to leave, have usually found me another tenant, a friend who has paid them a small tenant-right premium for what is left in the garden, or for the privilege of getting a house with such a garden.

A small garden is one of the best rivals to the fascinations of the tap-room; the strongest argument in favor of my canvas conservatories, and that which I reserve as the last, is that they are likely to become the poor man's drawing-room, where he may spend his summer evenings, smoke his pipe, contemplate his growing plants, and show them in rivalry to his friends, rather than slink away from an unattractive home to seek the sensual excitements that ruin so many of our industrious fellow-countrymen.

As above stated, I have not been able practically to test the filtering capabilities of the canvas, owing to my residence out of town, but since the above was written, i.e., on last Wednesday evening, I visited the Houses of Parliament, where, as I had been told, the ventilation arrangements include some devices for filtering the air by cotton, wool or otherwise.

I was much interested on finding that the long experience and many trials of Dr. Percy and his assistant engineer, Mr. Prim, have resulted in the selection of the identical material which I have chosen, and with which the above-described experiments have been made. A wall of such canvas surrounds a lower region of the Houses, and all the air that is destined to have the privilege of being breathed by British legislators is passed through this vertical screen, for the purpose of separating from it the sooty impurities that constitute the special abomination of our metropolitan atmosphere, and that of our great manufacturing towns. The quantity of sooty matter thus arrested is shown by the fact that it is found necessary to take the screens down once a week and wash them, the wash water coming away in a semi-inky condition.

I anticipate that the conservatory filters will rapidly clog, and, therefore, require washing. This may easily be done by means of a jet from a hand-syringe directed from within outwards, especially if the slope of the roof is considerable, which is to be recommended. The filtering screen of the Houses of Parliament is made by sewing the canvas edges together, to form a large continuous area, then edging the borders of this with tape, and stretching it bodily on to a stout frame. This method may be found preferable to that which I proposed above, and cheaper than I have estimated, as only very light intermediate cross-pieces would thus be required, merely to prevent bagging, the parliamentary quartering above described being nine feet apart instead of three. This would reduce the cost of timber to about one half of the above estimate.[33] The perpendicular walls of a conservatory, where such are required, may certainly be made thus, and I think the roof also, if the slope is considerable. Or, if in demand, the material may be made of greater width than the three feet.

So far, I have only mentioned back-yards; but, besides these, there are many very melancholy front areas, called "gardens," attached to good houses in some of the once suburban, but now internal regions of London, where the houses stand some distance back from the formerly rural highway. These spaces might be cheaply enclosed with canvas, and cultivated as kitchen gardens, orchard houses, flower gardens, or ferneries, thus forming elegant, refreshing, and profitable vestibules between the highway and the house-door, and also serve as luxurious summer drawing-rooms. The only objection I foresee to these bright enclosures will be their tendency to encourage the consumption of tobacco.

THE DISCUSSION WHICH FOLLOWED THE READING OF THE PRECEDING PAPER AT THE SOCIETY OF ARTS

A member asked if Mr. Williams had observed the effect of wind and rain on this material?

Mr. W. P. B. Shepheard said he was interested in a large square in London, and he had hoped to hear something about the cultivation of flowers in such places. Last year, they tried the experiment with several varieties of flower seeds, and they came up and bloomed well in the open ground without any protection whatever. In most London squares, the difficulty was to find anyone bold enough to try the experiment at all, and nothing but experience would prove what flowers would succeed and what would not. They were so successful last year that

---

[33] Subsequent experiments induce me not to recommend this economy, on account of the bagging which results from excessive width between the frames; 3 feet should not be exceeded.

several fine bouquets were gathered in July and August, and sent to some of the gardening magazines, who expressed their astonishment that such good results were possible in the circumstances. If flowers would answer, there would, of course, be more encouragement to try vegetables. One of the practical difficulties which occurred to him, with regard to this plan, was that the screens would be somewhat unsightly, and then again they might shrink, from alteration in the temperature and getting wet and dry. He would repeat, however, that, for a very small expense in seeds, a very good show of hardy annuals and perennials might be obtained in July and August even in London.

Mr. C. Cooke said a flower-garden had recently been opened in Drury Lane, on the site of an old churchyard, to which children were admitted; and he wished a similar arrangement might be made in some of the squares in crowded neighborhoods, such as Golden Square, and especially in Lincoln's Inn Fields. There were lots of children playing about in the streets, and he wished the good example set by the Templars might be followed.

Mr. Liggins, as an old member of the Royal Horticultural Society, felt a great interest in this subject. Among his poorer neighbors in the district of Kensington, cottage and window gardening had been encouraged for some years past, prizes having been awarded to those who were most successful, much to their gratification. This was a novel idea, but he felt quite sure that it would enable those who adopted it to obtain the crops which had been described. There were many collateral advantages which it would bestow on the working classes if largely followed by them, especially the one mentioned by Mr. Williams, that those who devoted their spare time to the cultivation of fruit and flowers would not be so open to the attractions of the public-house. When traveling through the United States some years ago, he was much struck with the difference in appearance of the houses in districts where the Maine liquor law was in force, and soon learned to distinguish where it was adopted by the clean, cheerful look of the workmen's dwellings, the neatness of the gardens, and the presence of trees and flowers which, in other districts, were wanting. He was not a teetotaler himself, and was not advocating such restrictions, but he could not help noticing the contrast; and he felt sure that in all our large towns great progress in civilization and morals would be effected if such an attraction were offered to the working classes. He believed there was so much intelligence and good sense among them, that if they only knew what could be done in this way they would attempt it; and when an Englishman attempted anything, he generally succeeded.

Mr. William Botly said they were much indebted to Mr. Williams for having called attention to this important subject. He quite agreed with the observations of the last speaker, for his own experience in building cottages showed him that the addition of a piece of garden ground had an excellent effect on the social, moral, and religious welfare of the inmates. It kept them from the public-house, and the children who were brought up to hoe and weed their parents' gardens turned out the most industrious laborers on his property. He had known of instances where houses had been built with flat concrete roofs, and covered in with glass, so as to form a conservatory, in which vegetables and salads grow very well, and he believed the cost was little, if any, more than ordinary slating.

The Chairman (Lord Alfred Churchill) in moving a vote of thanks to Mr. Williams, said there could be no doubt that if his suggestion were adopted it would lead to great economy, and have many other attractions for the working classes. During the last few years they had heard a good deal about floriculture in windows, and no doubt it was an excellent proposal, but if they could add to this the growth of vegetables it would have economical advantages also. The proposal to erect temporary conservatories on the roofs of some of these small houses was an admirable one. He saw no reason why you should not have a peach tree growing against many a tall chimney; you would only want a metal-lined tub filled with a good mold; the warmth of the chimney would aid in promoting the growth of the tree, and it could be protected from the smoke and frost by this canvas. One point he should like to know was, whether the fabric would not become rotted by the weather, and perhaps it might be protected by tanning, or some chemical preparation. The effect of the canvas in maintaining an equable temperature was a great consideration; the difference stated by Mr. Williams, of about five degrees in winter, in many cases would be just enough to save the life of a plant. Practical gardeners knew the value of placing a covering over a peach tree in early spring to keep off the frosts, and also to protect it from the attacks of birds. It was also a curious fact that even a slip of wood or slate a few inches wide, put on the top of a wall to which a fruit tree was nailed, acted as a protection from frost. He trusted that Mr. Williams' idea would find favor among the working classes, and thought it was a subject the Royal Horticultural Society might well take up and offer prizes for. He hoped in a short time, when that Society had passed through a crisis which was impending, it might emerge in a condition to devote attention to this matter. It already offered prizes for small suburban flower-shows, but had not yet turned its attention to the larger class aimed at by Mr. Williams.

Mr. Botly said he had forgotten to mention that he had a friend, a very excellent gardener, who always loosened his fruit trees from the wall for about three weeks before the time of blooming. The consequence was, they did not get so much heat from the wall, and the bloom was two or three weeks later in forming. After the spring frosts, the trees were again nailed up close, and he never failed in getting an excellent crop, when his neighbors often had none.

Mr. Trewby wished to caution those who read the paper against using what was commonly known as paperhangers' canvas, because it was made of two materials, hemp and jute, and if a piece of it were put into water it would soon be nothing but a lot of strings, the jute being all dissolved. It did very well for paper-hanging, but would be quite unsuitable for this purpose.[34]

The vote of thanks having been passed—

Mr. Williams, in reply, said he had had a piece of this canvas stretched on a frame exposed all the winter, and the only result was to make it rather dirty. He stretched it as tightly as he could in putting it on, but when it got wet it became still more tight, and gave a little again on becoming dry. It bore the weight of the snow which had fallen very well, and two or three spadefuls had been added to try it. He had a note from Mr. Prim, saying that at the Houses of Parliament the screens last about two sessions, being washed once a week, and the destruction is due to the wringing. But there is really no occasion for this, for if you syringe the stuff well from the inside, you make it sufficiently clear to allow the air and light to pass through, and it would probably last many years. He had tried the experiment of dipping it in a very weak solution of tar, but this had the effect of matting together the fine filaments, so that it did not act so effectually as a strainer. It acted best when wet, because the fine particles of soot adhered to it, and moist weather was just the time when the greatest quantity of soot fell. It might be easily tried in London squares to aid in the growth of flowers; he found that the cabbage plants which were so protected throve remarkably well,

---

[34] I have followed up Mr. Trewby's hint, and find that more than one quality of scrim is made. The best, made entirely of flax, costs rather more than the 2¼d. stated in the estimate, but it is the cheapest practically. The best I have seen is that used in the Houses of Parliament.

and he had no doubt that if flowers were planted and a screen put over them until they were ready to bloom, it would be a great advantage. The action of a little peat on the top of a wall to protect fruit trees is very simple, and the explanation was afforded by the experiments of Dr. Wells on dew. The frosts which did the greatest mischief, were due to radiation from the ground on clear nights; and it would be found that if one thermometer were placed in a garden under an umbrella, and another on the open ground near it, the differences of temperature would be very considerable; on cloudy nights there was very little difference. Last night there was only a difference of 2°, but a few nights before it was 6°. The period of greatest cold might not probably be more than hour, but it would be sufficient to do a great deal of mischief, and anything which would check the radiation would have the required effect. In the case of loosening the fruit trees from the wall there was, probably, a double action; it prevented the tree being forced on by the warmth or the wall in the daytime, and also avoided the chilling effect at night, a rough wall being a good radiator, and sinking to a low temperature. He did not think there was much danger to be apprehended from wind, because the canvas being so open, the wind would pass freely through it; but he had not seen it subjected to any violent gale.

Chapter 43

SOLIDS, LIQUIDS, AND GASES

The growth of accurate knowledge is continually narrowing, and often obliterating, the broad lines of distinction that have been drawn between different classes of things. I well remember when our best naturalists regarded their "species" of plants and animals as fundamental and inviolable institutions, separated by well-defined boundaries that could not be crossed. Darwin has upset all this, and now we cannot even draw a clear, sharp line between the animal and vegetable kingdoms. The chemist is even crossing the boundary between these and the mineral kingdom, by refuting the once positive dictum that organic substances (i.e., the compounds ordinarily formed in the course of vegetable or animal growth) cannot be produced directly from dead matter by any chemical device. Many of such organic compounds are now made in the laboratory from mineral materials.

We all know, broadly, what are the differences between solids, liquids, and gases, and, until lately, they have been very positively described as the three distinct states or modes of existence of matter. Mr. Crookes suggests a fourth. I will not discuss this at present, but merely consider the three old-established claimants to distinctive existence.

A solid is usually defined as a body made up of particles which hold together rigidly or immovably, in contradistinction to a fluid, of which the particles move freely over each other. "Fluids" is the general term including both gases and liquids, both being alike as regards the mobility of their particles. At present, let us confine our attention to liquids and solids.

The theoretical or perfect fluid which is imagined by the mathematician as the basis of certain abstract reasonings has no real existence. He assumes (and the assumption is legitimate and desirable, provided its imaginary character is always remembered) that the supposed particles move upon each other with perfect freedom, without any friction or other impediment; but, as a matter of fact, all liquids exert some amount of resistance to their own flowing; they are more or less *viscous*, have more or less of that sluggishness in their obedience to the law of finding their own level which we see so plainly displayed by treacle or castor oil.

This viscosity, added to the friction of the liquid against the solid on which it rests, or in which it is enclosed, may become, even in the case of water, a formidable obstacle to its flow. Thus, if we make a hole in the side of a tank at a depth of 16 feet below the surface, the water will spout from that hole at the rate of 32 feet per second, but if we connect with this hole a long horizontal pipe of the same internal diameter as the hole, and then observe the flow from the outlet of the pipe, we shall find its velocity visibly diminished, and we shall be greatly deceived if we make arrangements for carrying swift-flowing water thus to any great distances.

Three or four years ago an attempt was made to supersede the water-carts of London by laying down on each side of the road a horizontal pipe, perforated with a row of holes opening towards the horse-way. The water was to be turned on, and from these holes it was to jet out to the middle of the road from each side, and thus water it all. I watched the experiment made near the Bank of England.

Instead of spouting across the road from all these holes, as it would have done from any *one* of them, it merely dribbled; the reason being that, in order to supply them all, the water must run through the whole of the long pipe with considerable velocity, and the viscosity and friction to be overcome in doing this nearly exhausted the whole force of water-head pressure. Many other similar blunders have been made by those who have sought to convey water-power to a distance by means of a pipe of such diameter as should demand a rapid flow through a long pipe.

The resistance which water offers to the stroke of the swimmer or the pull of the rower is partly due to its viscosity, and partly to the uplifting or displacement of some of the water. If it were perfectly fluid, our movements within it, and those of fishes, etc., would be curiously different; the whole face of this globe would be strangely altered in many respects.

I will not now follow up this idea, but leave it as a suggestion for the reader to work out for himself, by considering what would remain undone upon the earth if water flowed perfectly, without any internal resistance, or friction upon the earth's surface.

The degrees of approach to perfect fluidity vary greatly with different liquids.

Is there any such a thing as an absolute solid, or a body that has no degree of fluidity, the particles or parts of which will admit of no change of their relative positions, no movement upon each other without fracture of the mass? This would constitute perfect *rigidity*, or the opposite to *fluidity*.

Take a piece of copper or soft iron wire, about one eighth of an inch in diameter, or thereabouts, and bend it backwards and forwards a few times as rapidly as possible, but without breaking it; then, without loss of time, feel the portion that has been bent. It is hot—painfully so—if the experiment is smartly made. How may this be explained?

It is evident that in the act of bending there must have been a displacement of the relative positions of the particles of the metal, and the force demanded for the bending indicated their resistance to this movement upon each other; or, in other words, that there was friction between them, or something equivalent to such internal friction, and thus the mechanical force exerted in the bending was converted into heat-force.

Here, then, was fluidity, according to the above definition; not perfect fluidity, but fluidity attended with resistance to flow, or what we have agreed to call viscosity. But water also offers such resistance to flow, or viscosity, therefore the difference between iron or copper wire and liquid water as regards their fluidity is only a difference of degree, and not of kind; the demarcation between solids and liquids is not a broad, clearly-defined line, but a band of blending shade, the depths of tint representing varying degrees of viscosity.

Multitudes of examples may be cited illustrating the viscosity of bodies that we usually regard as types of solidity, such, for example, as the rocks forming the earth's crust. In the "Black Country" of South Staffordshire, which is undermined by the great ten-yard coal-seam, cottages, chimney-shafts, and other buildings may be seen leaning over most grotesquely, houses split down the middle by the subsidence or inclination of one side, great hollows in fields or across roads that were once flat, and a variety of other distortions, due to the gradual sinking of the rock-strata that have been undermined by the colliery workings. In some cases the rocks are split, but usually the subsidence is a bending or flowing down of the rocks to fill up the vacuity, as water fills a hollow, or "finds its own level."

I have seen many cases of the downward curvature of the roof of a coal-pit, and have been told that in some cases the surrounding pressure causes the floor to curve upwards, but have not seen this.

Earthquakes afford another example. The so-called solid crust of the earth is upheaved, and cast into positive billows that wave away on all sides from the centre of disturbance. The earth-billows of the great Lisbon earthquake of 1755 traveled to this country, and when they reached Loch Lomond, were still of sufficient magnitude to raise and lower its banks through a perpendicular range of two feet four inches.

It is quite possible, or, I may say, probable, that there are tides of the earth as well as of the waters, and the subject has occupied much attention and raised some discussion among mathematicians. If the earth has a fluid centre, and only a comparatively thin crust, as some suppose, there must be such tides, produced by the gravitation of the moon and sun.

Ice presents some interesting results of this viscosity. At a certain height, varying with latitude, aspect, etc., we reach the "snow line" of mountain slopes, above which the snow of winter remains unmelted during summer, and, in most cases, goes on accumulating. It soon loses its flocculent, flaky character, and becomes coherent, clear blue ice by the pressure of its own weight.

A rather complex theory has been propounded to explain this change—the theory of *regelation*—i.e., re-freezing; a theory which assumes that the pressure first thaws a film of ice at the surface of contact, and that presently this re-freezes, and thus effects a healing or general solidification. Faraday found that two pieces of ice with moistened surfaces united if pressed together when at just about the temperature of freezing, but not if much colder. Tyndall has further illustrated this by taking fragments of ice and squeezing them in a mould, whereby they became a clear, transparent ball, or cake. Schoolboys did the like long before, when snowballing with snow at about the thawing point. Such snow, as we all remember, became converted into stony lumps when firmly pressed together. We also remember that in much colder weather no such cohesion occurred, but our snowballs remained powdery in spite of all our squeezing.

I am a sceptic as regards this theory of regelation. I believe that the true explanation is much simpler; that the crystals of snow or fragments of ice in these experiments are simply welded, as the smith unites two pieces of iron, by merely pressing them together when they are near their melting point. Other metals and other fusible substances may be similarly welded, provided they soften or become sufficiently viscous before fusing.

Platinum is a good example of this. It is infusible in ordinary furnaces, but becomes pasty before melting, and therefore, one method adopted in the manufacture of platinum ingots or bars from the ore, is to precipitate a sort of platinum snow (spongy platinum) from its solution in acid, and then compress this metallic snow in red-hot steel moulds by means of pistons driven with great force. The flocculent metal thus becomes a solid, coherent mass, just as the flocculent ice became coherent ice in Tyndall's experiment or in making hard snowballs.

Wax, pitch, resin, and all other solid that fuse *gradually*, cohere, are weldable, or, in very plain language, "stick together," when near their fusing point.

I have made the following experiment to prove that when this so-called regulation of snow or ice-fragments occurs, the ice is viscous or plastic, like wax or pitch. A strong iron squirt, with a cylindrical bore of half an inch in diameter, is fitted with an iron piston. This piston is driven forth by a screw working in a collar at one end of the squirt. Into the other end is screwed a brass nozzle with an aperature about one twentieth of an inch diameter, tapering or opening inwards gradually to the half-inch bore.

Into this bore I place snow or fragments of ice, then, holding the body of the squirt firmly in a vice, I work the lever of the screw, and thus drive forward the piston and crush down the snow or ice-fragments, which presently become coherent and form a half-inch solid cylinder of clear ice. Applying still more pressure, this cylinder is forced like a liquid through the small orifice of the nozzle of the squirt, and it jets or spouts out as a thin stick of ice like vermicelli, or the "leads" of ever-pointed pencils, for the moulding of which the squirt was originally constructed.

I find that ice at 32° can thus be squirted more easily than beeswax of the same temperature, and such being the case, I see no reason for imagining any complex operation of regulation in the case of the ice, but merely regard the adhesion of two pieces of ice when pressed together as similar to the sticking together of two pieces of cobblers'-wax, or softened sealing-wax, or beeswax, or the welding of iron or glass when heated to their welding temperatures, i.e., to a certain degree of incipient fluidity or viscosity.

If a leaden bullet be cut in half, and the two fresh-cut faces pressed forcibly together, they cohere at ordinary atmospheric temperatures, but we have no occasion for a regulation theory here. The viscosity of the lead accounts for all. At Woolwich Arsenal there is a monster squirt, similar to my little one. This is charged with lead, and, by means of hydraulic pressure, the lead is squired out of the nozzle as a cylindrical jet of any required diameter. This jet or stick of lead is the material of which the elongated cylindrical rifle bullets are now made.

But returning to the point at which we started, on the subject of ice, viz., its Alpine accumulation above the snow-line. If the snow-fall there exceeds the amount that is thawed and evaporated, it must either go on growing upward until it reaches the highest atmospheric region from which it falls, or is formed, or it must descend somehow.

If ice can be squirted through a syringe by mere hand-pressure, we are justified in expecting that it would be forced down a hill slope, or through a gully, or across a plain, by the pressure of its own weight when the accumulation is great. Such is the case, and thus are glaciers formed.

They are, strictly speaking, rivers or torrents of ice; they flow as liquid water does, and down the same channels as would carry the liquid surface drainage of the hills, were rain to take the place of snow. Like rivers, they flow with varying speed, according to the slope; like rivers, their current is more rapid in the middle than the sides; like rivers, they exert their greatest tearing force when squeezed narrow through gullies; and, like rivers, they spread out into lakes when they come upon an open basin-like valley, with narrow outlet.

The Justedalsbrae of Norway is a great ice-lake of this character, covering a surface of about 500 square miles, and pouring down its ice-torrents on every side, wherever there is a notch or valley descending from the table-land it covers. The rate of flow of such downpouring glaciers varies from two or three inches to as many feet per day, and they present magnificent examples of the actual fluidity or viscosity of an apparently solid mass. This viscosity has been disputed, and attempts have been made to otherwise explain the motion of glaciers; but while it is possible that it may be assisted by varying expansion and contraction, the downflow due to viscosity is now recognized as unquestionably the main factor of glacier motion.

Cascades of ice may be sometimes seen. In the course of my first visit to Norway, I wandered alone over a very desolate mountain region towards the head of the Justedal, and unexpectedly came upon a gloomy lake, the Styggevand, which lies at the foot of a precipice-boundary of the great ice-field above named. Here, the ice having no sloping valley-trough by which to descend, poured over the edge of the precipice as a great overhanging sheet or cornice, which bent down as it was pushed forward, and presented on the convex side of the sheet some fine blue cracks, or "crevasses" as they are called. These gradually widened and deepened, until the overhanging mass broke off and fell into the lake, on the surface of which I saw the result, in the form of several floating icebergs that had previously fallen.

Something like this, on a small scale, may be seen at home on the edge of a house roof, on which there has been an accumulation of snow; but, in this case, it is rather sliding than flowing that has made the cornice; but its *down-bending* is a result of viscosity.

These and a multitude of other facts that might be stated, many of which will occur to the reader, prove clearly enough that the solid and liquid states of matter are not distinctly and broadly separable, but are connected by an intermediate condition of viscosity.

We now come to the question whether there is any similar continuity between liquids and gases. Ordinary experience decidedly suggests a negative answer. We can point to nothing within easy reach that has the properties of a liquid and gaseous half-and-half; that stands between gases and liquids as pitch and treacle stand between solids and liquids.

Some, perhaps, may suggest that cloud-matter—London fog, for example—is in such an intermediate state. This, however, is not the case. White country fog, ordinary clouds, or the so-called "steam" that is seen assuming cloud forms as it issues from the spout of a tea-kettle or funnel of a locomotive, consists of minute particles of water suspended in air, as solid particles of dust are also suspended. It has been called "vesicular vapor," on the supposition that it has the form of minute vesicles, like soap-bubbles on a very small scale, but this hypothesis remains unproven. London fog consists of similar particles, varnished with a delicate film of coal-tar, and intersprinkled with particles of soot.

In order to clearly comprehend the above-stated question, we must define the difference between liquids and gases. In the first place, they are both fluids, as already agreed. What, then, is the essential difference between liquid fluidity and gaseous fluidity? The expert in molecular mathematics, discoursing to his kinematical brethren, would produce a tremendous reply to this question. He would describe the oscillations, gyrations, collisions, mean free paths, and mutual obstructions of atoms and molecules, and, by the aid of a maddening array of symbols, arrive at the conclusion that gases, unless restrained, expand of their own accord, while liquids retain definite limits or dimensions.

The matter-of-fact experimentalist demonstrates the same by methods that are easily understood by anybody. I shall, therefore, both for my own sake and my readers', describe some of the latter.

In the first place, we all see plainly that liquids have a surface, i.e., a well-defined boundary, and also that gases, unless enclosed, have not. But as this may be due to the invisibility of the gas, we must question it further. The air we breathe may be taken as a type of gases, as water may of liquids. It has weight, as we may prove by weighing a bottle full of air, then pumping out the contents, weighing the empty bottle, and noting the difference.

Having weight, it presses towards the earth, and is squeezed by all that rests above it; thus the air around us is constrained air. It is very compressible, and is accordingly compressed by the weight of all the air above it.

This being understood, let us take a bottle full of water and another full of air, and carry them both to the summit of Mont Blanc, or to a similar height in a balloon. We shall then have left nearly half of the atmosphere below, and thus both liquid and gas will be under little more than half of the ordinary pressure. What will happen if we uncork them both? The liquid will still display its definite surface, and remain in the bottle, but not so the gas. It will overflow upwards, downwards, or sideways, no matter how the bottle is held, and if we had tied an empty bladder over the neck before uncorking, we should find this overflow or expansion of the gas exactly proportionate to the removal of pressure, provided the temperature remained unaltered. Thus, at just half the pressure under which a pint bottle was corked, the air would measure exactly one quart, at one-eighth of the pressure one gallon, and so on.

We cannot get high enough for the latter expansion, but can easily imitate the effect of further elevation by means of an air-pump. Thus, we may put one cubic inch of air into a bladder of 100 cubic inches capacity, then place this under the receiver of an air-pump, and reduce the pressure outside the bladder to 1/100th of its original amount. With such atmospheric surrounding, the one cubic inch of air will plump out the flaccid bladder, and completely fill it. The pumpability of the air from the receiver shows that it goes on overflowing from it into the piston of the pump as fast as its own elastic pressure on itself is diminished.

Numberless other experiments may be made, all proving that all gases are composed of matter which is not merely incohesive, but is energetically self-repulsive; so much so, that it can only be retained within any bounds whatever by means of some external pressure or constraint. For aught we know *experimentally*, the gaseous contents of one of Mr. Glaisher's baloons would outstretch itself sufficiently to occupy the whole sphere of space that is spanned by the earth's orbit, provided that space were perfectly vacuous, and the baloon were burst in the midst of it, the temperature of the expanding gas being maintained.

Here, then, in this self-repulsiveness, instead of self-cohesion, this absence of self-imposed boundary or dimensions, we have a very broad and well-marked distinction between gases and liquids, so broad that there seems no bridge that can possibly cross it. This was believed to be the case until recently. Such a bridge has, however, been built, and rendered visible, by the experimental researches of Dr. Andrews; but further explanation is required to render this generally intelligible.

Until quite lately it was customary to divide gases into two classes—"permanent gases" and "condensable gases," or "vapors." Gaseous water or steam was usually described as typical of the latter; oxygen, hydrogen, or nitrogen of the former. Earlier than this, many other gases were included in the permanent list; but Faraday made a serious inroad upon this classification when he liquefied chlorine by cooling and compressing it. Long after this, the gaseous elements of water, and the chief constituents of air, oxygen, hydrogen, and nitrogen, resisted all efforts to condense them; but now they have succumbed to great pressure and extreme cooling.

We thus arrive at a very broad generalization, viz., that all gases are physically similar to steam (I mean, of course, "dry steam," i.e., true invisible steam, and not the cloudy matter to which the name of steam is popularly given), that they are all formed by raising liquids above their boiling point, just as steam is formed when we boil water and maintain the steam above the boiling-point of the water.

But some liquids boil at temperatures far below that at which others freeze; liquid chlorine boils at a temperature below that of freezing water, and liquid carbonic acid below even that of freezing mercury, and liquid hydrogen far lower still. These are cases of boiling, nevertheless, though it seems a paradox according to the ideas we commonly attach to this word. But such ideas are based on our common experience of the properties of our commonest of liquids, viz., water.

When water boils under the conditions of our ordinary experience, the passage from the liquid to the gaseous state is a sudden leap, with no intermediate state of existence that we are able to perceive; and the conditions upon which water is converted into steam—the liquid into the gas—while both are at the bottom of our atmospheric ocean, are such as to render an intermediate condition rationally, as well as practically, impossible.

We find that the expansive energy by which the steam is enabled to resist atmospheric pressure is conferred upon it by its taking into itself, and utilizing for its expansive efforts a large amount of calorific energy. When any given quantity of water is converted into steam, under ordinary circumstances, its bulk *suddenly* becomes above 1700 times greater—a cubic inch of water forms about a cubic foot of steam, and nearly 1000 degrees of heat (966·6) disappears *as temperature*. Otherwise stated, we must give to the cubic inch of water at 212° as much heat as would raise it to a temperature of 212 plus 966·6, or 1,178·6°, if it remained liquid. This is about the temperature of the glowing coals of a common fire; but the steam that has thus taken enough heat to make the water red-hot is still at 212°—no *hotter* than the water was while boiling.

This heat, which thus ceases to exhibit itself as *temperature*, is otherwise occupied. Its energy is partly devoted to the work of increasing the bulk of the water to the above-named extent, and partly in conferring on the steam its gaseous specialty—that is, in overcoming liquid cohesion, and substituting for it the opposite property of internal repulsive energy which is characteristic of gases. My reasons for thus defining and separating these two functions of the so-called "latent" heat will be seen when we come to the philosophy of the interesting researches of Dr. Andrews.

As already explained, all gases are now proved to be analogous to steam, they are matter expanded and rendered self-repulsive by heat. All *elementary* matter may exist in either of the three forms—solid, liquid, or gas, according to the amount of heat and pressure to which it is subjected. I limit this wide generalization to *elementary* substances for the following reasons:

Many compounds are made up of elements so feebly held together that they become "dissociated" when heated to a temperature below their boiling-point; or, their condition maybe otherwise defined by stating that the bonds of chemical energy, which hold their elements together, are weaker than the cohesion which binds and holds them in the condition of solid or liquid, and are more easily broken by the expansive energy of heat.

To illustrate this, let us take two common and well-known oils—olive oil and turpentine. The first belongs to the class of "fixed oils," and second to the "volatile oils." If we apply heat to liquid turpentine, it boils, passes into the state of gaseous turpentine, which is easily condensible by cooling it. If the liquid result of this condensation is examined, we find it to be turpentine as before. Not so with the olive oil. Just as this reaches its boiling point, the heat, which would otherwise convert it into olive-oil vapor, begins to dissociate its constituents, and if the temperature be raised a little higher, we obtain some gases, but these are the products of decomposition, not gaseous olive oil. This is called "destructive" distillation.

In olive oil, the boiling-point and dissociation point are near to each other. In the case of glycerine, these points so nearly approximate that, although we cannot distil it unbroken under ordinary atmospheric pressure, we may do so if some of this pressure is removed. Under such diminished pressure, the boiling-point is brought down below the dissociation point, and condensible glycerine gas comes over without decomposition.

Sugar affords a very interesting example of dissociation, commencing far below the boiling-point, and going on gradually and visibly, with increasing rapidity as the temperature is raised. Put some white sugar into a spoon, and heat the spoon gradually over the smokeless gas-flame or spirit-lamp. At first the sugar melts, then becomes yellow (barley sugar); this color deepens to orange, then red, then chestnut-brown, then dark brown, then nearly black (caramel), then quite black, and finally it becomes a mere cinder. Sugar is composed of carbon and water; the heat dissociates this compound, separates the water, which passes off as vapor, and leaves the carbon behind. The gradual deepening of the color indicates the gradual carbonization, which is completed when only the dry insoluble cinder remains. An appearance of boiling is seen, but this is the boiling of the dissociated water, not of the sugar.

The dissociation temperature of water is far above its boiling-point. It is 5072° Fahr., under conditions corresponding to those which make its boiling-point 212°. If we examine the variations of the boiling-point of water, as the atmospheric pressure on its surface varies, some curious results follow. To do this the reader must endure some figures. They are extremely simple, and perfectly intelligible, but demand just a little attention.

Following are three columns of figures. The first represents atmospheres of pressure—i.e., taking our atmospheric pressure when it supports 30 inches of mercury in the barometer tube as a unit, that pressure is doubled, trebled, etc., up to twenty times in the first column. The second column states the temperature at which water boils when under the different pressures thus indicated. The third column, which is the subject for special study just now, shows how much we must rise the temperature of the water in order to make it boil as we go on adding atmospheres of pressure; or, in other words, the increase of temperature due to each increase of one atmosphere of pressure. The figures are founded on the experiments of Regnault.

It may be seen from the above that, with the exception of one irregularity, there is a continual diminution of the additional temperature which is required to overcome an additional atmosphere of pressure, and if this goes on as the pressure and temperatures advance, we may ultimately reach a curious condition—a temperature at which additional pressure will demand no additional temperature to maintain the gaseous state; or, in other words, a temperature may be reached at which no amount of pressure can condense steam into water, or at which the gaseous and liquid states merge or become indifferent.

But we must not push this mere numerical reasoning too far, seeing that it is quite possible to be continually approaching a given point, without ever reaching it, as when we go on continually halving the remaining distance. The figures in the above do not appear to follow according to such a law—nor, indeed, any other regularity. This probably arises from experimental error, as there are discrepancies in the results of different investigators. They all agree, however, in the broad fact of the gradation above stated. Dulong and Arago, who directed the experiments of the French Government Commission for investigating this subject, state the pressure at 20 atmospheres to be 418·4, at 21 = 422·9, at 22 = 427·3, at

23 = 431·4, and at 24 atmospheres, their highest *experimental* limit, 435·5, thus reducing the rise of temperature between the 23d and 24th atmospheres to 4·1.

Pressure in Atmospheres	Temperature, F.	Rise of Temperature for each additional Atmosphere
1	212	
2	249·5	37·5
3	273·3	23·8
4	291·2	17·9
5	306·0	14·8
6	318·2	12·2
7	329·6	11·4
8	339·5	9·9
9	348·4	8·9
10	356·6	8·2
11	364·2	7·6
12	371·1	6·9
13	377·8	6·7
14	384·0	6·2
15	390·0	6·0
16	395·4	5·4
17	400·8	5·4
18	405·9	5·1
19	410·8	4·9
20	415·4	4·6

If we could go on heating water in a transparent vessel until this difference became a vanishing quantity, we should probably recognize a visible physical change coincident with this cessation of condensibility by pressure; but this is not possible, as glass would become red-hot and softened, and thus incapable of bearing the great pressure demanded. Besides this, glass is soluble in water at these high temperatures.

If, however, we can find some liquid with a lower boiling-point, we may go on piling atmosphere upon atmosphere of elastic expansive pressure, as the temperature is raised, without reaching an unmanageable degree of heat. Liquid carbonic acid, which, under a single atmosphere of pressure, boils at 112° below the zero of our thermometer, may thus be raised to a temperature having the same relation to its boiling-point that a red-heat has to that of water, and may be still confined within a glass vessel, provided the walls of the vessel are sufficiently thick to bear the strain of the elastic outstriving pressure. In spite of its brittleness glass is capable of bearing an enormous strain *steadily applied*, as may be proved by trying to break even a mere thread of glass by direct pull.

Dr. Andrews thus treated carbonic acid, and the experiment, as I have witnessed its repetition, is very curious. A liquid occupies the lower part of a very strong glass tube, which appears empty above. But this apparent void is occupied by invisible carbonic acid gas, evolved by the previous boiling of the liquid carbonic acid below. We start at a low temperature—say 40° Fahr. Then the temperature is raised; the liquid boils until it has given off sufficient gas or vapor to exert the full expansive pressure or tension due to that temperature. This pressure stops the boiling, and again the surface of the liquid is becalmed.

This is repeated at a higher temperature, and thus continued until we approach nearly to 88° Fahr., when the surface of the liquid loses some of its sharp outline. Then 88° is reached, and the boundary between liquid and gas vanishes; liquid and gas have blended into one mysterious intermediate fluid; an indefinite fluctuating something is there filling the whole of the tube—an etherealized liquid or a visible gas. Hold a red-hot poker between your eye and the light; you will see an upflowing wavy movement of what appears like liquid air. The appearance of the hybrid fluid in the tube resembles this, but is sensibly denser, and evidently stands between the liquid and gaseous states of matter, as pitch or treacle stands between solid and liquid.

The temperature at which this occurs has been named by Dr. Andrews the "*critical temperature*"; here the gaseous and liquid states are "*continuous*," and it is probable that all other substances capable of existing in both states have their own particular critical temperatures.

Having thus stated the facts in popular outline, I shall conclude the subject by indulging in some speculations of my own on the philosophy of these general facts or natural laws, and on some of their possible consequences.

As already stated, the conversion of water into steam under ordinary atmospheric pressure demands 966·6° of heat over and above that which does the work of raising the water to 212°, or, otherwise stated, as much heat is at work in a given weight of steam at 212°, as would raise the same quantity of water to 1178·6° if it remained liquid.

James Watt concluded from his experiments that a given weight of steam, whatever may be its density, or, in other words, under whatever pressure it may exist, contains the same quantity of heat. According to this, if we reduced the pressure sufficiently to bring down the boiling-point to 112°, instead of 212°, the latent heat of the steam thus formed would be 1066·6° instead of 966·6°, or if, on the other hand, we placed it under sufficient pressure to raise the boiling-point to 312°, the latent heat of the steam would be reduced to 866·6°, i.e., only 866·6° more would be required to convert the water into steam. If the boiling-point were 412°, as it is between 19 and 20 atmospheres of pressure, only 766·6° more heat would be required, and so on, till we reached a pressure which raised the boiling-point to 1178·6°; the water would then become steam without further heating, i.e., the critical point would be reached, and thus, if Watt is right, we can easily determine, theoretically, the critical temperature of water.[35]

Mr. Perkins, who made some remarkable experiments upon very high pressure steam many years ago, and exhibited a steam gun at the Adelaide Gallery, stated that red-hot water does not boil; that if the generator be sufficiently strong to stand a pressure of 60,000 lbs. load on the safety-valve, the water may be made to exert a pressure of 56,000 lbs. on the square inch at a cherry-red heat without boiling. He made a number of rather dangerous experiments in thus raising water to a red-heat, and his assertion that red-hot water does not boil is curious when viewed in connection with Dr. Andrews' experiments.

I cannot tell how he arrived at this conclusion, having been unable to obtain the original record of his experiments, and only quote the above second hand. It is worthy of remark that the temperature he names is about 1170°, or that which, if Watt is right, must be the critical temperature of the water. Perkins' red-hot water would not boil, being then in the intermediate condition.

---

[35] Watt's own figure for the latent heat of steam at 212° was 950°, but I adopt that which is now generally accepted.

So far, we have a nice little theory, which not only shows how the critical state of water must be reached, but also its precise temperature; but all this is based on the assumption that Watt made no mistake.

Unfortunately for the simplicity of this theory, Regnault states that *his* experiments contradict those of Watt, and prove that the latent heat of steam does not diminish just in the same degree as the boiling-point is raised, but that instead of this the diminution of the latent heat progresses 30½ per cent more slowly than the rise of temperature, so that, instead of the latent heat of steam between boiling-points of 212° and 312° falling from 966·6° to 866·6° it would only fall to 895·1° or 69·5° of latent heat for every 100° of temperature.

If this is correct, the temperature at which the latent heat of steam is reduced to zero is much higher than 1178·6°, and is, in fact, a continually receding quantity never absolutely reached; but I am not prepared to accept these figures of Regnault as implicitly as is now done in text-books (I was nearly saying "as is now the fashion"), seeing that they are not the actual figures obtained by his experiments, but those of his "empirical formulæ" based upon them. His actual experimental figures are very irregular; thus, between steam temperature of 171·6° and 183·2° a difference of 11·6°, the experimental difference in the latent heat came out as 4·7°; between steam temperature of 183·2° and 194·8°, or 11·6° again, the latent heat difference is tabulated as 8·0°.

Regnault's experiments were not carried to very high temperatures and pressures, and indicate that as these advance the deviation from Watt's law diminishes, and may finally vanish at about 1500° or 1600°, where the latent heat would reach zero, and there, according to the above, the critical temperature would be reached. Any additional heat applied after this will have but one function to perform, viz., the ordinary work of increasing the bulk of the heated body without doing anything further in the way of conferring upon it any new self-repulsive properties.

Our notions of solids, liquids, and gases are derived from our experiences of the state of matter here upon this earth. Could we be removed to another planet, they would be curiously changed. On Mercury water would rank as one of the condensible gases; on Mars, as a fusible solid; but what on Jupiter?

Recent observations justify us in regarding this as a miniature sun, with an external envelope of cloudy matter, apparently of partially condensed water, but red-hot, or probably still hotter within. His vaporous atmosphere is evidently of enormous depth, and the force of gravitation being on his visible outer surface two and a half times greater than that on our earth's surface, the atmospheric pressure in descending below this visible surface must soon reach that at which the vapor of water would be brought to its critical condition. Therefore we may infer that the oceans of Jupiter are neither of frozen liquid nor gaseous water, but are oceans or atmospheres of critical water. If any fish-birds swim or fly therein they must be very critically organized.

As the whole mass of Jupiter is three hundred times greater than that of the earth, and its compressing energy towards the centre proportional to this, its materials, if similar to those of the earth and no hotter, would be considerably more dense, and the whole planet would have a higher specific gravity; but we know by the movement of its satellites that, instead of this, its specific gravity is less than a fourth of that of the earth. This justifies the conclusion that it is intensely hot, for even hydrogen, if cold, would become denser than Jupiter under such pressure.

As all elementary substances may exist as solids, liquids, or gases, or critically, according to the conditions of temperature and pressure, I am justified in hypothetically concluding that Jupiter is neither a solid, a liquid, nor a gaseous planet, *but a critical planet*, or an orb composed internally of dissociated elements in the critical state, and surrounded by a dense atmosphere of their vapors, and those of some of their compounds, such as water. The same reasoning applies to Saturn and the other large and rarefied planets.

The critical temperature of the dissociated elements of the sun is probably reached at the base of the photosphere, or that region revealed to us by the sun-spots. When I wrote "The Fuel of the Sun," thirteen or fourteen years ago, I suggested, on the above grounds, the then heretical idea of the red-heat of Jupiter, Saturn, Uranus, and Neptune, and showed that all such compounds as water must be dissociated at the base of the sun's atmosphere; but being then unacquainted with the existence of this critical state of matter, I supposed the dissociated elements to exist as gases with a small solid nucleus or kernel in the centre.

Applying now the researches of Dr. Andrews to the conditions of solar existence, as I formerly applied the dissociation researches of Deville, I conclude that the sun has no nucleus, either solid, liquid, or gaseous, but is composed of dissociated matter in the critical state, surrounded, first, by a flaming envelope due to the re-combination of the dissociated matter, and outside of this another envelope of vapors due to this combination.

Chapter 44

MURCHISON AND BABBAGE

The curious contrast of character presented by these two eminent men, and the very different course of their lives, conveys a striking lesson to all those superficial thinkers and unthinking talkers who make sweeping generalizations concerning human character; who assume as a matter of course that any man who writes poetry must be merely a dreamer of day-dreams, incapable of transacting any practical daily business, and not at all reliable in money matters; whose eyes are always "in a fine frenzy rolling"; that he is, in short, a sort of amiable, harmless lunatic. All actors, according to such people, are dissipated spendthrifts; and if Sims Reeves, or any other public performer, is prevented by delicate larynx or other indisposition from appearing, they look knowing, shrug their shoulders, wink wisely, and assume, without the faintest shadow of evidence, that he is drunk.

In like manner they set up a typical philosopher of their own manufacture, and attribute his imaginary character to all who devote themselves to science. Their philosopher is a musty, dried-up, absent-minded pedant, whose ordinary conversation is conducted in words of seven syllables, who is always lost in profound abstractions; takes no interest in common things; regards music, dancing, play-acting, poetry, and every cheerful pursuit as frivolous and contemptible—a creature who never makes a joke, seldom laughs, and who in matters of business is even more incapable than the poet.

The singular contrast of character presented by Babbage and Murchison affords at once a most complete refutation of such generalizations. Here were two men, both philosophers, one the very type of amiability, suavity, and all conceivable polish, the very perfection of a courtier, but differing from the vulgar courtier of the Court in this respect, that his high-toned courtesy was not bestowed upon kings only, but also upon all his human brethren, and with especial gracefulness upon those whose rank was below his own.

I doubt whether there is any man now living, or has lived during this generation, that could equal Sir Roderick Murchison in the art of distributing showers of compliments upon a large number of different people in succession, and making each recipient delightfully satisfied with himself. In his position as Chairman to the Geological Section of the British Association, he did this with marvelous tact, without the least fulsomeness or repetition, or any display of patronizing. Every man who read a paper before that section was better than ever satisfied with the great merits and vast importance of his communication, after hearing the Chairman's comments upon it. None but a most detestably strong-minded and logical brute could resist the insinuating flattery of Sir Roderick.

How different was poor Babbage! Who that attends any sort of scientific gatherings has not seen Sir Roderick? but who in the world, excepting the organ-grinders and the police magistrate has ever seen Babbage, or even his portrait? What a contrast between the seclusion and the public existence; between the hedgehog bristles and the velvet softness, of the one and the other!

Those who were on intimate terms with Babbage (I have never met or heard of such a person) could probably tell us that all his irritability and roughness were outside, and that, in the absence of organ-grinders, he was a kind and amiable gentleman; but, even admitting this, the contrast between the two philosophers is as great as could well be found between any two men following the most widely divergent studies or professions.

Those who would reply that mathematics and geology are such different studies have only to go a little further back on the death-roll, and they will find the name of De Morgan, a pure mathematician, like Babbage. He was a man of exuberant fun and humor, and so far from hating music of either a humble or pretentious character, was a highly accomplished musician, both theoretical and practical, and if we are to believe confidential communications, one of his favorite instruments was the penny whistle, on which he was a most original and peculiar performer.

I had not intended to reprint the above, which was written just after the death of Murchison and Babbage, but the comments that have recently followed the death of Darwin induce me to do so.

Many have expressed their surprise at the unanimous expressions of Darwin's friends concerning the geniality of his disposition, his gentleness, cheerfulness; his *genuine* humility and simplicity of character.

A third type of character is here presented, and that which corresponds most correctly with the true ideal of a modern philosopher, also represented by that great master of experimental science, Faraday. In both of these there was the full measure of Murchison's amiability, but without the courtly polish of the ex-soldier. Philosophic meditation and close application to original research may, and often does, induce a certain degree of shyness due to a consciousness of the social disqualification which arises from that inability to fulfil all the demands for small attentions which constitute conventional politeness; a disability due to habits of consecutive thought and mental abstraction.

A sensitive and amiable man would suffer much pain on finding that he had neglected to supply the small wants of the lady sitting next to him at a dinner party, and would withdraw himself from the risk of repeating such unwitting rudeness. This holding back from ordinary society, though really due to a conscientious sense of social duty and tender regard for the feelings of others, is too often referred to a churlish unsociality or arrogant assumption of superiority.

If Newton really did mistake the lady's finger for a tobacco-stopper, depend upon it the pain he suffered was far more acute than that which he inflicted, and was suffered over and over again whenever the incident was recollected.

Chapter 45

ATMOSPHERE *VERSUS* ETHER

One of the most remarkable meteors of which we have a reliable record appeared on February 6, 1818. Several accounts of it were published, the fullest being that in *The Gentleman's Magazine* of the time. (I may here add, parenthetically, that one reason why I have especial pleasure in writing these notes is that they contribute something towards the restoration of the ancient status of this magazine, which was at one time the only English serial that ventured upon any notable degree of exposition of *popular* science.)

Upon the data supplied by this account, Mr. Joule has calculated the height of the meteor to have been 61 miles above the surface of the earth, and he states that "this meteor is one of the few that have been seen in the daytime, and is also interesting as having been one of the first whose observation afforded materials for the estimation of its altitude." It was seen in the neighborhood of Cambridge at 2 P.M., also at Swaffham in Norfolk, and at Middleton Cheney near Banbury. The distance between this and Cambridge is sufficient to afford a measurement of its height, provided its position above the horizon at both places was determined with tolerable accuracy.

According to the orthodox text-books, the atmosphere of this earth terminates at a height of about 45 or 50 miles, or, if not absolutely ended there, it ceases to be of appreciable density anywhere above this elevation.

But here we have a fact which flatly contradicts the calculation. At 61 miles above the earth's surface there must be atmospheric matter of sufficient density to offer to the passage of this meteor through it an amount of resistance which produced an intense white heat, visible by its luminosity in broad daylight.

In the above-quoted paper, read by Mr. Joule before the Manchester Literary and Philosophical Society on December 1, 1863, he refers to subsequent observations and estimates 116 miles as "the elevation at which meteors in general are first observed"—i.e., where our atmosphere is sufficiently dense to generate a white-heat by the resistance it offers to the rapidly flying meteor.

It is curious to observe how, in dealing with actual physical facts, a mathematician of the solid practical character of Joule becomes compelled to practically throw overboard the orthodox theory of limited atmospheric extension. Here, in making his calculations of the resistance of atmospheric matter at this elevation, he bases them on the assumption of a decrease of density at the rate of "one quarter for every seven miles," and indicates no limit at which this rate shall vary. Very simple arithmetic is sufficient to show that this leads us to the unlimited atmospheric extension, for which I have contended we may go on for ever taking off a quarter at every seven miles, and there will still remain the three quarters of the quantity upon which we last operated, or, more practically stated, we shall thus go on seven after seven until we reach the boundaries of the atmospheric grasp of the gravitation of some other sphere.

Surely the time has arrived for the full reconsideration of this fundamental question of whether the universe is filled with atmospheric matter or is the vacuum of the molecular mathematicians plus the imaginary "ether," which has been invented by its mathematical creators only to extricate them from the absurd dilemma into which they are plunged when they attempt to explain the transmission of light and heat by undulations traveling through space containing nothing to undulate.

They have filled it with immaterial matter evolved entirely from their own consciousness, which they have gratuitously endowed with whatever properties are required for the fitting of their theories—properties that are self-contradictory and without any counterpart in anything seen or known outside of the fertile imagination of these reckless theorists.

We know of nothing that can penetrate every form of matter without adding either to its weight or its bulk; we know of nothing that can communicate motion to ponderable matter without itself being ponderable—i.e., having the primary property of matter, viz., mass, or weight, and consequent *vis viva* when moving; we know of nothing that can set bodies in motion without proportionally resisting the motion of bodies through it; and if the waving of the ether is (as Tyndall describes it) "as real and as truly mechanical as the breaking of sea-waves upon the shore," the material of the breakers must be like the "jelly" to which he compares it, and have some viscosity, or resistance to penetration, or pushing aside.

We have not a shadow of direct evidence of the existence of the "interatomic" spaces occupied by the other, and in the midst of which the atoms are made to theoretically swing, nor even of the existence of the atoms themselves.

The "ether" of to-day, with its imaginary penetration and its material action without material properties, has merely taken the place of the equally imaginary phlogiston, caloric, electric, and magnetic fluids, the "imponderables" of the past. I have little doubt that ere long the modern modification of these physical superstitions will share their fate, and we shall all adopt the simple conception that heat, light, end electricity are, like sound, merely transmissible states or affections of matter itself regarded bodily, as it is seen and felt to exist.

This may possibly throw a good many mathematicians out of work—or into more useful work; but, however that may be, it will certainly aid the general diffusion of science as the intellectual inheritance of every human being. At present the explanations of the simple phenomena of light and heat are incomparably more difficult to understand and to account for than the facts which they attempt to elucidate.

Chapter 46

A Neglected Disinfectant

When the household of our grandmothers was threatened with infection, the common practice was to sprinkle brimstone on a hot shovel or on hot coals on a shovel, and carry the burning result through the house. But now this simple method of disinfecting has gone out of fashion without any good and sufficient reason. The principal reason is neither good nor sufficient, viz., that nobody can patent it and sell it in shilling and half-crown bottles.

On September 18th last, M. d'Abbadie read a paper at the Academy of Sciences on "Marsh Fevers," and stated that in the dangerous regions of African river mouths immunity from such-fevers is often secured by sulphur fumigations on the naked body. Also that the Sicilian workers in low ground sulphur mines suffer much less than the rest of the surrounding population from intermittent fevers. M. Fouqué has shown that Zephyria (on the volcanic island of Milo or Melos, the most westerly of the Cyclades), which had a population of 40,000 when it was the centre of sulphur-mining operations, became nearly depopulated by marsh fever when the sulphur-mining was moved farther east, and the emanations prevented by a mountain from reaching the town. Other similar cases were stated.

It is well understood by chemists that bleaching agents are usually good disinfectants; that which can so disturb an organic compound as to destroy its color, is capable of either arresting or completing the decompositions that produce vile odors and nourish the organic germs or ferments which usual accompany, or, as some affirm, cause them. Sulphurous acid is, next to hypochlorous acid, one of the most effective bleaching agents within easy reach.

I should add that sulphurous acid is the gas that is *directly* formed by burning sulphur. By taking up another dose of oxygen it becomes sulphuric acid, which, combined with water, is oil of vitriol. The bleaching and disinfecting action of the sulphurous acid is connected with its activity in appropriating the oxygen which is loosely held or being given off by organic matter. Chlorine and hypochlorous acid (which is still more effective than chlorine itself) act in the opposite way, so do the permanganates, such as Condy's fluid, etc. They supply oxygen in the presence of water. It is curious that opposite actions should produce like results. A disquisition on this and its suggestions would carry me beyond the limits of a note.

Chapter 47

ANOTHER DISINFECTANT

The above-named disinfectants are objectionable on account of their own odors and their corrosive action. Both sulphurous acid and hypochlorous acid (the active principle of the so-called "chloride of lime") have a disagreeable habit of rusting iron and suggesting antique green bronzes by their action on brass ornaments. Under serious conditions this should be endured, but in many cases where the danger is not already developed, the desired end may be attained without these annoyances.

Sulphate of copper, which is not patented or "brought out" by a limited company, may be bought at its fair retail value of 6*d.* or less per lb. (the oil-shop name for it is "blue vitriol"), in crystals, readily soluble in water.

I have lately used it in the case of a trouble to which English households are too commonly liable, and one that has in many cases done serious mischief. The stoppage of a soil-pipe caused the overflow of a closet, and a consequent saturation of floor boards, that in time would probably have developed danger by nourishing and developing those germs of bacteria, bacilli, etc., which abound in the air, and are ready to increase and multiply wherever their unsavory food abounds.

By simply mopping the floor with a solution of these green crystals, and allowing it to soak well into the pores of the wood, they cease to become a habitat for such microscopic abominations. The copper-salt poisons the poisoners.

Dr. Burg goes so far as to recommend that building materials, articles of furniture, and clothing, etc., should be injected with sulphate of copper, in order to avert infection, and in support of this refers to the immunity of workers in copper from cholera, typhoid fever, and infectious diseases generally.

I agree with him to the extent of suggesting the desirability of occasionally mopping house floors with this solution. Its visible effects on the wood are first to stain it with a faint green tinge which gradually tones down to a brown stain, giving to deal the appearance of oak, a change which has no disadvantage from an artistic point of view. If the wood is already tainted with organic matter capable of giving off sulphureted hydrogen, the darkening change is more rapid and decided, owing to the formation of sulphide of copper.

The solution of sulphate should not be put into iron or zinc vessels, as it rapidly corrodes them, and deposits a non-adherent film of copper. It will even disintegrate common earthenware, by penetrating the glaze, and crystallizing within the pores of the ware, but this is a work of time (weeks or months). Stoneware resists this, and wooden buckets may be used safely. It is better to keep the crystals and dissolve when required. Ordinary earthenware may be used with impunity if washed immediately afterwards.

Chapter 48

ENSILAGE

This subject has been largely expounded and discussed lately in the *Times* and other newspapers. As most of my readers are doubtless aware, it is simply a substitute for haymaking, by digging pits, paving and building them round with stone or concrete, then placing the green fodder therein and covering it over with sufficient earth to exclude the air.

We are told that very inferior material (such as coarse maize grass mixed with chaff) when thus preserved gives better feeding and milking results than good English hay.

I may mention a very humble experience of my own that bears upon this. When a boy, I was devoted to silkworms, and my very small supply of pocket-money was over-taxed in the purchase of exorbitantly small pennyworths of mulberry leaves at Covent Garden. But a friend in the country had a mulberry tree, and at rather long intervals I obtained large supplies, which, in spite of all my careful wrapping in damp cloths, became rotted in about ten days. I finally tried digging a hole and burying them. They remained fresh and green until all my silkworms commenced the working and fasting stage of their existence. This was ensilage on a small scale.

The correspondence in the newspapers has suggested a number of reasons why English farmers do not follow the example of their continental neighbors in this respect; climate, difference of grasses, etc., etc., are named, but the real reason why this is commercially impossible, and farming, properly so called, is becoming a lost art in England (mere meadow or prairie grazing gradually superseding it) is not named in any part of the discussion that I have read.

I refer to the cause which is abolishing the English dairy, which drives us to the commercial absurdity of importing fragile eggs from France, Italy, Spain, etc., apples from the other side of the Atlantic, tame house-fed rabbits from Belgium, and so on, with all other agricultural products which are precisely those we are *naturally* best able to produce at home; I mean *those demanding a small area of land and a proportionately large amount of capital and labor*. A poultry or rabbit farm, acre for acre, demands fully ten times the capital, ten times the labor, and yields ten times the produce obtained by our big-field beef and mutton graziers.

The scientific and economic merits of ensilage are probably all that is claimed for it, and it is especially adapted for our uncertain haymaking climate, but what farmer who is merely a lodger on the land, holding it as an annual tenant-at-will or under a stinted beggarly lease of

21 years, would expend his capital in building a costly *silo*, which becomes by our feudal laws and usages the absolute property of the landlord?

Our tenant farmers employ the latest and best achievements of engineering science in the form of implements, but take care that they shall be *upon wheels*, or otherwise non-fixtures, and use rich chemically prepared manures, provided they are not permanent, while they abstain from improvements which involve any serious outlay in the form of fixtures on the land. Those who lecture them about their want of enterprise should always remember that their condition is merely a form of feudal serfdom, tempered by the possession of capital, and that all their agricultural operations are influenced by a continual struggle to prevent their capital from falling into the hands of the feudal lord. Anybody who has ever read an ordinary form of English farm-lease, with its prohibitions concerning the sale of hay and straw, and restrictions to "four-course," or other mode of cultivation, must see the hopelessness of any development of British agriculture comparable to that of British commerce and manufactures.

Imagine the condition of a London shopkeeper or Midland manufacturer holding his business premises as a yearly tenant, liable at six months' notice to quit, with confiscation of all his business fixtures.

Chapter 49

THE FRACTURE OF COMETS

The view of the constitution of comets expounded in one of my notes of April last, viz., that they are meteoric systems consisting of a central mass, or masses, round which a multitude of minor bodies are revolving like satellites around their primary, is strongly confirmed by the curious proceedings of the present comet, which proceedings also justify my last note of last month pointing out the omission of our astronomers, who have neglected the positive and irregular repulsive action of the sun upon comets, that, like the great comets of 1843, 1880, and 1882, come within a few hundred thousand miles of the visible solar surface.

The solar prominences are stupendous eruptions from the sun, consisting, as the spectroscope demonstrates, of hydrogen flames and incandescent metallic vapors ejected with furious violence to visible distances ranging from ten or twenty to above three hundred thousand miles, but this flame shown by the spectroscope is but the flash of the gun, the actual ejection proceeding vastly farther, far beyond the limits of the corona, as described in last month's notes. These eruptions are so abundant that Secchi alone observed and recorded 2767 in one year (1871). Speaking generally, the sun is never free from them, and they proceed from all parts of the sun, but most abundantly from the sun-spot zones.

A system of meteoric bodies such as I suppose to form a comet (I mean the comet as it exists in space before the generation of its tail, which is only formed as it approaches the sun) could not approach so near to the sun as did the present comet at perihelion, without encountering more or less of these furious blasts the flash of some of which have been seen to move with a measurable mean velocity of above 300 miles per second, and a probable maximum velocity sufficient to eject solid matter beyond the reclaiming grasp of solar gravitation.

It is evident that such a meteoric system as I suppose to constitute a comet would, in the course of a rapid perihelion flight crossing these outblasts, be liable to various degrees of ejection in different parts, that would disturb its original structure by blowing some of its constituents out of their orbits, or even quite away from the control of the feeble gravitation of the general meteoric mass, and thus effecting a rupture of the comet.

Now such a disintegration or dispersion of the present comet has been actually observed. Several able observers have described a breaking of the head of this comet shortly after its perihelion passage. Commander Sampson's observations with the great 26-inch equatorial telescope of the Washington Naval Observatory are very explicit. On October 25 he saw the nucleus as a single well-defined globular body. On November 3, with the same telescope, he saw a triple nucleus, due to the formation of two additional minor bodies. These were more distinctly seen on November 6. Mr. W. R. Brooks, of New York, saw a detached fragment of the comet which afterwards faded out of view. Professor Schmidt observed another and similar fragment which has likewise disappeared.

All these observations indicate disruption due to some disturbing force, acting with different degrees of violence upon different portions of the comet.

Minor disturbances of this kind will, I think, account for the trail of meteoric bodies which Schiaparelli has shown to follow the paths of other comets. A great disturbance might give quite a new orbit to the meteoric fragments.

These considerations suggest another and a curious view of the question of possible cometary collision with the sun, viz., that a comet might be traveling in such an orbit as to make it mathematically due to plunge obliquely beneath the solar surface at its next perihelion; but on its approach to the surface of the sun it might encounter so violent an outrush of solar-prominence matter as to drive it bodily out of its course, and avert the threatened peril to its existence.

Chapter 50

THE ORIGIN OF COMETS

We read in story-books of uncomfortable people who have cherished a guilty secret in their bosoms, that it has "gnawed their vitals," until at last they have carried it to the grave. I have such a secret that does the gnawing business whenever I write or speak of comets, concerning the origin of which I am guilty of an hypothesis that has hitherto been cherished as aforesaid from the very shame of adding another to an already exaggerated heap of speculations on celestial physics.

It assumes, in the first place, that all the other suns which we see as stars are constituted like our own sun; that they eject great eruptions similar to the prominences above described, and even of vastly greater magnitude, as in the case of the flashing stars that have excited so much wonderment among astronomers, but which I regard simply as suns like ours, subject, like ours, to periodic maximum and minimum activities, but of greater magnitude.

If such is the case, some of the prominence matter or vaporous constituents of these suns must be ejected with much greater proportional violence than are those from our sun. But those from our sun have been proved to rush out on some occasions with a velocity so great that the solar gravitation cannot bring them back. If such is ever the case with the explosions of our sun, it must be of frequent occurrence with the greater explosions of certain stars. and therefore vast quantities of meteoric matter are continually ejected into space, and traveling there until they come within the gravitation domain of some other sun like ours, when they will necessarily be bent into such orbits as those of comets.

But what will be the nature of this meteoric matter?

If from our sun, it would be a multitude of metallic hailstones, due to the condensation of the metallic vapor by cooling as it leaves the sun, and such meteoric hail would correspond to the meteoric stones that fall upon our earth, and which, for reasons stated in "The Fuel of the Sun," I believe to be of solar origin. Besides these, there would be ice-hail, such as Schevedorf claims to be meteoric.

A star mainly composed of hydrogen and carbon, or densely enveloped in these gases (as the spectroscope indicates to be the case in some of these flashing stars), would eject hydrocarbon vapors, condensible by cooling into solids similar to those we obtain by the condensation of terrestrial hydrocarbon vapors (paraffin, camphor, turpentine, and all the essential oils, for example), and thus we should have the meteoric systems composed of these particles circulating about their own common centre of mass as above stated, and displaying the spectrum which Dr. Huggins has found common to comets.

If this is correct, the present comet comes from a sun that contains metallic sodium in addition to the hydrocarbons, as the spectrum of this metal was seen when this comet was near enough to the sun to render its vapor incandescent.

INDEX

A

abatement, 217, 224, 226, 227
acid, 7, 9, 10, 29, 30, 41, 43, 44, 49, 50, 91, 117, 137, 148, 177, 180, 181, 185, 190, 191, 202, 211, 212, 213, 225, 255, 259, 261, 269, 271
acoustics, 151, 152
agricultural chemistry, 173
air temperature, 128, 130
Albion, 177
alternative hypothesis, 56
aluminium, 11, 44, 63, 180
ammonia, 52, 104, 107, 227
architect, 67, 127, 152, 212, 213
Arctic Expedition, 109, 118
Aristotle, 191
asteroids, 2, 15, 18, 37, 59, 68
atmosphere, 1, 2, 3, 5, 6, 7, 8, 9, 11, 12, 14, 17, 20, 31, 33, 34, 37, 40, 41, 49, 50, 63, 64, 65, 89, 90, 91, 92, 94, 98, 99, 113, 121, 127, 129, 136, 137, 152, 170, 171, 211, 212, 219, 223, 229, 231, 234, 238, 240, 241, 243, 248, 249, 258, 260, 261, 263, 264, 267
atmospheric pressure, 3, 4, 8, 37, 92, 259, 260, 262, 263
atoms, 13, 35, 38, 165, 166, 167, 169, 180, 257, 268
authorities, 1, 26, 123, 126

B

barium, 11, 63, 179, 180
barometer, v, 48, 89, 90, 91, 92, 93, 94, 95, 96, 97, 98, 99, 100, 101, 102, 260
barriers, 42, 125, 131
beef, 123, 127, 222, 273
beetles, 23
Belgium, 273
bending, 32, 77, 78, 96, 254, 255, 257
beneficial effect, 217
Bible, 40
birds, 83, 84, 246, 247, 251, 263
bog-land, 103, 107, 108
boilers, 127, 177, 187, 202, 223, 224, 225
boils, 103, 107, 209, 259, 260, 261
brain, 23, 26, 27, 50, 127, 133, 158
brass, 212, 215, 256, 271
breathing, 9, 234, 237
Britain, 56, 92, 133, 191, 200, 208, 233
bronchitis, 136, 137, 229
burn, 40, 42, 52, 105, 121, 174, 180, 216, 226, 230

C

cabbage, 23, 104, 243, 245, 252
cabbage butterfly, 23
calcium, 11, 30, 43, 44, 63, 179, 180, 202
carbon, 44, 47, 48, 49, 50, 51, 104, 119, 137, 173, 175, 218, 225, 226, 244, 260, 277
carbonization, 260
caricature, 136, 234
cascades, 234, 235
catastrophes, 171
Central Asia, 243
certificate, 193, 222
chemical, 4, 27, 31, 38, 51, 55, 57, 119, 132, 137, 147, 157, 162, 166, 178, 191, 202, 208, 209, 212, 218, 226, 251, 253, 259
chemistry, v, 12, 20, 27, 51, 103, 167, 173, 177, 190
chimneys, 136, 138, 235, 242
China, 42, 132, 134, 143, 147, 148
chlorine, 258, 269
circumpolar region, 110
civilization, 131, 133, 179, 191, 243, 250
classes, 40, 76, 134, 173, 212, 218, 250, 251, 253, 258

climate, 33, 56, 76, 77, 104, 107, 112, 115, 116, 138, 203, 208, 231, 235, 237, 240, 241, 242, 245, 246, 247, 273
climates, 76, 88, 235, 247
coal, 10, 51, 52, 55, 56, 57, 83, 117, 119, 121, 122, 123, 124, 125, 126, 127, 128, 130, 131, 132, 133, 134, 135, 136, 139, 152, 183, 184, 185, 186, 211, 212, 216, 217, 218, 219, 220, 222, 224, 225, 226, 227, 229, 233, 237, 238, 239, 240, 242, 254, 255, 257
cobalt, 11, 12, 63
coke, 51, 52, 185, 218, 227
collateral, 61, 88, 248, 250
collisions, 13, 35, 38, 169, 257
color, 53, 144, 148, 149, 201, 211, 260, 269
combustion, 5, 6, 7, 9, 10, 12, 40, 42, 104, 105, 137, 138, 174, 175, 180, 211, 217, 225, 226, 241
commercial, 39, 51, 133, 135, 143, 144, 148, 149, 181, 203, 223, 273
commodity, 123, 131, 147, 148, 217
communication, 5, 27, 48, 50, 61, 70, 79, 117, 138, 163, 178, 235, 239, 240, 241
communities, 191
competition, 51, 131, 132, 213, 227
composition, 27, 29, 30, 42, 55, 56, 62, 80, 105, 119, 125, 180, 187, 198, 202, 208, 211, 218, 227, 241
compound eye, 25
compounds, 4, 17, 29, 43, 44, 119, 144, 148, 155, 173, 179, 180, 253, 259, 264
compression, 3, 5, 8, 9, 14
conception, 2, 15, 27, 37, 60, 82, 268
condensation, 4, 5, 6, 8, 38, 62, 92, 230, 259, 277
conductor, 26, 49, 128, 208, 215, 217, 221
conductors, 47, 138
conference, 119
configuration, 31, 56, 85, 110, 116, 117
confinement, 159
constituents, 5, 44, 169, 174, 208, 227, 258, 259, 275, 277
construction, 35, 41, 47, 90, 96, 127, 135, 136, 137, 156, 205, 215, 222, 224, 227, 229, 233, 234, 244
consumption, 121, 123, 124, 136, 137, 191, 225, 227, 249
cooking, 123, 221, 222, 224
cooling, 29, 31, 32, 43, 44, 45, 128, 129, 130, 170, 205, 216, 226, 230, 231, 240, 241, 258, 259, 277
copper, 11, 12, 48, 49, 63, 124, 133, 134, 144, 215, 254, 271
cornea, 23, 24
cost, 42, 51, 52, 61, 103, 106, 112, 123, 125, 130, 131, 132, 135, 139, 144, 177, 178, 181, 194, 195, 212, 222, 224, 225, 226, 239, 242, 243, 244, 249, 250

cotton, 11, 104, 133, 149, 178, 179, 189, 191, 249
covering, 75, 81, 90, 132, 234, 243, 245, 251, 256, 273
crops, 103, 104, 107, 108, 174, 175, 184, 247, 248, 250
crude oil, 183, 184, 186
crust, 16, 29, 30, 31, 32, 43, 44, 45, 121, 124, 170, 171, 202, 207, 254, 255
crystalline, 109, 201, 207
crystals, 30, 109, 207, 209, 255, 271, 272
cultivation, 108, 245, 247, 249, 250, 274
customers, 143, 149, 158, 183, 221, 223

D

danger, 74, 112, 114, 248, 252, 271
decomposition, 4, 53, 106, 208, 260
deduction, 16, 121, 123
degradation, 147, 158
delusion, 115, 133, 136, 159, 225
delusions, 23, 158, 159, 160
deposition, 49, 55, 57, 79, 80, 83, 87, 187, 202
deposits, 55, 56, 73, 74, 75, 76, 80, 81, 82, 87, 184, 199, 202, 230, 271
depth, 9, 11, 12, 30, 33, 44, 54, 56, 77, 79, 81, 82, 83, 84, 86, 87, 91, 96, 103, 107, 123, 124, 125, 126, 127, 128, 129, 130, 131, 132, 141, 162, 171, 208, 216, 223, 234, 237, 239, 253, 263
derivatives, 51
destruction, 41, 57, 170, 251
detonation, 60
deviation, 171, 238, 263
diffusion, 35, 37, 65, 153, 211, 268
discharges, 12, 13, 181
diseases, 40, 41, 271
dissociation, 4, 5, 6, 8, 14, 16, 33, 43, 49, 260, 264
dissociation temperature, 260
distillation, 42, 51, 57, 183, 184, 187, 226, 227, 260
distilled water, 104, 209
distribution, 3, 17, 81, 100, 116
division of labor, 61, 62
dogs, 27, 119, 161
Donegal, 103, 105
dragon-fly, 23
drainage, 75, 103, 104, 107, 238, 256
draught, 23, 136, 217, 219, 235, 240, 241
drawing, 14, 87, 156, 237, 248, 249

E

earthquakes, 45, 170
electricity, 2, 25, 47, 48, 49, 50, 52, 181, 268

electro-magnet, 47
emission, 138, 180, 215
employment, 100, 127, 143
enamel, 127, 180
encouragement, 250
energy, 6, 7, 20, 26, 36, 37, 38, 51, 52, 94, 130, 131, 133, 165, 259, 263
engineering, 223, 229, 274
England, 69, 73, 96, 113, 114, 132, 133, 134, 137, 143, 165, 190, 203, 219, 222, 242, 273
epidemic, 41
equilibrium, 37, 80, 96
equipment, 118
erosion, 80, 87
ethers, 13, 17, 35
Europe, 41, 73, 75, 77, 82, 84, 86, 110, 111, 113, 116, 137, 139, 208, 221
evaporation, 4, 91, 127, 129, 231, 241
evidence, 2, 11, 13, 16, 20, 31, 34, 75, 78, 84, 87, 149, 156, 157, 166, 217, 233, 265, 268
evil, 41, 105, 231, 241
evolution, 5, 6, 7, 29, 30, 32, 212
excitation, 18
exercise, 35, 61, 67, 127, 158, 224
exposure, 142, 207, 212
extinction, 6, 224, 231
extrusion, 27, 76

F

famine, 121, 122, 123, 124, 135, 139, 226
farmers, 273, 274
fat, 42, 177, 178, 190, 223
fear, 39, 125, 126, 134, 143, 178, 187, 220
fermentation, 230
fertility, 55, 83, 84, 133, 174, 175
fever, 40, 41, 126, 269
filters, 241, 249
filtration, 197, 244
Finland, 86, 111, 113, 116, 117
fires, 17, 121, 127, 130, 190, 212, 218, 222, 224, 225, 226, 227, 231, 235, 242
fish, 9, 26, 83, 84, 117, 156, 180, 263
fishing, 83, 84, 100, 117
flame, 5, 7, 9, 10, 11, 20, 30, 39, 60, 64, 179, 215, 218, 223, 224, 225, 226, 230, 260, 275
flotation, 43, 77, 87
flowers, 61, 160, 245, 249, 250, 252
fluctuations, 76, 90, 91
fluid, 20, 24, 38, 43, 89, 171, 184, 253, 254, 255, 262, 269
food, 39, 41, 83, 108, 112, 135, 139, 174, 190, 221, 246, 271

force, 4, 6, 7, 15, 20, 30, 35, 36, 37, 38, 49, 51, 60, 70, 81, 89, 95, 98, 129, 132, 133, 135, 138, 155, 156, 157, 159, 160, 186, 206, 207, 209, 223, 250, 254, 255, 256, 263, 276
formation, 12, 32, 34, 45, 49, 56, 57, 75, 78, 79, 80, 81, 84, 88, 125, 179, 194, 198, 199, 202, 203, 225, 271, 276
Fountain of Cyane, 53
fragments, 11, 13, 23, 53, 54, 55, 61, 68, 70, 71, 76, 79, 80, 82, 109, 110, 111, 166, 167, 171, 198, 206, 208, 209, 255, 256, 276
France, 97, 177, 273
freezing, 88, 109, 114, 115, 138, 165, 169, 205, 206, 207, 208, 209, 230, 231, 255, 259
friction, 77, 79, 206, 207, 253, 254
frost, 84, 92, 93, 205, 207, 208, 209, 245, 251
fruits, 189, 245, 246
fuel of the sun, 20
fusion, 29, 30, 31, 44, 48, 199, 217

G

Galileo, 141
general knowledge, 73
geological history, 6
Germany, 137, 138
gill, 84
glacial deposits, 73, 74, 75
glaciation, 73, 74, 75, 76, 77, 81, 82, 85, 86, 88
glasses, 23
grass, 82, 107, 174, 175, 202, 273
grasses, 108, 273
gravitation, 3, 6, 17, 20, 35, 36, 37, 60, 70, 71, 155, 156, 233, 255, 263, 268, 275, 277
gravity, 7, 12, 263
Great Britain, 56, 68, 121, 123, 125, 132, 134
Great Ice Age, v, 73, 74, 75
Greece, 69
Greeks, 189
greenhouse, 244
Grove, Mr., 2, 6, 8, 9, 68
Grove, Sir William, 1, 68
growth, 1, 56, 73, 88, 104, 106, 158, 202, 224, 245, 246, 251, 252, 253

H

health, 135, 149, 237
height, 15, 30, 31, 43, 49, 60, 64, 77, 80, 84, 85, 86, 87, 90, 91, 92, 93, 96, 98, 106, 132, 151, 153, 224, 233, 255, 258, 267
helplessness, 112

hemisphere, 95, 109, 116
hemp, 251
herbage, 103, 107
highways, 109, 162
history, 3, 15, 19, 56, 69, 73, 94, 95, 96, 143, 144, 162, 166, 167, 184, 185, 189, 200
House, 173, 221, 224, 241
house-fly, 23
huge cavities, 141
human agency, 160
human body, 127, 231
human existence, 130
human health, 40, 41
human nature, 40
humidity, 92, 104, 128
humus, 104
hunting, 178
hybrid, 262
hydrocarbons, 51, 52, 57, 119, 185, 278
hydrogen, 5, 12, 14, 16, 51, 60, 61, 64, 65, 117, 119, 165, 181, 202, 218, 226, 258, 259, 263, 271, 275, 277
hydrogen gas, 65, 117, 119

I

Iceland, 83, 84, 110, 113, 191
illumination, 9, 14, 41, 47, 48, 49, 50, 51, 52, 64, 180
imagination, 1, 13, 24, 28, 54, 60, 67, 158, 166, 169, 178, 268
impurities, 143, 149, 211, 249
industry, 106, 122, 132, 133, 134, 143, 185, 229
infection, 137, 269, 271
inheritance, 134, 233, 268
insect, 23, 24, 25, 26, 27
insects, 24, 25, 26, 27
institutions, 136, 222, 253
intellect, 6, 13, 26, 35, 67, 134, 158, 167, 169
intellectual attainment, 157
intelligence, 23, 61, 160, 169, 222, 250
interference, 158, 224
inventors, 14, 47, 180, 181
Ireland, vi, 69, 73, 87, 92, 96, 99, 103, 105, 106, 107, 108, 185, 197, 199, 200, 203
iron, v, 10, 11, 12, 29, 30, 32, 44, 45, 49, 63, 65, 117, 122, 124, 125, 127, 133, 134, 136, 137, 138, 143, 144, 147, 148, 149, 152, 153, 177, 178, 186, 187, 195, 206, 207, 209, 211, 215, 216, 217, 218, 221, 222, 223, 224, 229, 231, 233, 254, 255, 256, 271
irradiation, 141
islands, 74, 75, 83, 84, 85, 87, 96, 99, 100, 113, 116, 122, 193, 202
Italy, 63, 69, 166, 190, 203, 273

J

Jupiter, 3, 12, 15, 17, 39, 59, 70, 71, 263, 264

K

Kensington Gardens, vi, 173, 174, 175
kerosene, 186
kinetics, 35, 36
Kylemore, 103, 105, 106, 108

L

lakes, 43, 53, 82, 87, 106, 109, 131, 202, 256
lava, 29, 30, 31, 32, 43, 44, 45, 189, 198, 199
laws, 2, 3, 5, 6, 14, 15, 16, 17, 18, 38, 39, 57, 131, 233, 274
leakage, 112, 136, 195, 234
learning, 14, 42, 167
light, 2, 6, 8, 9, 10, 11, 13, 14, 15, 17, 18, 25, 32, 36, 37, 38, 39, 40, 41, 42, 48, 50, 51, 52, 59, 62, 63, 68, 70, 81, 98, 105, 106, 107, 112, 113, 141, 142, 151, 153, 158, 179, 180, 181, 186, 187, 216, 219, 227, 241, 244, 249, 251, 262, 268
limestone, 30, 105, 106, 183, 184, 185, 187, 198, 201, 202, 203, 212
liquids, 63, 119, 170, 205, 206, 216, 253, 254, 257, 258, 259, 263, 264
Lofoden Islands, 83, 85, 109
luminosity, 9, 11, 13, 25, 33, 153, 181, 267

M

machinery, 8, 20, 49, 50, 127, 130, 158, 177, 178, 181, 184
magnesium, 11, 43, 63, 180
magnitude, 7, 8, 9, 10, 11, 14, 18, 26, 29, 30, 31, 36, 44, 48, 54, 55, 60, 67, 68, 75, 78, 81, 82, 83, 85, 154, 159, 171, 184, 185, 193, 239, 255, 277
manufacturing, 13, 122, 132, 191, 211, 223, 249
manure, 56, 103, 104, 107, 108
mapping, 170
Mars, 15, 17, 59, 117, 263
marsh, 269
mass, 3, 6, 11, 17, 20, 30, 31, 32, 37, 38, 44, 45, 60, 65, 77, 78, 80, 82, 84, 114, 197, 199, 203, 254, 255, 256, 257, 263, 268, 275, 277
materials, 1, 4, 17, 19, 29, 31, 33, 36, 43, 44, 61, 64, 67, 119, 169, 177, 187, 201, 202, 208, 211, 212, 213, 215, 231, 251, 253, 263, 267, 271
matrix, 80, 82, 198, 199

melted rock, 43
melting, 44, 76, 77, 198, 206, 255
Mercury, 3, 15, 17, 31, 263
metals, 5, 9, 12, 43, 44, 47, 63, 64, 187, 216, 255
meteor, 65, 68, 267
meteorites, 2, 16, 37
microscope, 60, 151, 167
miniature, 17, 29, 30, 31, 45, 198, 263
modifications, 9, 35, 47, 96
moisture, 103, 104, 107, 223, 230, 231
molecular structure, 203
molecules, 25, 35, 37, 133, 165, 167, 180, 257
monopoly, 39, 51, 221, 227

N

natural resources, 105
natural selection, 131
nebulæ, 2, 17, 18
negotiating, 180
neuralgia, 229
nickel, 11, 12, 63, 133
North Tyrol, 53
Norway, 54, 55, 56, 73, 74, 75, 76, 78, 81, 83, 85, 86, 87, 88, 96, 109, 110, 111, 113, 115, 116, 137, 138, 246, 256, 257
Norwegian fjords, 55, 56, 109
nucleus, 6, 7, 9, 10, 12, 71, 264, 275

O

oil, 39, 40, 41, 42, 179, 181, 183, 184, 185, 186, 187, 189, 190, 193, 194, 195, 211, 259, 269, 271
old age, 171, 237
olfactory nerve, 26, 27
olive oil, 41, 42, 190, 193, 259, 260
operations, 10, 41, 103, 125, 127, 130, 149, 160, 269, 274
opportunities, 57, 74, 106, 114, 199
optic nerve, 23, 24, 180
orbit, 15, 16, 68, 69, 70, 116, 258, 276
organic compounds, 253
organic matter, 137, 231, 269, 271
organs, 25, 26, 27, 152
oxidation, 12, 44, 104, 180
oxygen, 5, 7, 10, 48, 49, 50, 51, 91, 211, 222, 226, 248, 258, 269
oyster, 45, 179
ozone, 243, 248

P

pasture, 83, 108
peat, 56, 57, 106, 108, 122, 131, 132, 184, 252
penumbra, 10, 11, 33
percolation, 184, 185
petroleum, 42, 183, 184, 185, 186, 187
phalanx, 86
phosphates, 175
phosphorescence, 180
phosphorus, 179, 180
photographs, 11, 14, 59, 61, 65, 142
pitch, 24, 25, 83, 244, 255, 256, 257, 262
pitch pine, 244
planets, 2, 6, 7, 12, 15, 17, 18, 20, 31, 37, 59, 68, 70, 71, 170, 264
plants, 42, 56, 104, 107, 167, 190, 243, 245, 248, 252, 253
platinum, 44, 48, 52, 255
practical knowledge, 111
precipitation, 77, 92, 202
principles, 49, 51, 96, 128, 160, 167, 215, 222, 239, 241, 242
probability, 6, 17, 68, 87, 125, 166
profit, 121, 177, 221, 227, 243
project, 10, 12, 111, 119, 198, 242, 245, 247
prosperity, 106, 132, 133, 134, 135, 191, 218
protection, 245, 246, 249, 251

R

radiation, 6, 7, 9, 10, 11, 16, 20, 31, 35, 37, 38, 39, 43, 86, 128, 170, 215, 216, 217, 218, 245, 252
radiometer, v, 35, 36, 37, 38
radius, 15, 78, 114
resistance, 7, 30, 48, 78, 105, 126, 206, 207, 208, 253, 254, 267, 268
resources, 87, 130, 133
restoration, 20, 174, 175, 211, 267
restrictions, 221, 250, 274
rings of Saturn, 2, 18
risk, 23, 111, 112, 170, 266
river flows, 107

S

safety, 24, 118, 193, 194, 212, 262
salts, 30, 31, 51, 52, 104, 105, 174, 175, 209
saturation, 55, 209, 230, 271
Saturn, 2, 12, 17, 18, 39, 70, 71, 264
Scandinavia, 73, 76, 77, 81, 82, 83, 86, 87, 109, 116
scientific knowledge, 133

Scotch firths, 109
sea-level, 31, 77, 81, 82, 85, 86, 87, 90, 94
seed, 107, 167, 169, 170, 171, 247, 248
semicircle, 122, 135
shape, 7, 35, 88, 105, 106, 153, 154, 180, 195, 198, 199
sheep, 103, 106, 107, 175
shores, 42, 55, 76, 87, 92, 111, 114, 177, 191
showing, 9, 18, 38, 61, 62, 77, 94, 95, 98, 108, 128, 161, 178, 206, 243, 244
Siberia, 111, 113, 115, 116, 200
Siemens, Dr. C. W., v, 19, 20, 229, 230, 231, 232
signs, 26, 85, 92, 96, 114, 212
silicon, 43, 44
silk, 50, 114, 149
silver, 128, 133, 215
skeleton, 31, 43, 145, 244, 245
skin, 44, 127, 130, 136, 189
smoking, 115, 226, 235
smoothing, 85, 193, 201
smoothness, 144
social change, 132
sodium, 11, 30, 63, 166, 278
solar activity, 2
solar eclipse, v, 16, 59
solar eruptions, 12, 59, 60, 71
solar heat, 2, 18, 34, 68, 116
solar system, 2, 7, 37, 60, 68, 69, 170
solid rock, 43, 87, 125
solution, 17, 49, 59, 61, 75, 118, 119, 128, 167, 190, 193, 209, 211, 217, 237, 251, 255, 271
South America, 112, 200
specific gravity, 11, 43, 77, 138, 263
specific heat, 4, 12, 130
speculation, 14, 19, 27, 29, 35, 63, 71, 187
sponge, 103, 106, 201
stars, 6, 20, 21, 37, 60, 61, 71, 171, 277
stellar heat, 2
stock, 6, 108, 112, 122, 123, 124, 181
stoves, 136, 137, 139, 218, 221, 222, 230, 231, 234, 237, 240
structure, 2, 24, 26, 27, 53, 57, 77, 86, 152, 198, 201, 202, 203, 207, 208, 211, 221, 244, 275
succession, 50, 53, 65, 93, 114, 243, 248, 265
sulphur, 44, 179, 180, 181, 191, 211, 269
sun-spots, v, 2, 8, 10, 12, 18, 33, 141, 142, 264
Sweden, 86, 110, 137, 138

T

tar, 183, 227, 251, 257
temperature, 3, 4, 5, 7, 9, 10, 12, 20, 30, 33, 34, 36, 37, 43, 48, 90, 91, 92, 98, 107, 114, 115, 124, 125, 126, 127, 128, 129, 130, 132, 137, 138, 165, 169, 186, 205, 215, 216, 223, 226, 230, 233, 235, 237, 238, 240, 241, 245, 246, 247, 250, 251, 252, 255, 256, 258, 259, 260, 261, 262, 263, 264
testing, 2, 34, 104, 156, 208, 212, 213
Thames, vi, 177, 178
thinning, 76, 79, 86, 173, 174
Thomson, Sir William, 43, 44, 165, 167
tides, 7, 11, 91, 255
tin, 92, 133, 216, 244
titanium, 11, 63
tobacco, 175, 249, 266
Torricelli, 17, 89
trade, 40, 108, 112, 131, 132, 133, 143, 147
trade policy, 131
traditions, 119, 122, 134
training, 62, 67, 133, 134, 154, 162
transmission, 2, 36, 38, 268
transparent medium, 230
treatment, 30, 74, 108
tremor, 24, 25, 180
trial, 68, 117, 244
typhoid fever, 271
typhus, 40, 41

U

United Kingdom, 40, 42, 100, 122
United States, 100, 131, 231, 250
universe, 2, 3, 6, 18, 20, 36, 37, 38, 67, 69, 73, 171, 268

V

vapor, 4, 5, 6, 7, 12, 13, 16, 17, 18, 34, 49, 62, 63, 64, 65, 92, 109, 117, 130, 183, 185, 186, 212, 213, 223, 230, 231, 257, 259, 260, 261, 263, 277, 278
vegetables, 104, 169, 173, 245, 247, 250, 251
vegetation, 56, 57, 75, 103, 104, 132, 167, 241, 243
velocity, 8, 12, 15, 16, 20, 60, 68, 70, 95, 115, 165, 253, 254, 275, 277
ventilation, 122, 128, 129, 136, 138, 229, 233, 234, 235, 237, 238, 239, 240, 241, 242, 243, 248, 249
vessels, 35, 38, 82, 89, 271
victims, 23, 84, 112, 136, 158, 159
violence, 12, 16, 29, 156, 194, 205, 275, 276, 277
viscosity, 32, 126, 206, 207, 253, 254, 255, 256, 257, 268
vision, 23, 25, 26, 27, 32, 35, 63, 64, 121
volcanic activity, 29, 199
volcanoes, v, 29, 30, 44, 45, 170, 200, 203

W

waste, 51, 52, 104, 105, 121, 122, 123, 135, 189, 219, 224, 227
wells, 184, 185, 186, 187, 202
Wollaston, Dr., 1, 2, 14
wood, 38, 53, 54, 122, 131, 132, 139, 190, 221, 222, 224, 251, 271
workers, 41, 107, 177, 269, 271

Y

yield, 32, 53, 78, 103, 105, 107, 139, 207, 246, 248
Yorkshire, 177, 178

Z

zinc, 11, 48, 63, 133, 240, 244, 271

Planetary Influence on the Sun and the Earth, and a Modern Book-Burning

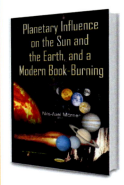

Author: Nils-Axel Mörner (Paleogeophysics & Geodynamics, Stockholm, Sweden)

Series: Space Science, Exploration and Policies

Book Description: This book is primarily concerned with fundamental components of solar physics, terrestrial geophysics and general climate issues.

Hardcover ISBN: 978-1-63482-837-6
Retail Price: $230

U.S. and World Coal Production: Developments and Projections

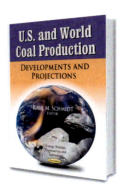

Editor: Raul M. Schmidt

Series: Energy Science, Engineering and Technology

Book Description: This book serves as a primer on U.S. and world coal resources and production and highlights some of the congressional interest related to coal production on U.S. federal lands.

Softcover ISBN: 978-1-62618-976-8
Retail Price: $69

The Science and Engineering of Sustainable Petroleum

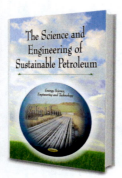

Editor: Rafiq Islam

Series: Energy Science, Engineering and Technology

Book Description: We are facing a crisis that threatens the sustainability of the entire planet. Civilization has been defined up to now by how efficiently we handle our energy needs; nevertheless, today we are bombarded by proposals for alternative technologies that are more energy-intensive than whatever preceded.

Hardcover ISBN: 978-1-62618-601-9
Retail Price: $135